AF665583

ICIAM2023 Springer Series

Volume 1

Editor-in-Chief

Hisashi Okamoto, Department of Mathematics, Gakushuin University, Tokyo, Japan

Series Editors

Pingwen Zhang, School of Mathematical Sciences, Peking University, Beijing, China

Ricardo H. Nochetto, Department of Mathematics, University of Maryland, College Park, USA

Carlos Parés, Mathematical Analysis, Statistics, and Applied Mathematics, University of Malaga, Málaga, Spain

Giulia Di Nunno, Department of Mathematics, University of Oslo, Oslo, Norway

This series aims to publish some of the most relevant results presented at the ICIAM 2023 held in Tokyo in August 2023.

The series is managed by an independent Editorial Board, and will include peer-reviewed content only, including the Invited Speakers volume as well as books resulting from mini-symposia and collateral workshops.

The series is aimed at providing useful reference material to academic and researchers at an international level.

Shin'ichi Oishi · Hisashi Okamoto · Ken Hayami
Editors

Recent Developments in Industrial and Applied Mathematics

Springer

Editors
Shin'ichi Oishi
Department of Applied Mathematics
Waseda University
Tokyo, Japan

Hisashi Okamoto
Department of Mathematics
Gakushuin University
Toshima City, Tokyo, Japan

Ken Hayami
National Institute of Informatics
Tokyo, Japan

ISSN 3091-3098 ISSN 3091-3101 (electronic)
ICIAM2023 Springer Series
ISBN 978-981-95-1445-8 ISBN 978-981-95-1446-5 (eBook)
https://doi.org/10.1007/978-981-95-1446-5

Mathematics Subject Classification: 00-02, 00B20, 00A69

This work was supported by The Japan Society for Industrial and Applied Mathematics.

© The Editor(s) (if applicable) and The Author(s) 2026. This book is an open access publication.

Open Access This book is licensed under the terms of the Creative Commons Attribution 4.0 International License (http://creativecommons.org/licenses/by/4.0/), which permits use, sharing, adaptation, distribution and reproduction in any medium or format, as long as you give appropriate credit to the original author(s) and the source, provide a link to the Creative Commons license and indicate if changes were made.
The images or other third party material in this book are included in the book's Creative Commons license, unless indicated otherwise in a credit line to the material. If material is not included in the book's Creative Commons license and your intended use is not permitted by statutory regulation or exceeds the permitted use, you will need to obtain permission directly from the copyright holder.
The use of general descriptive names, registered names, trademarks, service marks, etc. in this publication does not imply, even in the absence of a specific statement, that such names are exempt from the relevant protective laws and regulations and therefore free for general use.
The publisher, the authors and the editors are safe to assume that the advice and information in this book are believed to be true and accurate at the date of publication. Neither the publisher nor the authors or the editors give a warranty, expressed or implied, with respect to the material contained herein or for any errors or omissions that may have been made. The publisher remains neutral with regard to jurisdictional claims in published maps and institutional affiliations.

This Springer imprint is published by the registered company Springer Nature Singapore Pte Ltd.
The registered company address is: 152 Beach Road, #21-01/04 Gateway East, Singapore 189721, Singapore

If disposing of this product, please recycle the paper.

Preface

This volume is the record of ICIAM 2023 (International Congress of Industrial and Applied Mathematics 2023, the tenth ICIAM). It was held at Waseda University, Tokyo, Japan in August 20th–25th, 2023, with 5594 participants (4329 in-person, 1247 online. Student participants 1566) and 233 accompanying persons. It consisted of four public lectures, 34 plenary and invited talks, three SIAM prize lectures, 476 mini-symposia, 809 contributed talks, and 242 posters. The papers in this volume are authored by some of the invited speakers of ICIAM2023. This volume is a part of a series dedicated to ICIAM 2023. Other volumes will appear soon from Springer Nature.

We are sure that this volume will contribute greatly to the applied mathematics community in the world. This volume was made open access by the courtesy of the organizing committee of the congress and the Japan Society for Industrial and Applied Mathematics. We thank very much for all who took pains to make it open access.

The congress, hosted by the Japan Society for Industrial and Applied Mathematics (JSIAM) and the Mathematical Society of Japan (MSJ), was held at Waseda University in Tokyo Japan, under the auspices of the International Council for Industrial and Applied Mathematics (ICIAM). With the greatest ever number of participants as indicated above, we may safely say that ICIAM 2023 was a great success. The number of the participants demonstrates that the enthusiasm of industrial and applied mathematics communities is getting stronger despite the pandemic in 2020–2022.

The industrial aspect of the congress was enriched by, following the traditions of ICIAM, organizing a specific activity: 'The Industry Day.' Fourteen speakers, selected from a broad representation of different sectors, presented the results of ongoing collaborations with academia and the benefits derived from it. In addition, 19 industrial mini-symposia were scheduled during the congress. Fifteen satellite events took place before and after ICIAM2023. They covered a broad range of topics within industrial and applied mathematics. We thank greatly the efforts of the organizers of all satellite events.

From the announcement of the candidacy of JSIAM in 2016, it started, with the help of MSJ, the long process involved in the planning of this complex event. Our deepest gratitude and heartiest thanks go to all the people who helped to organize this congress. A list of all the committees and people involved in this task is given below.

The congress could not have been possible without the support of a large set of sponsors. The venue of the Congress was offered, free of charge, by Waseda University. The support of TCVB (Tokyo Convention and Visitors Bureau) is highly

Fig. 1 The Logo of ICIAM2023

10th International Congress on Industrial and Applied Mathematics

ICIAM 2023 TOKYO

Fig. 2 At the opening ceremony at the Okuma auditorium on August 20th, 2023, Ms. Nagaoka Keiko (center) delivers her message. On the left sits Dr. Tanaka Aiji, the President of Waseda University. Dr. Yuan Ya-xiang, the president of ICIAM, on the right

acknowledged. A thankful recognition is also due to the Ministry of Education, Culture, Sports, Science and Technology. Last but not least, on behalf of ICIAM 2023, we would like to express our most sincere gratitude to the invited speakers who have contributed to this volume for taking the time to provide their valuable contributions. The reader may grasp the atmosphere of the opening ceremony in Figs. 1–6.

Tokyo, Japan — Shin'ichi Oishi
Tokyo, Japan — Hisashi Okamoto
Tokyo, Japan — Ken Hayami

Congress Director's Speech

On behalf of the Japanese scientific community, the organizers of ICIAM 2023 warmly welcome participants from over 87 countries to Tokyo, Japan.

Fortunately, ICIAM 2023 can be held in person. In addition, all sessions are available for online participation. ICIAM 2023 features 27 invited talks, one Olga Taussky-Todd plenary lecture, six plenary talks from ICIAM prize recipients, and four public lectures. Furthermore, 477 mini-symposia composed of 4000 talks are organized

Fig. 3 Japanese traditional music and dance were performed at the Opening Ceremony in the Okuma Auditorium

Fig. 4 A scene of the opening ceremony at Okuma Auditorium during the Congress Director Oishi's welcoming speech

together with 1000 contributed talks, and 500 poster presentations. Approximately 20 exhibitions are held.

The organizers would like to extend their sincere thanks to the presenters of talks, as well as to the ICIAM officers led by President Yuan Ya-xiang, the members of the scientific program committee chaired by Prof. Yasumasa Nishiura, and the members of the local programming committee headed by Prof. Takeshi Ogita. Additionally, they express their gratitude to SIAM for managing the SIAM prize lectures, and organizing several mini-symposia based on SIAM activities.

ICIAM 2023's venue at Waseda University has been provided free of charge. This has enabled us to set the registration fee at a reasonable price. We would like to extend their gratitude to the Tokyo Convention and Visitors Bureau (TCVB) for their support in various aspects related to ICIAM 2023, especially we are able to provide financial support for approximately 500 attendees from around the world.

The organizers finally would like to thank everyone who supported ICIAM 2023, with special recognition to the Secretariat Group led by Prof. Naoya Yamanaka, for their contribution in making ICIAM 2023 a vibrant platform for researchers in industrial and applied mathematics. The organizers of ICIAM 2023 eagerly anticipate the engagement of participants in fruitful exchanges of research ideas and declare now the opening of ICIAM 2023.

Shin'ichi Oishi The Congress Director

ICIAM President's Speech

Distinguished Guests, Friends, Ladies and Gentlemen, Good morning!

I'm delighted to welcome you all to the 10th ICIAM Congress, on behalf of the International Council for Industrial and Applied Mathematics.

The International Council for Industrial and Applied Mathematics has consistently promoted the growth of mathematics and its applications globally, even amid the challenges posed by COVID-19. In these testing times, we've remained resilient, achieving notable progress. We have established a new ICIAM prize: the ICIAM Industry prize. We have upgraded to a Full Member of the International Science Council. ICIAM is active and represented in the Standing Committee for Gender Equality in Science, and the Steering Committee of International Year of Basic Sciences for Sustainable Development.

The ICIAM congresses, starting 36 years ago in Paris, are known for their rich programme. Today, we continue this tradition with a diverse and robust agenda. Immediately after this ceremony, we'll honor our six ICIAM prize winners. Although their victory was announced last year, today we officially appreciate their immense contribution.

Over the course of the congress, you will have the opportunity to hear from these prize winners, along with 27 top mathematicians from diverse fields, providing a rich blend of knowledge and experience. In addition, we'll host mini-symposia, contributed talks, and poster sessions, showcasing the breadth of industrial and applied mathematics.

The Congress also provides a platform to highlight the importance of mathematics to the wider public through the OTT lecture and three public lectures.

Finally, I would like to thank the local organizing committee, chaired by Prof. Oishi, for their tremendous efforts in assembling and coordinating the programme, and their kind hospitality in these wonderful conference facilities.

Ladies and Gentlemen, it's my great pleasure to officially open the Congress. Welcome to ICIAM 2023!

Ya-xiang Yuan
The ICIAM President

International Congress of Industrial and Applied Mathematics in the Past

1987 Paris
1991 Washington, D.C.
1995 Hamburg
1999 Edinburgh
2003 Sydney
2007 Zurich

2011 Vancouver
2015 Beijing
2019 Valencia
2023 Tokyo

The next ICIAM will be held in Hague, the Netherlands in 2027.

ICIAM Prize Winners

ICIAM Collatz Prize

1999 Stefan Müller, Max Planck Institut für Mathematik Leipzig, Germany.
2003 Weinan E., Princeton University, USA.
2007 Felix Otto, Universität Bonn, Germany.
2011 Emmanuel J. Candès, Stanford University and CALTECH, USA.
2015 Annalisa Buffa, CNR-IMATI, Italy.
2019 Siddharta Mishra, ETH Zürich, Switzerland.
2023 Maria Colombo, École polytechnique fédérale de Lausanne, Switzerland.

ICIAM Lagrange Prize

1999 Jacques-L. Lions, Collège de France and Académie des Sciences de Paris, France.
2003 Enrico Magenes, Università di Pavia, Italy.
2007 Joseph Keller, Stanford University, USA.
2011 Alexandre J. Chorin, UC Berkeley and LBNL, USA.
2015 Andrew J. Majda, New York University, USA.
2019 George Papanicolaou, Stanford University, USA.
2023 Alfio Quarteroni, Polytechnic University of Milan, Italy.

ICIAM Maxwell Prize

1999 Grigory I. Barenblatt, UC Berkeley, USA, and University of Cambridge, UK.
2003 Martin D. Kruskal, Rutgers University, USA.
2007 Peter Deuflhard, Zuse Institute Berlin, Germany.
2011 Vladimir Rokhlin, Yale University, USA.

2015 Jean-Michel Coron, Université Pierre et Marie Curie, France.
2019 Claude Bardos, Université Paris Diderot (Paris II), France.
2023 Weinan E, Peking University, China.

ICIAM Pioneer Prize

1999 Ronald R. Coifman, Yale University, USA, Helmut Neunzert, University of Kaiserslautern, Germany.
2003 Stanley Osher, University of California, Los Angeles, USA.
2007 Ingrid Daubechies, Princeton University, USA, Heinz Engl, Johannes Kepler University, and Austrian Academy of Sciences, Austria.
2011 James Albert Sethian, UC Berkeley and LBNL, USA.
2015 Björn Engquist, The University of Texas at Austin, USA.
2019 Yvon Maday, Sorbonne University, Paris, France.
2023 Leslie Greengard, New York University, USA.

ICIAM Su Buchin Prize

2007 Gilbert Strang, Massachusetts Institute of Technology, USA.
2011 Edward Lungu, University of Botswana, Botswana.
2015 Li Tatsien, Fudan University, P.R. China.
2019 Giulia Di Nunno, University of Oslo, Norway.
2023 José Mário Martinez Perez, Brazilian Academy of Sciences, Brazil.

ICIAM Industry Prize

2023 Cleve Moler, Mathworks, USA.

Organization of ICIAM 2023

Congress Director: Shin'ichi Oishi, Waseda University, Japan

Fig. 5 The prize winners of ICIAM2023 and the president of ICIAM at the opening ceremony

Organizing Committee

President: Shin'ichi Oishi, Waseda University

Members

Keiko Imai, Chuo University, Japan
Kenji Kajiwara, Kyushu University, Japan
Hiroshi Kokubu, Kyoto University, Japan
Motoko Kotani, Tohoku University, Japan
Kenichi Maruno, Waseda University, Japan
Hisashi Okamoto, Gakushuin University, Japan
Osamu Saeki, Kyushu University, Japan
Kazue Sako, NEC Corporation, Japan
Katsuyuki Takashima, Mitsubishi Electric Corporation, Japan
Shao Liang Zhang, Nagoya University, Japan

The Organizing Committee prepared the Bid for ICIAM2023 and decided the framework of the congress.

Scientific Program Committee (Appointed by ICIAM)

Chair: Yasumasa Nishiura, Tohoku University, Japan

Members

Martine Labbé, Université de Libre Bruxelles, Belgium
Helena Judith Nussenzveig Lopes, Federal University of Rio de Janeiro, Brazil
Alejandro Jofre, Universidad de Chile, Chile
Tao Tang, Southern University of Science and Technology, China
Xiaoshan Gao, Institute of Systems Science, Chinese Academy, China
Gabriel Peyré, Centre National de la Recherche Scientifique, France
Sylvie Méléard, Centre de Mathématiques Appliquées, France
Barbara Wohlmuth, Technische Universität München, Germany
Alex Martin, Friedrich-Alexander University Erlangen-Nuremberg, Germany
Dirk Hartmann, Siemens, Germany
Roberto Natalini, Consiglio Nazionale delle Ricerche, Italy
Hiroshi Kokubu, Kyoto University, Japan
Hinke Osinga, University of Auckland, New Zealand
Jin Keun Seo, Yonsei University, Korea
Alfredo Bermudez De Castro, Universidad de Santiago de Compostela, Spain
Carola-Bibiane Schónlieb, University of Cambridge, UK
Des Higham, University of Strathclyde, UK
Tammy Kolda, Sandia National Laboratories, USA
Wolfgang Dahmen, University of South Carolina, USA
Karen Wilcox, University of Texas at Austin, USA
Katya Scheinberg, Cornell University, USA

Executive Committee

Chair: Shin'ichi Oishi, Waseda University

Members

Daisuke Furihata, Osaka University, Japan
Elliott Ginder, Meiji University, Japan
Kenji Kajiwara, Kyushu University, Japan
Kenichi Maruno, Waseda University, Japan

Masaharu Nagayama, Hokkaido University, Japan
Takeshi Ogita, Waseda University, Japan
Hisashi Okamoto, Gakushuin University, Japan
Katsuhisa Ozaki, Sibaura Institute of Technology, Japan
Takashi Sakajo, Kyoto University, Japan
Kouta Sekine, Chiba Institute of Technology, Japan
Hiroshi Suito, Tohoku University, Japan
Naoya Yamanaka, Meisei University, Japan

ICIAM2023 in Numbers

- 41 Distinguished Lectures (27 Invited, six ICIAM Prize, three SIAM Prize, one Olga Taussky-Todd, four Public Lectures).
- Industry Day (14 invited, industry-driven lectures).
- 4880 scientific contributions from participants (mini-symposia: 3829 talks, 809 contributed talks, 242 posters displayed). 93 simultaneous sessions.
- Scholarships (financial support) awarded to 427 persons from the developing countries.
- 1 embedded meeting and 15 satellite meetings.

Fig. 6 The ICIAM flag was handed from Tokyo, Japan, to Hague, the Netherlands

Acknowledgements

Special support by TCVB (Tokyo Convention and Visitors Bureau) is greatly acknowledged.It included the preparation of the Bid and travel support of various sorts. Its website is: https://www.tcvb.or.jp/jp/

ICIAM2023 was supported by

Science Council of Japan
Waseda University
National Institute of Advanced Industrial Science and Technology
The Institute of Statistical Mathematics
The Institute of Electronics, Information and Communication Engineers
The Institute of Electrical Engineers of Japan
The Japan Society of Mechanical Engineers
The Physical Society of Japan
The Operations Research Society of Japan
The Japan Statistical Society
The Japan Society for Simulation Technology
The Japan Society for Computational Engineering and Science
The Japan Society of Fluid Mechanics
The Gas Turbine Society of Japan
Mathematical Education Society of Japan
Research Institute for Mathematical Sciences, Kyoto University
Institute of Mathematics for Industry, Kyushu University
Meiji Institute for Advanced Study of Mathematical Sciences, Meiji University

Contents

Exploring Matrices

Cleve Moler

Abstract I have spent much of my career working to bring abstract linear algebra and practical matrix computation closer together. This project is my latest effort.

Exploring Matrices

Exploring Matrices is a multi-media project that shows matrices in action. Short videos, blog posts, interactive MATLAB software and self-study exercises investigate applications of matrices. The material is intended for students in an undergraduate course in linear algebra or computational science. However, anyone using matrices should find topics that interest them.

The first release of Exploring Matrices has six modules. All of the modules use animated MATLAB displays. The modules are:

- Matrix Multiplication
- Rotation and Scaling
- Computer Graphics
- Matrices and Cubes
- Simulink
- AI and Gorillas.

The first four of these modules describe animations produced by MATLAB apps–`Multiply`, `Rotate`, `Grafix`, and `Qube`. The other two modules describe applications–simulation of control systems and neural networks for facial recognition (of gorillas).

C. Moler (✉)
MathWorks Inc., Natick, MA, USA
e-mail: moler@mathworks.com

© The Author(s) 2026

S. Oishi et al. (eds.), *Recent Developments in Industrial and Applied Mathematics*, ICIAM2023 Springer Series 1, https://doi.org/10.1007/978-981-95-1446-5_1

1 Matrix Multiplication

Some may just be learning the mechanics of matrix multiplication. Others encountered it years ago. The traditional algorithm for computing the product of two matrices involves inner products between the rows of the first matrix and the columns of the second. A less familiar algorithm, which involves linear combinations of the columns of the first matrix, is often more efficient and informative. The two approaches produce the same final result from intermediate terms in different orders.

Here is one frame from the animation of these two algorithms generated by our `Multiply` app. The highlighted element in the first matrix either moves across the rows or goes down the columns.

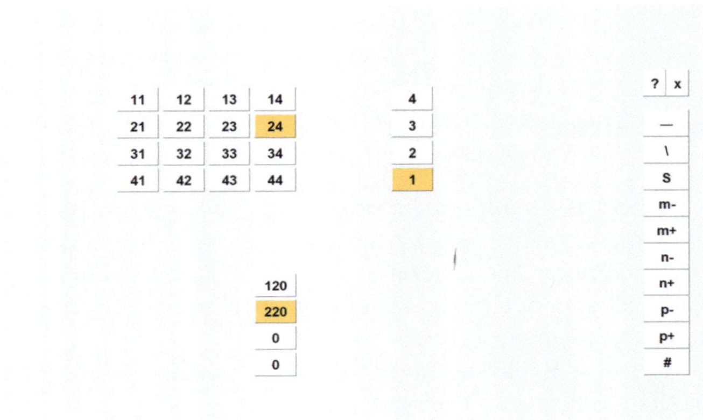

2 Rotation and Scaling

Our first matrices are 2-by-2. We see how the matrix

$$R = \begin{pmatrix} \cos\theta & \sin\theta \\ -\sin\theta & \cos\theta \end{pmatrix}$$

rotates points by the angle θ, measured in degrees.

We also see how the matrix

$$S = \begin{pmatrix} \sigma & 0 \\ 0 & \sigma \end{pmatrix}$$

makes objects larger and smaller.

The two can be combined with matrix multiplication. For more operations in higher dimensions, matrix multiplication provides a unifying framework.

Here is one frame from the animation of rotation and scaling generated by the Rotate app. The first panel displays a 2-by-2 rotation matrix, the second panel displays a 2-by-2 diagonal scaling matrix, and the third panel displays their product.

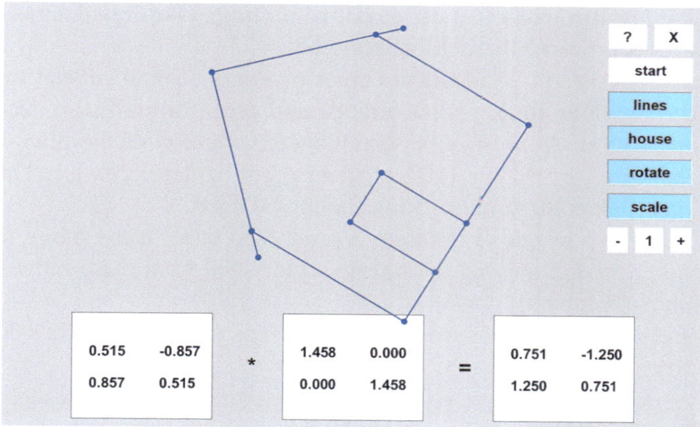

3 Computer Graphics

Operations with the 4-by-4 matrices that are at the heart of modern computer graphics employ a system known as “homogeneous coordinates”. The leading 3-by-3 submatrix produces rotation and scaling in three dimensions. The fourth column produces translations.

Here is one frame from an animation of rotation about the x-axis generated by the Grafix app. This is often called “pitch”. Rotation about the y- and z-axes are “roll” and “yaw”,

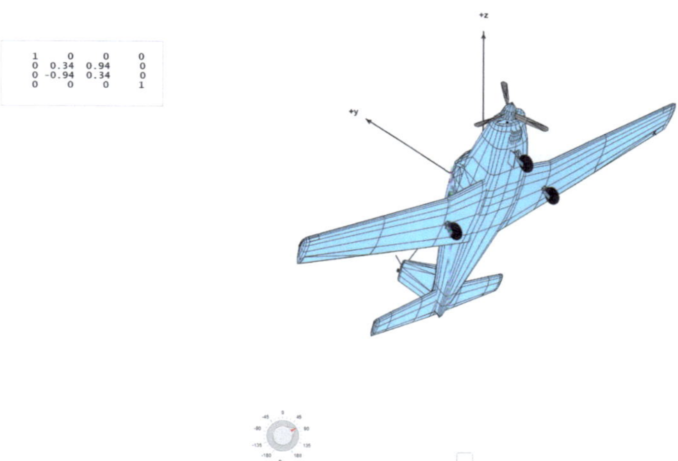

4 Matrices and Cubes

Rubik's Cube, named for its inventor, Ernő Rubik, a Hungarian professor of architecture, is the greatest mathematical puzzle of all time. Our digital simulator of the puzzle, `Qube`, is powered by rotation matrices.

The model consists of 27 identical copies of a single small cubelet whose sides are colored red, white, blue, yellow, orange and green. Initially, all cubelets have the same orientation. A move is the simultaneous rotation of the nine cubelets in one of the six faces, by 90° or 180°, clockwise or counterclockwise. This leads to 4.3×10^{19} possible configurations for a scrambled cube.

The object of the puzzle is to return a scrambled cube to the initial state. Most people are interested in solving the puzzle rapidly, but I am more interested in the number of moves required.

`Qube` offers animations of many mathematical properties of Rubik's cubes. Here is a frame from one of them.

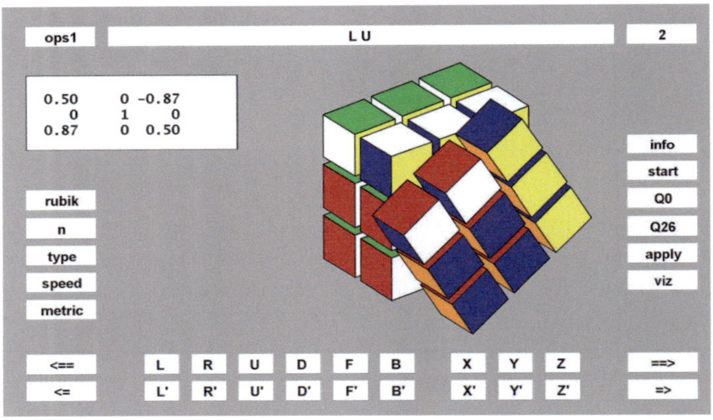

5 Simulink

MATLAB's companion product, Simulink, is a block diagram programming environment used to design and simulate systems with multidomain models and to automatically generate the code required to operate embedded processors.

Matrices are involved in dozens of different ways by Simulink, although most users rarely see operations at that detailed level. Our Simulink module shows a model of an automobile being driven on a test track and displays the pitch, roll and yaw recorded by the matrix connecting the coordinate system for the automobile to the coordinate system for the track.

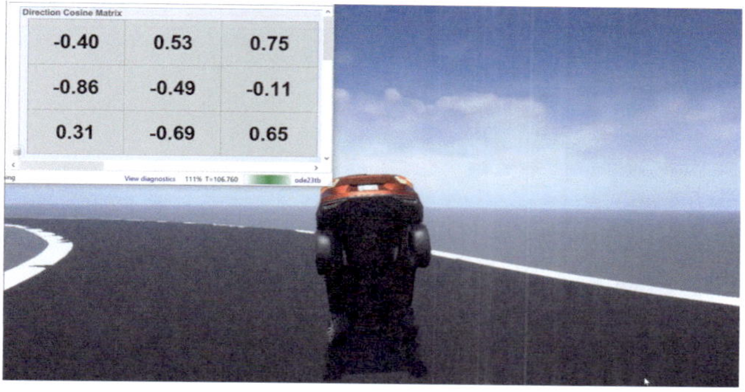

6 AI and Gorillas

This is a personal story about a project in the early stages of development.

My wife and I first visited the mountain gorillas in the Volcano National Park of Rwanda twelve years ago. An American primatologist named Dian Fossey had studied the mountain gorillas between 1966 and her murder by poachers in 1985. Her book Gorillas in the Mist was very popular and was the basis for a critically acclaimed 1988 Hollywood movie starring Sigourney Weaver.

We have become good friends with the people in the Gorilla Doctors organization. These African and American veterinarians attend to the health of the roughly 1,000 gorillas in the park. Most of the gorillas have African names like "Indundwa" and "Maisha". We envision a gorilla facial recognition system that is available on cell phones and tablets so that new guides and doctors can learn the names of their patients.

Inception-v3 is a convolutional neural network (CNN) that is widely used for image processing. We have a version of the network pretrained on more than a million images from the ImageNet database. This publicly available system knows nothing about gorillas. We must do additional training using photos of our unique subjects.

This is where matrices are applied. Training a CNN involves determining the values of thousands of weights and coefficients. The digital photos, regarded as vectors, are repeatedly multiplied by circulant matrices where each row is a shifted copy of the other rows. Importantly, a modern CNN also contains some nonlinear layers.

Here is one photo from a small test collection. Indundwa appears to have his own selfie stick.

Software and Videos

A self-extracting MATLAB source archive of the four apps is available at

`https://blogs.mathworks.com/cleve/files/Apps_mzip.m`

An introduction and six videos ranging in length from one to six minutes are available on YouTube at

```
https://www.youtube.com/playlist?
        list=PLn8PRpmsu08oGNmtBfFOmgVC0TlXDaLDJ
```

Dedication

We dedicate the Gorillas project to the memory of Mike Cranfield, DVM. Mike was Executive Director of the Mountain Gorillas Veterinary Project in Rwanda from 1999 until 2014. Before Rwanda, he held various positions at the Maryland Zoo in Baltimore.

Last year, Mike sent us a disc drive containing over 14,000 photographs of mountain gorillas he had taken in Rwanda. Shortly after sending the disc, Mike was hospitalized from an apparent attack of West Nile Virus. He passed away on August 27, 2023. Ironically, after years of working safely in the mountain jungles of Central Africa, it is likely that he acquired the virus from a mosquito bite at his family's cabin in Canada.

We plan to sort and organize his photos to provide specialized training of the facial recognition neural net.

Open Access This chapter is licensed under the terms of the Creative Commons Attribution 4.0 International License (http://creativecommons.org/licenses/by/4.0/), which permits use, sharing, adaptation, distribution and reproduction in any medium or format, as long as you give appropriate credit to the original author(s) and the source, provide a link to the Creative Commons license and indicate if changes were made.

The images or other third party material in this chapter are included in the chapter's Creative Commons license, unless indicated otherwise in a credit line to the material. If material is not included in the chapter's Creative Commons license and your intended use is not permitted by statutory regulation or exceeds the permitted use, you will need to obtain permission directly from the copyright holder.

Fast and Numerically Stable Implementation of Rate Constant Matrix Contraction Method

Shinichi Hemmi, Satoru Iwata, and Taihei Oki

Abstract The rate constant matrix contraction (RCMC) method, proposed by Sumiya et al. (2015, 2017), enables efficient and stable simulations of chemical kinetics in large-scale reaction networks. Iwata et al. (2023) reformulated the RCMC method as an algorithm for solving master equations with rate constant matrices satisfying detailed balance. This study focuses on accelerating RCMC by addressing its bottleneck: the greedy selection of steady states, which is equivalent to MAP inference in determinantal point processes (DPPs) under cardinality constraints. LazyFastGreedy, introduced by Hemmi et al. (2022), combines the practically efficient lazy greedy method with the DPP MAP inference but suffers from numerical instability due to wide-ranging reaction time scales in chemical kinetics. We propose modified LazyFastGreedy that mitigates numerical instability by avoiding subtraction of like-sign numbers, leveraging properties of rate constant matrices and connections to Cholesky decomposition. Numerical experiments on real chemical reaction instances demonstrate that the proposed method achieves significant speed improvements while maintaining numerical stability.

S. Hemmi (✉)
Preferred Networks, Inc., Tokyo, Japan
e-mail: hemmi.shinichi@gmail.com

S. Iwata (✉)
Department of Mathematical Informatics, Graduate School of Information Science and Technology, The University of Tokyo, Hongo 7-3-1, Bunkyo-ku, Tokyo 113-8654, Japan

Chemical Reaction Design and Discovery (WPI-ICReDD), Hokkaido University, Kita 21, Nishi 10, Kita-ku, Sapporo, Hokkaido 001-0021, Japan
e-mail: iwata@mist.i.u-tokyo.ac.jp

T. Oki
Chemical Reaction Design and Discovery (WPI-ICReDD), Hokkaido University, Kita 21, Nishi 10, Kita-ku, Sapporo, Hokkaido 001-0021, Japan
e-mail: oki@icredd.hokudai.ac.jp

Center for Advanced Intelligence Project, RIKEN, Nihombashi 1-4-1, Chuo-ku, Tokyo 103-0027, Japan

© The Author(s) 2026

S. Oishi et al. (eds.), *Recent Developments in Industrial and Applied Mathematics*, ICIAM2023 Springer Series 1, https://doi.org/10.1007/978-981-95-1446-5_2

1 Introduction

1.1 Background

Simulation of chemical kinetics is a pivotal tool for predicting chemical reactions on reaction path networks. A reaction path network consists of n equilibrium states (EQs) and paths connecting two EQs through transient states (TSs). The state of a chemical reaction at time $t \geq 0$ is represented as the *yield vector* $x(t)$, which is an n-dimensional vector whose vth ($v \in V := \{1, \ldots, n\}$) component $x_v(t)$ corresponds to the yield (amount) of the vth EQ. The dynamics of the yield vector follow the *master equation*

$$\dot{x}(t) = Kx(t). \tag{1}$$

In (1), $K = (K_{uv})_{u,v \in V}$ is an $n \times n$ *rate constant matrix*, which is determined from the potential energies of EQs and TSs if the system is described by the microcanonical ensemble (Sumiya and Maeda 2020; Sumiya et al. 2015). The rate constant matrix K satisfies the following conditions (Iwata et al. 2023):

(RCM1) $K_{uv} \geq 0$ for $u \neq v$,
(RCM2) $K_{vv} = -\sum_{u \neq v} K_{uv}$ for $v \in V$,
(RCM3) there exists a positive vector $\pi = (\pi_v)_{v \in V}$ such that $K_{uv}\pi_v = K_{vu}\pi_u$ for $u, v \in V$.

The (u, v) off-diagonal entry K_{uv} of K represents the reaction rate constant from the vth EQ to the uth EQ, and is non-negative by (RCM1). The second condition (RCM2) comes from the conservation law of the sum of yields, i.e., $\sum_{v \in V} x_v(t)$ is constant for all t. The last condition (RCM3) and the vector π are referred to as the *detailed balance condition* and a *stationary* (or *equilibrium*) *distribution*, respectively. The yield vector $x(t)$ converges to π as $t \to \infty$ regardless of the initial yield $x(0)$ if the network is connected (Anderson 1991). The conditions (RCM1–3) can be rephrased to state that $L = -K\Pi$ is a (weighted) *graph Laplacian matrix*, where Π is the diagonal matrix with diagonal entries $\pi_1, \ldots, \pi_n$. In probability theory, this system of $x(t)$ is known as the *continuous-time Markov chain*, and the master Eq. (1) is called the *Kolmogorov forward equation*.

Rate constant matrices are typically ill-conditioned due to the wide range of reaction time scales. Furthermore, recent advancements in automatic chemical reaction pathway search techniques, such as the anharmonic downward distortion following (ADDF) method (Ohno and Maeda 2004) and the single-component artificial force induced reaction (SC-AFIR) method (Maeda et al. 2014), have resulted in large-scale ($n \gtrsim 10,000$), sparse ($\mathrm{nnz}(K) \lesssim 30,000$), and extremely ill-conditioned rate constant matrices; the ratio of the largest to smallest non-zero off-diagonal entries of K can exceed 10^{200} (see Table 3). Numerically solving such a stiff differential equation has been recognized as a challenging task. While one can obtain $x(t)$ for every t from the eigendecomposition of K via the explicit formulation of the analytic solution $x(t) = e^{tK}x(0)$, computing the eigendecomposition of ill-conditioned matrices requires time-consuming multi-precision arithmetic as the double-precision arithmetic suffers from heavy catastrophic cancellations.

To address these issues, Sumiya et al. (2015, 2017) presented the *rate constant matrix contraction* (RCMC) method for approximately solving master equations described by the canonical or microcanonical ensemble. The RCMC method relies on neither finite difference, multi-precision arithmetic, nor subtraction of like-sign numbers. The original description (Sumiya et al. 2015; 2017) of the RCMC method was based on chemical intuitions. Subsequently, Iwata et al. (2023) provided a mathematical reformulation of the RCMC method, showing that it works for the master Eq. (1) with any coefficient matrix K satisfying (RCM1–3) and performing a theoretical error analysis. Recent work (Harabuchi et al. 2024) applies the reformulated RCMC method to the sensitivity analysis of complicated reaction path networks.

The reformulated procedure of the RCMC method is outlined as follows (Iwata et al. 2023, Sect. 3). The input of the RCMC method is a rate constant matrix K, an initial yield vector p, and the end time $t_{\max}$ of the simulation. The RCMC method runs in two steps: Steps 1 and 2. Step 1 constructs a sequence of growing sets $\emptyset = S^{(0)} \subsetneq S^{(1)} \subsetneq \cdots \subsetneq S^{(k)} \subsetneq V$ of EQs with $0 \leq k \leq n$ in a greedy fashion. This procedure is essentially the pivoted Gaussian elimination for Cholesky decomposition on $L = -K\Pi$ with a minor modification to avoid subtraction of like-sign numbers. Intuitively, every EQ $v \in S^{(j)}$ is regarded as "steady" after some time $t^{(j)}$, i.e., $\dot{x}_v(t) \approx 0$ for $t \geq t^{(j)}$. On the other hand, EQs in $T^{(j)} := V \setminus S^{(j)}$ are said to be "transient". The value k is the number of EQs that are regarded as steady at $t_{\max}$. In typical use such as $t_{\max} = 86,400$ sec (= one day), k roughly satisfies $0.4n \lesssim k \lesssim 0.7n$. The idea of bipartitioning EQs by steadiness comes from the classical *quasi-steady state approximation* (QSSA).

Step 2 of the RCMC method computes a vector $q^{(j)}$ and the time $t^{(j)}$ for $j = 1, \ldots, k$ by explicit matrix calculations. It is guaranteed that $0 < t^{(1)} < \cdots < t^{(k)} \leq t_{\max}$ and each $q^{(j)}$ is a non-negative vector with $\sum_{v \in V} q_v^{(j)} = \sum_{v \in V} p_v$. The value $t^{(j)}$ is a reference time when the EQs in $S^{(j)}$ and $T^{(j)} := V \setminus S^{(j)}$ are regarded as steady and transient, respectively, and the vector $q^{(j)}$ serves as an approximation to $x(t^{(j)})$. This is a "full" scheme to obtain a rough sketch of the entire trajectory of the yield vector $x(t)$ through $t \in [0, t_{\max}]$. If only the yield vector $x(t_{\max})$ at the last moment $t_{\max}$ is needed, it suffices to compute only the last vector $q^{(k)}$. We call the former and the latter way the `full` and `last` options, respectively. See Sect. 2.1 for more on the RCMC method.

The advantages of the RCMC method lie in its numerical stability and running time. Regarding numerical stability, the RCMC method is free from subtractions of like-sign numbers, and thus, catastrophic cancellation never occurs. In fact, it is reported in (Iwata et al. 2023, Sect. 6) that using double-precision floating-point numbers is enough for obtaining a good approximation to $x(t)$. The paper (Iwata et al. 2023) also observed that the RCMC method with `full` option runs within seconds for rate constant matrices with $n \approx 1700$.[1]

[1] Several versions of the RCMC methods are presented in (Iwata et al. 2023). In this paper, we employ the standard Type A method with the "diag" option, in which the RCMC method coincides with the original description in (Sumiya et al. 2015, 2017).

1.2 Aim of This Paper

The aim of this paper is to speed up the (reformulated) RCMC method. This is motivated by the fact that large-scale rate constant matrices with $n \gtrsim 10,000$ are recently obtained by GRRM and AFIR. For such large-scale matrices, the RCMC method took more than 5 min, even with the faster `last` option (see Sect. 6). Since the RCMC method is repeatedly applied in the kinetics-based navigation in AFIR (Sumiya and Maeda 2020), its speeding up is crucial for accelerating the entire procedure of AFIR.

Step 1 of the RCMC method runs in $\Theta(kn^2)$ time. Step 2 takes $\Theta(k^3)$ and $\Theta(k^2)$ times for `full` and `last` options, respectively, assuming that $k = \tilde{\Omega}(\sqrt{n})$ and K is a sparse matrix, i.e., the number of non-zero entries in K is $\tilde{\Theta}(n)$ (see Sect. 2.2). Here, $\tilde{\Omega}$ and $\tilde{\Theta}$ hides logarithmic factors. This means that Step 1 is the bottleneck of the RCMC method. We have also confirmed in numerical experiments that Step 1 actually takes a large portion of the running time for small k; see Sect. 6.

Our starting point for accelerating Step 1 lies in its connection to the greedy algorithm for *maximum a posteriori* (MAP) *inference* for *determinantal point processes* (DPPs) under cardinality constraints. A DPP is a probability distribution on the power set 2^V of a finite set $V = \{1, \dots, n\}$ determined from an $n \times n$ positive semi-definite matrix L (Macchi 1975). The probability of observing $X \subseteq V$ is proportional to $\det L_{XX}$, where L_{XX} denotes the principal submatrix of L with rows and columns indexed by X. The MAP inference for DPPs under the cardinality constraint, which is to compute X with maximum $\det L_{XX}$ subject to $|X| = k$, has many applications in machine learning, such as recommendation systems (Chen et al. 2018), document summarization (Kulesza and Taskar 2012), and diverse molecular selection (Nakamura et al. 2022). While exact solving the MAP inference is NP-hard (Ko et al. 1995), a greedy algorithm provides a good approximation due to the submodularity of the function $X \mapsto \log\det L_{XX}$; see Sect. 2.3 for detail.

We then observe in Sect. 2.4 that Step 1 of the RCMC method is essentially equivalent to a slight generalization of the greedy MAP inference for DPPs. Therefore, any implementation of the greedy MAP inference for DPPs can be used as that of Step 1. We refer to the original implementation (Iwata et al. 2023) of Step 1 as GREEDY. Chen et al. (2018) presented an algorithm, called FASTGREEDY in (Hemmi et al. 2022), for faster implementation of greedy MAP inference, improving the running time from $\Theta(kn^2)$ to $\Theta(k^2n)$. FASTGREEDY is based on Doolittle's algorithm (Doolittle 1878) for Cholesky decomposition, which computes entries of the Cholesky factor directly. Combining FASTGREEDY with the lazy greedy algorithm (Minoux 1978) for cardinality-constrained submodular function maximization, Hemmi et al. (2022) proposed a practically faster implementation, called LAZYFASTGREEDY. Intuitively, LAZYFASTGREEDY reduces the running time by delaying the computation of entries in the Cholesky factor that are not immediately needed. The running time of LAZYFASTGREEDY depends on how well this lazy heuristic works. In the best case, LAZYFASTGREEDY runs in $\Theta(k^3)$ time and is not slower than FASTGREEDY even in the worst case (up to logarithmic terms). In our numerical experiment, FASTGREEDY ran faster than GREEDY, and LAZYFASTGREEDY outperformed FASTGREEDY in real instances. See Sect. 3 for details on FASTGREEDY and LAZYFASTGREEDY.

Unfortunately, however, both FASTGREEDY and LAZYFASTGREEDY suffer from numerical instability. In fact, our numerical experiment reported in Sect. 6 indicates that the set $S^{(j)}$ computed by these two algorithms with double-precision floating-point numbers was incorrect at some j because of the accumulated cancellation errors.

1.3 Our Contributions

To achieve fast and numerically stable computation, we present a modification of LAZYFASTGREEDY, termed STABLELAZYFASTGREEDY. As stated above, Step 1 of the RCMC method is essentially the Cholesky decomposition of the graph Laplacian matrix $L = -K\Pi$. The only point where LAZYFASTGREEDY may subtract like-sign numbers is in the computation of diagonal entries of the Cholesky factor C of L. Due to the conditions (RCM1–3) on K, the Cholesky factor C has the property that its off-diagonal entries are non-positive, and every diagonal entry is the minus of the sum of the off-diagonal entries with the same column. Therefore, we can retrieve each vth diagonal entry of C by summing all the non-zero off-diagonal entries in the same column, i.e., $C_{vv} = -\sum_{v \in T'} C_{uv}$ with $T' := T^{(j)} \setminus \{v\}$. As $C_{uv} \leq 0$, this way of computing C_{vv} does not involve additions of opposite-sign numbers. This is, however, incompatible with the approach of LAZYFASTGREEDY, which benefits from delaying the computation of off-diagonal entries of C.

Our idea for overcoming this issue is to consider the *compression* $L^{[j,v]}$ of L by T', a matrix obtained by aggregating the rows (columns) of L indexed by T' into a single row (column). The compression $L^{[j,v]}$ has the properties that (i) the Cholesky factor $C^{[j,v]}$ of $L^{[j,v]}$ is the compression of C by T' and (ii) $L^{[j,v]}$ is again a graph Laplacian matrix. The first property implies that we can obtain C_{vv} as the negative of an off-diagonal entry $C^{[j,v]}_{\star v}$, where $\star$ denotes the index of the compressed row. Furthermore, the second property enables us to regard the compression $L^{[j,v]}$ as an input matrix instead of L and obtain $C^{[j,v]}_{\star v}$ in the same way as other off-diagonal entries of the Cholesky factor. If the lazy heuristics work well, this reduces the computational cost compared to computing C_{uv} for all $u \in T'$.

In implementing STABLELAZYFASTGREEDY, the values of compressed entries $L^{[j,v]}_{\star w} = \sum_{u \in T'} L_{uw}$ for all $w \in S^{(j)} \cup \{v\}$ are required for computing $C^{[j,v]}_{\star v}$. To calculate the compressed entries only with additions of non-negative numbers, we employ a data structure called a *segment tree* (or a *range tree*) (Chazelle 1988). A segment tree manages a one-dimensional array of n entries of a semigroup and allows $\mathrm{O}(\log n)$-time queries for executing the semigroup operation on a consecutive subarray and updating an array element. Using segment trees regarding the set of non-negative reals with the addition as a semigroup, we can compute the compressed entries for successive $j = 0, 1, 2, \ldots$ efficiently.

Table 1 Comparison of the running time and the numerical stability. In running time analyses, $k = \Omega(\mathrm{nnz}(K)^{1/3})$ is assumed, where $\mathrm{nnz}(K) = \Omega(n)$ denotes the number of non-zero entries in K. Check marks on the stability column mean that the method does not cause catastrophic cancellations by avoiding subtracting reals close to each other

Method	Lazy	Relax	Running time	Stability
GREEDY (Iwata et al. 2023)	–	–	$\Theta(kn^2)$	✓
FASTGREEDY (Chen et al. 2018)	–	–	$\Theta(k^2n)$	
LAZYFASTGREEDY (Hemmi et al. 2022)	Best	–	$\Theta(k^3)$	
	Worst	–	$\Theta(k^2n)$	
STABLELAZYFASTGREEDY	best	–	$\Theta(k^3)$	✓
	Worst	–	$\Theta(k^3n)$	✓
RELAXEDSTABLELAZYFASTGREEDY	Best	Best	$\Theta(k^3)$	✓
	Best	Worst	$\Theta(k^3)$	✓
	Worst	Best	$\Theta(k^2n)$	✓
	Worst	Worst	$\Theta(k^3n)$	✓

If the lazy heuristics works the best, STABLELAZYFASTGREEDY runs in $\Theta(k^3)$ time (ignoring logarithmic factors), which is as asymptotically fast as LAZYFASTGREEDY. However, if the lazy heuristics works poorly, the worst-case time complexity swells up to $\Theta(k^3n)$. To reduce this complexity in practice, we introduce RELAXEDSTABLELAZYFASTGREEDY, which implements additional heuristics called the *relaxing heuristics*. Relaxing heuristics aims to accelerate the algorithm by allowing the subtraction of like-sign numbers if their ratio is smaller than a prescribed threshold. To bound errors that may occur by subtractions, for two positive numbers $\hat{a}$ and $\hat{b}$ containing errors from their true values a and b, respectively, we give an upper bound on the relative error of $\hat{a} - \hat{b}$ to the true difference $a - b$. Through our error analysis, we can determine the threshold depending on the error tolerance. When the relaxing heuristics works well, i.e., much computation has been done with subtractions, RELAXEDSTABLELAZYFASTGREEDY runs in $\Theta(k^2n)$ time even if the lazy heuristics work the worst. We summarize the methods, their running times, and numerical stability in Table 1.

Finally, we numerically compare the methods with double-precision arithmetic for six rate constant matrices obtained from chemical reaction path networks. We observe that LAZYFASTGREEDY and FASTGREEDY fail before $t^{(j)}$ reaches $t_{\max} = 86,400$ sec due to catastrophic cancellations, whereas the other three methods succeed. For the largest data with $n = 12{,}215$ and $k = 8056$ corresponding to $t_{\max} = 86{,}400$ sec, GREEDY and RELAXEDSTABLELAZYFASTGREEDY (with a reasonable error tolerance) took 1041 seconds and 154 seconds, respectively. That is, RELAXEDSTABLELAZYFASTGREEDY ran about seven times as fast as GREEDY. We also observe that both the lazy and relaxing heuristics work well in practice by measuring the number of arithmetic operations omitted due to the heuristics.

1.4 Organization

The remainder of this paper is organized as follows. Section 2 describes the RCMC method and its connection to the greedy MAP inference for DPPs. Section 3 explains FastGreedy and LazyFastGreedy, providing their running time analyses. We then present StableLazyFastGreedy in Sect. 4 and RelaxedStableLazyFastGreedy in Sect. 5. Section 6 provides results on numerical experiments. Finally, Sect. 7 concludes this paper.

2 RCMC Method and Greedy MAP Inference for DPPs

This section introduces the RCMC method in Sect. 2.1 and discusses its running time and numerical stability in Sect. 2.2. Then, Sect. 2.3 describes the greedy MAP inference for DPPs, and Sect. 2.4 explains its relationship to the RCMC method.

Notations

Let $\mathbb{R}$, $\mathbb{R}_{\geq 0}$, and $\mathbb{R}_{>0}$ denote the set of reals, non-negative reals, and positive reals, respectively. For an non-negative integer n, we denote $[n] := \{1, \ldots, n\}$ and $[0, n] := \{0, 1, \ldots, n\}$. Note that $[0] := \varnothing$.

For a finite set U, let $\mathbb{R}^U$ denote the set of all $|U|$-dimensional real vectors whose components are indexed by U. Similarly, for finite sets U and V, we denote by $\mathbb{R}^{U\times V}$ the set of real matrices of size $|U| \times |V|$ with rows and columns indexed by U and V, respectively. We simply denote $\mathbb{R}^n := \mathbb{R}^{[n]}$, $\mathbb{R}^{U\times m} := \mathbb{R}^{U\times[m]}$, and $\mathbb{R}^{n\times m} := \mathbb{R}^{[n]\times[m]}$. For a matrix $A \in \mathbb{R}^{U\times V}$, let $A_{I,J}$ and A_{IJ} denote the submatrix of A whose rows and columns are indexed by $I \subseteq U$ and $J \subseteq V$, respectively. We simply denote $A_{\{i\}J}$, $A_{I\{j\}}$, and $A_{\{i\}\{j\}}$ as A_{iJ}, A_{Ij}, and A_{ij}, respectively. For a vector $a \in \mathbb{R}^U$, we denote by a_I its subvector indexed by $I \subseteq U$. In particular, if $I = \{i\}$, we denote it by a_i.

For two (row or column) vectors $a, b \in \mathbb{R}^U$, we mean their (standard) inner product by $\langle a, b\rangle := \sum_{i\in U} a_i b_i$. We denote the 2-norm of $a \in \mathbb{R}^U$ by $\|a\| := \sqrt{\langle a, a\rangle}$. Let $\mathbf{1}$ denote the all-one vector of an appropriate dimension. For a square matrix $L \in \mathbb{R}^{V\times V}$, let $\operatorname{diag}(L)$ denote the vector in $\mathbb{R}^V$ obtained by extracting diagonal entries of L. Conversely, for a vector $d \in \mathbb{R}^V$, let $\operatorname{Diag}(d)$ denote the $V \times V$ diagonal matrix obtained by arranging the components of d into the diagonals.

The number of non-zero entries of a matrix A is denoted by $\operatorname{nnz}(A)$. We say that a matrix $A \in \mathbb{R}^{U\times V}$ is *dense* if $\operatorname{nnz}(A_{IJ}) = \Theta(|I||J|)$ and *sparse* if $\operatorname{nnz}(A_{IJ}) = \tilde{\Theta}(|I| + |J|)$ for any $I \subseteq U$ and $J \subseteq V$, where $\tilde{\Theta}(\cdot)$ hides logarithmic factors.

Algorithm 1 RCMC method (Iwata et al. 2023; Sumiya and Maeda 2020)

Input: rate constant matrix $K \in \mathbb{R}^{V \times V}$, initial value $p \in \mathbb{R}^V$, $t_{\max} \in \mathbb{R}_{>0}$, opt $\in \{\texttt{full}, \texttt{last}\}$
Output: $t^{(1)}, \ldots, t^{(k)} \in \mathbb{R}_{>0}$ and $q^{(1)}, \ldots, q^{(k)} \in \mathbb{R}^V$ for some $k \in [0, n)$
1: $S^{(0)} := \varnothing$, $T^{(0)} := V$, $K^{(0)} := K$
2: **for** $j = 1, \ldots, n$ **do**
3: $s \leftarrow s^{(j)} := \arg\max \left\{ -K_{vv}^{(j-1)} \middle| v \in T^{(j-1)} \right\}$ ▷ Ties are consistently broken
4: **if** $-K_{ss}^{(j-1)} < 1/t_{\max}$ **then**
5: $k := j - 1$ and **break**
6: $S^{(j)} := S^{(j-1)} \cup \{s\}$, $T^{(j)} := T^{(j-1)} \setminus \{s\}$
7: $K_{uv}^{(j)} := K_{uv}^{(j-1)} - K_{us}^{(j-1)} K_{sv}^{(j-1)} / K_{ss}^{(j-1)}$ for $u, v \in T^{(j)}$ ▷ Use (4) for diagonals
8: Compute $t^{(j)}$ and $q^{(j)}$ by (3) for $j \in [k]$ if opt $=$ `full` and for $j = k$ if opt $=$ `last`

2.1 RCMC Method

Throughout this paper, V is a finite set of cardinality n. A *rate constant matrix* over V is an $n \times n$ matrix $K \in \mathbb{R}^{V \times V}$ satisfying 1–3. The vector $\pi = (\pi_v)_{v \in V}$ in (RCM3) is called a *stationary distribution*. If we set $\Pi := \mathrm{Diag}(\pi)$, we can express a rate constant matrix as $K = -L\Pi^{-1}$, where $L \in \mathbb{R}^{V \times V}$ is a symmetric matrix with $L_{uv} \leq 0$ for $u, v \in V$ where $u \neq v$ and $L_{vv} = -\sum_{u \neq v} L_{uv}$ for $v \in V$. Such a matrix L is positive semi-definite (PSD) and is called a (*weighted*) *graph Laplacian matrix*.

Consider the master Eq. (1) defined by a rate constant matrix K with an initial condition $x(0) = p \in \mathbb{R}^n$ over a time interval $t \in [0, t_{\max}]$. The *rate constant matrix contraction* (RCMC) *method* (Iwata et al. 2023; Sumiya and Maeda 2020), described in Algorithm 1, computes an approximate solution to the master equation. The RCMC method consists of the following two steps:

Step 1: greedy selection of steady-states (Lines 1–7),
Step 2: computation of approximate solutions (Line 8).

To distinguish Lines 1–7 in Algorithm 1 and other implementations of Step 1 (Algorithms 2–4) that will be presented later, we refer to Lines 1–7 as Greedy. Greedy constructs a growing sequence of subsets $\emptyset = S^{(0)} \subsetneq S^{(1)} \subsetneq \cdots \subsetneq S^{(k)} \subseteq V$ with $S^{(j)} = \{s^{(1)}, \ldots, s^{(j)}\}$, where k denotes the number of iterations. Each element $s = s^{(j)}$ is called a *steady-state* in expectation of $\dot{x}_s(t) \approx 0$ for $t \geq t^{(j)}$, where $t^{(j)}$ is the *reference time* that will be computed in Step 2 later. Conversely, elements in $T^{(j)} := V \setminus S^{(j)}$ are said to be *transient*. Starting with $S^{(0)} = \varnothing$, $T^{(0)} = V$ and $K^{(0)} := K$, in the jth iteration, the algorithm takes the element $s^{(j)} \in T^{(j-1)}$ with the largest diagonal entry of $K^{(j-1)}$ in the absolute value at Line 3. Here, we assume that ties are consistently broken by choosing the smallest element with respect to a fixed total order on V, including all the subsequent algorithms. Then, the algorithm moves $s^{(j)}$ from $T^{(j-1)}$ to $S^{(j-1)}$ (Line 6) and modifies $K^{(j-1)}$ into $K^{(j)}$ by a rank-one update (Line 7). The matrix $K^{(j)}$ is the Schur complement of $S^{(j)}$ in $K = K^{(0)}$ (the original rate constant matrix), i.e.,

$$K^{(j)} = K_{TT} - K_{TS}K_{SS}^{-1}K_{ST} = -(L_{TT} - L_{TS}L_{SS}^{-1}L_{ST})\Pi_{TT}^{-1} \tag{2}$$

holds with $S = S^{(j)}$ and $T = T^{(j)}$. Since the Schur complement $L_{TT} - L_{TS}L_{SS}^{-1}L_{ST}$ of S in L is again a graph Laplacian matrix, $K^{(j)}$ is also a rate constant matrix, which is formally stated as follows.

Proposition 1.1 (Iwata et al. 2023) *The matrix $K^{(j)} \in \mathbb{R}^{T^{(j)} \times T^{(j)}}$ satisfies the axioms (RCM1–3) of rate constant matrices on $T^{(j)}$.*

The iteration of GREEDY ends when all the diagonal entries in $-K^{(j)}$ are less than $1/t_{\max}$.

Step 2 computes the reference time $t^{(j)} \in \mathbb{R}_{>0}$ and a vector $q^{(j)} \in \mathbb{R}^V$ according to the following formulas[2]:

$$t^{(j)} := -\frac{1}{K_{ss}^{(j)}}, \quad q^{(j)} = \begin{pmatrix} q_S^{(j)} \\ q_T^{(j)} \end{pmatrix} := \begin{pmatrix} K_{SS}^{-1}K_{ST}UK_{TS}K_{SS}^{-1} & -K_{SS}^{-1}K_{ST}U \\ -UK_{TS}K_{SS}^{-1} & U \end{pmatrix} \begin{pmatrix} p_S \\ p_T \end{pmatrix}, \tag{3}$$

where $s = s^{(j)}$, $S = S^{(j)}$, $T = T^{(j)}$, $K = K^{(0)}$ is the original rate constant matrix, and $U := \mathrm{Diag}(\mathbf{1}^\top + \mathbf{1}^\top K_{TS}K_{SS}^{-2}K_{ST})$. The reference times $t^{(j)}$ satisfy $0 \le t^{(1)} \le \cdots \le t^{(k)} \le t_{\max}$ and capture times when the solution $x(t)$ changes significantly. Every $q^{(j)}$ serves as a rough approximation of $x(t^{(j)})$; for the accuracy analysis, see (Iwata et al. 2023, Sect. 4).

An argument "opt" controls the set of j for which $(t^{(j)}, q^{(j)})$ is computed in Step 2. The algorithm computes the entire sequence $(t^{(0)}, q^{(0)}), \ldots, (t^{(k)}, q^{(k)})$ if opt $=$ `full` and only $(t^{(k)}, q^{(k)})$ if opt $=$ `last`. The former is used when it is necessary to capture the entire trajectory $\{ (t, x(t)) \mid 0 \le t \le t_{\max} \}$, while the latter is used when it is enough only to obtain the last value $x(t_{\max})$, reducing the computational cost of Step 2.

2.2 *Efficiency and Numerical Stability of the RCMC Method*

Here, we analyze the running time of the RCMC method. For an efficient implementation, we hold $K^{(j)}$ as a dense matrix since we cannot expect $K^{(j)}$ to be sparse for larger j due to fill-ins even if K is sparse. Then, the time complexity of Algorithm 1 is analyzed as follows.

Theorem 1.1 Greedy *runs in $\Theta(kn^2)$ time and Step 2 runs in time summarized in Table 2.*

[2] This formula for $t^{(j)}$, called "diag" in (Iwata et al. 2023), is used in the original RCMC method by Sumiya et al. (2015). More accurate, albeit complicated, formulas are presented in (Iwata et al. 2023).

Table 2 Running time of Step 2

Sparsity of K	opt	Running time
Dense	`full`	$\Theta(k^2 n)$
Dense	`last`	$\Theta(kn)$
Sparse	`full`	$\Theta(k^3) + \tilde{\Theta}(kn)$
Sparse	`last`	$\Theta(k^2) + \tilde{\Theta}(n)$

Proof It is clear that the bottleneck of GREEDY is Line 7, which takes $\Theta(kn^2)$ time. We analyze the running time of Step 2. As described in (Iwata et al. 2023, Sect. 5.1), the LU decomposition of K_{SS} can be immediately obtained from the Schur complements of K. Since we can compute $q^{(j)}$ in $\Theta(j^2 + \mathrm{nnz}(K_{S^{(j)}T^{(j)}}))$ time for every $j \in [k]$, the computational costs of Step 2 listed in Table 2 are obtained. □

From Theorem 1.1 and Table 2, the total running time of the RCMC method is $\Theta(kn^2)$ regardless of the sparsity of K and the option.

Iwata et al. (2023, Sect. 5.2) gave the following observation on the numerical stability of the RCMC method. By Proposition 1.1, $K_{ss}^{(j-1)} < 0$ and $K_{us}^{(j-1)}, K_{sv}^{(j-1)} \geq 0$ hold for any $u, v \in T^{(j)}$. Thus, the fraction $K_{us}^{(j-1)} K_{sv}^{(j-1)} / K_{ss}^{(j-1)}$ is non-positive. As a result, the subtraction of $K_{us}^{(j-1)} K_{sv}^{(j-1)} / K_{ss}^{(j-1)}$ from $K_{uv}^{(j-1)}$ at Line 7 for $u \neq v$ is indeed the addition of two nonnegative numbers, ensuring the operation numerically stable. On the other hand, in the case of $u = v$, both $K_{vv}^{(j-1)}$ and $K_{vs}^{(j-1)} K_{sv}^{(j-1)} / K_{ss}^{(j-1)}$ become non-positive. Thus, their subtraction may cause a catastrophic cancellation. For this problem, the following straightforward remedy was proposed in (Iwata et al. 2023): first fill out the off-diagonal entries of $K^{(j)}$, and then, calculate every vth ($v \in T^{(j)}$) diagonal entry $K_{vv}^{(j)}$ as

$$K_{vv}^{(j)} \leftarrow - \sum_{u \in T^{(j)} \setminus \{v\}} K_{uv}^{(j)}. \tag{4}$$

Since $K^{(j)}$ satisfies (RCM2), (4) computes the same value as Line 7, without increasing the time complexity. In addition, all the summands in (4) are nonnegative by (RCM1). Hence, (4) provides a numerically stable implementation of Line 7. Line 8 is also free from subtractions of like-sign numbers (Iwata et al. 2023, Sect. 5.2).

2.3 Greedy MAP Inference for DPPs

Let $L \in \mathbb{R}^{V \times V}$ be a positive semi-definite (PSD) matrix. A *determinantal point process* (DPP) (Macchi 1975) with a kernel matrix L is a probability distribution on 2^V such that the probability of obtaining every $X \subseteq V$ is proportional to $\det L_{XX}$. The *maximum a posteriori* (MAP) *inference* for DPPs (under the cardinality constraint) is the problem of finding $X \subseteq V$ with $|X| = k$ that maximizes $\det L_{XX}$, where $k \in [0, n]$ is a prescribed parameter. This problem is equivalent to maximizing a function $f_L \colon 2^V \to \mathbb{R} \cup \{-\infty\}$ defined by

$$f_L(X) := \log\det L_{XX} \quad (X \subseteq V), \tag{5}$$

where $\log 0$ is set to $-\infty$, under $|X| \leq k$. The function f_L is known to be *submodular* (Fan 1968). Here, a function $f \colon 2^V \to \mathbb{R} \cup \{-\infty\}$ is called *submodular* if $f(v \mid X) \geq f(v \mid Y)$ for all $X, Y \subseteq V$ with $X \subseteq Y$ and $v \in V \setminus Y$, where $f(v \mid X) := f(X \cup \{v\}) - f(X)$ denotes the *marginal gain* of $v \in V \setminus X$ under $X \subseteq V$.

Maximizing a submodular function $f \colon 2^V \to \mathbb{R} \cup \{-\infty\}$ under the cardinality constraint $|X| \leq k$ requires exponentially many queries of function values in general (Nemhauser and Wolsey 1978), and even the MAP inference for DPPs is NP-hard (Ko et al. 1995). Thus, polynomial-time approximation algorithms are commonly used for this problem. The standard *greedy method* works as follows: starting from $S^{(0)} := \varnothing$, in each jth ($j \in [k]$) iteration, take $s^{(j)} := \arg\max \left\{ f(v \mid S^{(j-1)}) \,\middle|\, v \in T^{(j-1)} \right\}$ with $T^{(j-1)} := V \setminus S^{(j-1)}$ and let $S^{(j)} := S^{(j-1)} \cup \left\{ s^{(j)} \right\}$. The greedy method evaluates f a total of $\Theta(kn)$ times, and enjoys a $(1 - 1/\mathrm{e})$-approximation guarantee if f is *monotone*, i.e., $f(X) \leq f(Y)$ for $X \subseteq Y$ (Nemhauser et al. 1978). Even if not, the greedy method is commonly adopted as a practical heuristic.

2.4 Relationship Between the RCMC Method and the Greedy DPP MAP Inference

Iwata et al. (2023) indicated that Greedy can be regarded as a variant of the greedy DPP MAP inference as follows. Let $K = -L\Pi^{-1} \in \mathbb{R}^{V \times V}$ be a rate constant matrix and $K^{(j-1)}$, $S^{(j-1)}$, and $T^{(j-1)}$ the values computed in Algorithm 1. At Line 3, the vth ($v \in T^{(j-1)}$) diagonal entry of $K^{(j-1)}$ can be expressed as

$$K_{vv}^{(j-1)} = \frac{\det K_{S\cup\{v\}, S\cup\{v\}}}{\det K_{SS}} = -\frac{\det L_{S\cup\{v\}, S\cup\{v\}}}{\pi_v \det L_{SS}} \tag{6}$$

by (Iwata et al. 2023, Appendix E), where $S = S^{(j-1)}$. Define functions $f_{\Pi^{-1}}$ and f_{-K} in the same way as f_L, i.e.,

$$f_{\Pi^{-1}}(X) := \log\det(\Pi^{-1})[X] = -\sum_{v \in X} \log \pi_v \quad (X \subseteq V), \tag{7}$$

$$f_{-K}(X) := \log\det(-K)[X] = \log\det L_{XX} - \sum_{v \in X} \log \pi_v = f_L(X) + f_{\Pi^{-1}}(X) \quad (X \subseteq V). \tag{8}$$

Then, f_{-K} is submodular since f_L is submodular and $f_{\Pi^{-1}}$ is modular (linear). By (6) and (8), we have the following proposition.

Proposition 1.2 (Iwata et al. 2023, Appendix E) *For $j \in [k]$ and $v \in T^{(j-1)}$,*

$$f_{-K}(v \mid S^{(j-1)}) = \log(-K_{vv}^{(j-1)}) \tag{9}$$

holds.

Proposition 1.2 implies that Step 1 of the RCMC method is essentially the greedy algorithm for maximizing the submodular function f_{-K}. In particular, if $\pi = \mathbf{1}$, it coincides with the greedy DPP MAP inference for a kernel matrix $L = -K$, in which only the stopping criterion is changed from $|S| = k$ to $-K_{ss}^{(j-1)} < 1/t_{\max}$. Algorithm 1 for general K is the variant of the greedy DPP MAP inference with the objective submodular function modified by adding the modular function $f_{\Pi^{-1}}$.

Algorithm 2 pivoted Gaussian elimination

Input: A PSD matrix $L \in \mathbb{R}^{V \times V}$ and $\varepsilon \in \mathbb{R}_{\geq 0}$
Output: A (partial) Cholesky factor C of L

$S^{(0)} := \varnothing$, $T^{(0)} := V$, $L^{(0)} := L$
for $j = 1, \ldots, n$ **do**
  $s \leftarrow s^{(j)} := \arg\max_{v \in T^{(j-1)}} L_{vv}^{(j-1)}$
  if $L_{ss}^{(j-1)} < \varepsilon$ **then**
    $k := j - 1$ and **break**
  $S^{(j)} := S^{(j-1)} \cup \{s\}$, $T^{(j)} := T^{(j-1)} \setminus \{s\}$
  $C_{sj} := \sqrt{L_{ss}^{(j-1)}}$
  $C_{uj} := L_{us}^{(j-1)} / C_{sj}$ for $u \in T^{(j)}$
  $L_{uv}^{(j)} := L_{uv}^{(j-1)} - L_{us}^{(j-1)} L_{sv}^{(j-1)} / L_{ss}^{(j-1)}$ for $u, v \in T^{(j)}$

Algorithm 3 pivoted Doolittle's algorithm

Input: A PSD matrix $L \in \mathbb{R}^{V \times V}$ and $\varepsilon \in \mathbb{R}_{\geq 0}$
Output: A (partial) Cholesky factor C of L

$S^{(0)} := \varnothing$, $T^{(0)} := V$, $d^{(0)} := \mathrm{diag}(L)$
for $j = 1, \ldots, n$ **do**
  $s \leftarrow s^{(j)} := \arg\max_{v \in T^{(j-1)}} d_v^{(j-1)}$
  if $d_s^{(j-1)} < \varepsilon$ **then**
    $k := j - 1$ and **break**
  $S^{(j)} := S^{(j-1)} \cup \{s\}$, $T^{(j)} := T^{(j-1)} \setminus \{s\}$
  $C_{sj} := \sqrt{d_s^{(j-1)}}$
  $C_{uj} := (L_{us} - \langle C_{uJ}, C_{sJ} \rangle) / C_{sj}$ for $u \in T^{(j)}$, where $J := [j-1]$
  $d_v^{(j)} := d_v^{(j-1)} - {C_{vj}}^2$ for $v \in T^{(j)}$

3 Accelerating the RCMC Method

As we described in Sect. 2, the bottleneck of the RCMC method is GREEDY for Step 1, which is essentially the same as the greedy DPP MAP inference. This section describes how we can accelerate Step 1 via existing faster implementations of the greedy DPP MAP inference. In Sect. 3.1, we introduce FASTGREEDY, which is an adaptation of the algorithm of Chen et al. (2018), through its connection to the Cholesky the decomposition. In Sect. 3.2, we describe a faster algorithm, LAZYFASTGREEDY, presented by Hemmi et al. (2022). We consider the time complexity and the numerical stability of LAZYFASTGREEDY in Sect. 3.3.

3.1 FastGreedy via Cholesky Decomposition

The (pivoted) *Cholesky decomposition* of a PSD matrix $L \in \mathbb{R}^{V \times V}$ is a factorization of L in the form of $L = CC^\top$, where $C = PG \in \mathbb{R}^{V \times n}$ is the product of a permutation matrix $P \in \mathbb{R}^{V \times n}$ and a lower-triangular matrix $G \in \mathbb{R}^{n \times n}$. Note that rows and columns of C are indexed by V and $[n]$, respectively. Algorithms 2 and 3 detail two methods for the Cholesky decomposition with a tolerance parameter ε. Both algorithms fill entries of C column-wise from left to right, treating unassigned entries of C as 0. Algorithm 2 is the standard Gaussian elimination with full pivoting. We call C computed by Algorithm 2 with $\varepsilon = 0$ the *Cholesky factor* of L with the order of pivots $s^{(1)}, \dots, s^{(k)}$, where $k = \operatorname{rank} L$ holds here. In particular, the corresponding permutation matrix P is given by $P_{s^{(j)} j} = 1$ for $j \in [k]$. The intermediate matrix $L^{(j)}$ is the Schur complement of $S^{(j)}$ in L. Algorithm 3 keeps track of only the diagonal entries of $L^{(j)}$ as a vector $d^{(j)}$, without explicitly computing off-diagonals of $L^{(j)}$. Note that C_{uJ} and C_{sJ} at Line 8 are (row) vectors of dimension $|J| = j - 1$ as u and s are single elements in V; hence their inner product is defined. Whereas the unpivoted version of Algorithm 3 originates from Doolittle (1878) and Gauss (1809), the origin of the pivoted version is unclear; for a detailed history of the Cholesky decomposition, see (Grcar 2011). Properties of Algorithms 2 and 3 are summarized below.

Theorem 1.2 *In Algorithms 2 and 3, the following hold:*

1. *both algorithms yield the same values for k, $s^{(j)}$ ($j \in [k+1]$), and C_{uj} ($j \in [k]$, $u \in T^{(j-1)}$),*
2. *$d_v^{(j)} = L_{vv}^{(j)}$ holds for $j \in [0, k]$ and $v \in T^{(j)}$.*

Although we believe that Theorem 1.2 is well-known in the literature, we provide its proof as we could not find exact references. For this, we first present the following lemma, which will also be used in the proof of Lemma 1.3.

Lemma 1.1 *Let k, C, $L^{(j)}$, and $T^{(j)}$ ($j \in [0, k]$) be the values computed by Algorithm 2 with $\varepsilon = 0$. Then, for $j \in [0, k]$ and $u, v \in T^{(j)}$, we have $L_{uv}^{(j)} = L_{uv} - \langle C_{uJ}, C_{vJ} \rangle$ with $J := [j]$.*

Proof Let $T = T^{(j)}$ and $S = V \setminus T$. By $L = CC^\top$, we have

$$\begin{pmatrix} L_{SS} & L_{ST} \\ L_{TS} & L_{TT} \end{pmatrix} = \begin{pmatrix} C_{SJ} & O \\ C_{TJ} & * \end{pmatrix} \begin{pmatrix} C_{SJ}^\top & C_{TJ}^\top \\ O & * \end{pmatrix} = \begin{pmatrix} C_{SJ} C_{SJ}^\top & C_{SJ} C_{TJ}^\top \\ C_{TJ} C_{SJ}^\top & * \end{pmatrix}, \tag{10}$$

where O denotes a zero matrix of appropriate size and $*$ means some matrix. Since $L^{(j)}$ is the Schur complement of S in L, we have

$$L^{(j)} = L_{TT} - L_{TS} L_{SS}^{-1} L_{ST} = L_{TT} - C_{TJ} C_{SJ}^\top (C_{SJ})^{\top^{-1}} (C_{SJ})^{-1} C_{SJ} C_{TJ}^\top = L_{TT} - C_{TJ} C_{TJ}^\top \tag{11}$$

as desired. □

Proof of Theorem 1.2 Let k, C, and $s^{(j)}$ ($j \in [k+1]$) denote the values computed by Algorithm 2, and $\tilde{k}$, $\tilde{C}$, and $\tilde{s}^{(j)}$ ($j \in [\tilde{k}+1]$) those by Algorithm 3. Let $S^{(j)} := \{s^{(1)}, \dots, s^{(j)}\}$ and $T^{(j)} := V \setminus S^{(j)}$ for $j \in [0, k]$, $\tilde{S}^{(j)} := \{\tilde{s}^{(1)}, \dots, \tilde{s}^{(j)}\}$ and $\tilde{T}^{(j)} := V \setminus \tilde{S}^{(j)}$ for $j \in [0, \tilde{k}]$. By induction on $j \in [0, \min\{k, \tilde{k}\}]$, we show $S^{(j)} = \tilde{S}^{(j)}$ (hence $T^{(j)} = \tilde{T}^{(j)}$), $C_{ul} = \tilde{C}_{ul}$ for $l \in [j]$ and $u \in T^{(l-1)} = \tilde{T}^{(l-1)}$, and $d_v^{(j)} = L_{vv}^{(j)}$ for $v \in T^{(j)} = \tilde{T}^{(j)}$. The claims are clear when $j = 0$. For $j \geq 1$, suppose that the claims hold for $j-1$ and consider the case of j. By $d_v^{(j-1)} = L_{vv}^{(j-1)}$ for $v \in T^{(j-1)}$, we have $s := s^{(j)} = \tilde{s}^{(j)}$ and hence the conditions $L_{ss}^{(j-1)} < \varepsilon$ and $d_s^{(j)} < \varepsilon$ are equivalent, meaning $k = \tilde{k}$ if the induction works correctly. If these conditions are met, we also have $S^{(j)} = \tilde{S}^{(j)}$ and $C_{sj} = \tilde{C}_{sj}$. For $u \in T^{(j)}$, by Lemma 1.1 and the inductive assumptions, we have

$$C_{uj} = \frac{L_{us}^{(j-1)}}{C_{sj}} = \frac{L_{us} - \langle C_{uJ}, C_{sJ} \rangle}{C_{sj}} = \frac{L_{us} - \langle \tilde{C}_{uJ}, \tilde{C}_{sJ} \rangle}{\tilde{C}_{sj}} = \tilde{C}_{uj}, \tag{12}$$

where $J := [j-1]$. Furthermore, for $v \in T^{(j)}$, we have

$$L_{vv}^{(j)} = L_{vv}^{(j-1)} - \frac{L_{vs}^{(j-1)2}}{L_{ss}^{(j-1)}} = L_{vv}^{(j-1)} - C_{vj} = d_v^{(j-1)} - \tilde{C}_{vj} = d_v^{(j)}, \tag{13}$$

completing the proof. □

The matrix $L^{(j)}$ in Algorithm 2 is the Schur complement of $S^{(j)}$ in $L = L^{(0)}$. Hence, if L is a graph Laplacian matrix, the following holds similarly to Proposition 1.1.

Proposition 1.3 *If L is a graph Laplacian matrix, so is $L^{(j)}$ for any $j \in [0, k]$.*

Proposition 1.3 immediately implies the following properties on the Cholesky factor of a graph Laplacian matrix.

Proposition 1.4 *Let L be a graph Laplacian matrix and C the Cholesky factor of L computed by Algorithms 2 and 3. Then, $C_{uj} \leq 0$ and $C_{s^{(j)}j} \geq 0$ hold for $j \in [k]$ and $u \in T^{(j)}$.*

Proof The claims hold from Proposition 1.3 and the equality $C_{uj} = L_{us}^{(j)} / \sqrt{L_{ss}^{(j)}}$ with $s = s^{(j)}$ for $u \in T^{(j)}$ from Line 8 in Algorithm 2. □

We get back to the RCMC method. In case of $\pi = \mathbf{1}$, Greedy for Step 1 is the same as Algorithm 2 applied to $L = -K$ and $\varepsilon = 1/t_{\max}$ except Algorithms 7 and 8. Since Step 1 with $\pi = \mathbf{1}$ and the greedy DPP MAP inference are equivalent and Algorithms 2 and 3 yield the same output by Theorem 1.2, Algorithm 3 can be used as an alternative implementation of the greedy DPP MAP inference, which is nothing but the *fast greedy* presented by Chen et al. (2018). We can also employ Algorithm 3 as an implementation for Step 1 of the RCMC method with a general rate constant matrix $K = -L\Pi^{-1}$ by modifying Algorithm 3 as follows:

- replace Line 3 with $s \leftarrow s^{(j)} := \arg\max \{d_v^{(j-1)}/\pi_v \mid v \in T^{(j-1)}\}$, and
- change the stopping criterion at Line 4 to $d_s^{(j-1)}/\pi_s < 1/t_{\max}$.

This algorithm, which we refer to as FASTGREEDY, yields the same result as GREEDY because of the equivalence of GREEDY and Algorithm 2, and that of Algorithms 2 and 3 due to Theorem 1.2. This improves the running time of Step 1 from $\Theta(kn^2)$ to $\Theta(k^2n)$, summarized as follows.

Theorem 1.3 FastGreedy *computes the same values for* k *and* $s^{(j)}$ ($j \in [k+1]$) *as* Greedy *in* $\Theta(k^2n)$ *time.*

Proof The equivalence of the algorithms is explained above. As for the running time, we take $\Theta(j)$ time to compute C_{uj} for each $j \in [k]$ and $u \in T^{(j)}$ at Line 8 in Algorithm 3. Hence, Line 8 takes $\Theta(k^2n)$ time in total. □

Note that FASTGREEDY does not impose extra computational cost in Step 2 as the LU decomposition of K_{SS} with $S = S^{(j)}$ is obtained from the Cholesky factor C_{SJ} ($J := [j]$) of L_{SS} as $K_{SS} = -C_{SJ} \cdot C_{SJ}^{\top} \Pi_{SS}^{-1}$. Since Step 2 can be done in $O(k^2n)$ time by Table 2, the total running time of the RCMC method with FASTGREEDY is improved to $\Theta(k^2n)$ regardless of the sparsity of K and the option of `full` or `last`.

3.2 *LazyGreedy and LazyFastGreedy*

The *lazy greedy* (Minoux 1978) is a practically faster implementation of the greedy method for maximizing a general submodular function $f : 2^V \to \mathbb{R} \cup \{-\infty\}$ under a cardinality constraint. In addition to the current solution $S^{(j-1)} \subseteq V$, the lazy greedy maintains an upper bound ρ_v on the marginal gain $f(v \mid S^{(j-1)})$ for each $v \in T^{(j-1)} := V \setminus S^{(j-1)}$. The value of ρ_v represents an "old" marginal gain of $v \in T^{(j-1)}$ in a sense that $\rho_v = f(v \mid S^{(b_v)})$ holds for some $b_v \leq j-1$, and the submodularity of f guarantees $f(v \mid S^{(j-1)}) \leq \rho_v$. At each jth iteration, the lazy greedy picks $u \in T^{(j-1)}$ with the largest ρ value and updates ρ_u to the current marginal gain $f(u \mid S^{(j-1)})$. If the updated ρ_u is still no less than ρ_v for any $v \in T^{(j-1)}$, then we can conclude that u has the largest marginal gain under $S^{(j-1)}$ without updating the others, hence we add u to $S^{(j-1)}$ to construct $S^{(j)}$. Otherwise, this procedure is repeated until an element with the largest marginal gain is found. The upper bounds $(\rho_v)_{v \in T^{(j-1)}}$ are efficiently managed by using a priority queue implemented by a binary heap. If the lazy update works the best, the method evaluates the objective function only $\Theta(n)$ times. While the worst-case time complexity of the lazy greedy remains unchanged up to slight additional costs of heap operations, it empirically outperforms the naïve greedy method in many real instances (Hemmi et al. 2022).

We can apply the lazy greedy to the greedy DPP MAP inference and Step 1 of the RCMC method since they are equivalent to the greedy maximization of the submodular functions f_L and f_{-K}, respectively. FASTGREEDY (Algorithm 3) and the lazy greedy were combined as the *lazy fast greedy* by Hemmi et al. (2022), which

Algorithm 4 LAZYFASTGREEDY

Input: rate constant matrix $K = -L\Pi^{-1} \in \mathbb{R}^{V \times V}$, initial vector $p \in \mathbb{R}^V$, $t_{\max} \in \mathbb{R}_{>0}$
Output: $t^{(1)}, \dots, t^{(k)} \in \mathbb{R}_{>0}$ and $q^{(1)}, \dots, q^{(k)} \in \mathbb{R}^n$

1: $j \leftarrow 1$, $S^{(0)} := \varnothing$, $T^{(0)} := V$, $d^{(0)} := \mathrm{diag}(L)$, $b_v \leftarrow 0$, $\rho_v \leftarrow d_v^{(0)}/\pi_v$ $(v \in V)$
2: **while** true **do**
3: $\quad u \leftarrow \arg\max \left\{\rho_v \middle| v \in T^{(j-1)}\right\}$
4: $\quad$ **for** $l = b_u + 1, \dots, j-1$ **do**
5: $\quad\quad C_{ul} := (L_{us^{(l)}} - \langle C_{uJ}, C_{s^{(l)}J} \rangle)/C_{s^{(l)}l}$ with $J := [l-1]$
6: $\quad d_u^{(j-1)} := d_u^{(b_u)} - \sum_{l=b_u+1}^{j-1} C_{ul}{}^2$ $\quad$ ▷ STABLELAZYFASTGREEDY modifies this line
7: $\quad b_u \leftarrow j-1$ and $\rho_u \leftarrow d_u^{(j-1)}/\pi_u$
8: $\quad$ **if** $\rho_u \geq \rho_v$ for all $v \in T^{(j-1)}$ **then**
9: $\quad\quad s \leftarrow s^{(j)} := u$
10: $\quad\quad$ **if** $\rho_u < 1/t_{\max}$ **then**
11: $\quad\quad\quad k := j-1$ and **break**
12: $\quad\quad S^{(j)} := S^{(j-1)} \cup \{s\}$, $T^{(j)} := T^{(j-1)} \setminus \{s\}$
13: $\quad\quad C_{sj} := \sqrt{d_s^{(j-1)}}$
14: $\quad\quad j \leftarrow j+1$

can be easily modified to the setting of the RCMC method as we did for Algorithm 3. Algorithm 4 details this variant, which we call LAZYFASTGREEDY. The value ρ_v for $v \in V$ is the exponential of an upper bound on the marginal gain $f_{-K}(v \mid S^{(j-1)})$ and b_v keeps track of when ρ_v is last updated. Specifically,

$$\rho_v = \exp(f_{-K})(v \mid S^{(b_v)}) = -K_{vv}^{(b_v)} = \frac{L_{vv}^{(b_v)}}{\pi_v} = \frac{d_v^{(b_v)}}{\pi_v} \tag{14}$$

holds from Proposition 1.2 and Theorem 1.2; see also (Hemmi et al. 2022, Proposition 2.1). The validity of LAZYFASTGREEDY is summarized as follows.

Theorem 1.4 LazyFastGreedy *(Algorithm 4) computes the same values for k, $s^{(j)}$ $(j \in [k+1])$ and $d_u^{(j)}$, C_{uj} $(u \in V,\ j \in [b_u])$ as* Greedy *and* FastGreedy, *where b_u $(u \in V)$ denotes the value at the end of Algorithm 4.*

Proof The claim follows from Theorem 1.3 and the validity of the lazy greedy (Minoux 1978). □

3.3 Runtime and Numerical Stability of LazyFastGreedy

We perform the running time analysis of LAZYFASTGREEDY, modifying (Hemmi et al. 2022, Theorem 3.2) to the setting of the RCMC method. For $v \in V$, let b_v be the value in Algorithm 4 at the end of the algorithm. In addition, we let c_j be the count of the **while** iterations of Lines 2–14 with $j \in [k+1]$.

Lemma 1.2 *The values of b_v and c_j satisfy the following.*

1. *We have* $b_{s^{(j)}} = j - 1$ *for* $j \in [k+1]$, $0 \le b_v \le k$ *for* $v \in T^{(k)}$, $c_1 = 1$, *and* $1 \le c_j \le n - j + 1$ *for* $j = 2, \dots, k+1$. *These lower and upper bounds are attained if the lazy heuristics works the best and worst, respectively.*
2. *The inequality* $\sum_{j=1}^{k+1} c_j - 1 \le \sum_{v \in V} b_v \le \sum_{j=1}^{k} (j-1) c_j$ *holds.*

Proof The statement 1 is clear from the algorithm. The first inequality in 2 follows from the facts that $\sum_{j=1}^{k+1} c_j$ is the total count of the **while** iterations, and every iteration increases some b_v by at least one, except the first iteration with $j = 1$. Similarly, the second inequality in 2 is due to the fact that each iteration with j increases some b_v by at most $j - 1$. □

Let M_{offdiag} be the total dimension of vectors for which the inner products are computed in Line 5. We can represent M_{offdiag} as

$$M_{\text{offdiag}} = \sum_{v \in V} \sum_{l=1}^{b_v} (l-1) = \frac{1}{2} \sum_{v \in V} b_v (b_v - 1). \tag{15}$$

Substituting into (15) the lower and upper bounds on b_v provided by Lemma 1.2, we obtain the following exact bounds on M_{offdiag}:

$$\frac{k(k-1)(k+1)}{6} \le M_{\text{offdiag}} \le \frac{k(k-1)(3n-2k-2)}{6}. \tag{16}$$

Theorem 1.5 (see (Hemmi et al. 2022, Theorem 3.2)) LazyFastGreedy *(Algorithm 4) runs in* $\Theta(M_{\text{offdiag}} + n) + \mathrm{O}\big(\sum_{j=1}^{k+1} c_j \log n\big)$ *time, which is* $\Theta(k^3 + n) + \mathrm{O}(k \log n)$ *and* $\Theta(k^2 n) + \mathrm{O}(kn \log n)$ *if the lazy heuristic works the best and worst, respectively.*

Proof The initialization at Line 1 takes $\Theta(n)$ time, including the heap construction. Algorithms 3 and 5 take $\mathrm{O}\big(\sum_{j=1}^{k+1} c_j \log n\big)$ time, respectively, and the other operations can be done within these times. The lower and upper bounds are obtained from Lemma 1.2 and (16). □

We will observe in Sect. 6 that LazyFastGreedy runs substantially faster than FastGreedy for real data.

We now consider the possibility of numerical cancellations in LazyFastGreedy. Recall that $C_{uj} \le 0$ holds for $u \in V$ and $j \in [b_u]$ by Proposition 1.4, and we also have $d_v^{(j)} = L_{vv}^{(j)} \ge 0$ for $v \in V$ and $j \in [0, k]$ since $L^{(j)}$ is a graph Laplacian matrix. Thus, the operation at Line 5 in Algorithm 4, which is the subtraction of $\langle C_{uJ}, C_{s^{(l)}J} \rangle$ from $L_{us^{(l)}}$, is free from subtractions of like-sign numbers. Unfortunately, in contrast, Line 6 involves subtracting two non-negative numbers, making the algorithm numerically unstable. In fact, we have confirmed that LazyFastGreedy with double-precision floating numbers output incorrect $s^{(j)}$ at some j for real data due to catastrophic cancellations; see Sect. 6.

4 Stabilizing LazyFastGreedy

This section presents a numerically stable implementation of LazyFastGreedy. In Sect. 4.1, we propose a stable subroutine used instead of Line 6 in Algorithm 4. We then describe its faster implementation via segment trees in Sect. 4.2 and its running time analysis in Sect. 4.3.

4.1 *StableLazyFastGreedy*

In Sect. 2.1, we have stabilized the computation of $K_{vv}^{(j)} = -L_{vv}^{(j)}/\pi_j = -d_v^{(j)}/\pi_j$ in Greedy (Line 7 in Algorithm 1) using the identity $\sum_{u \neq v} K_{uv}^{(j)} = 0$, which is equivalent to

$$d_v^{(j)} = - \sum_{u \in T^{(j)} \setminus \{v\}} L_{uv}^{(j)}. \tag{17}$$

We cannot directly apply the same method for the computation of $d_v^{(j)}$ in FastGreedy (Line 9 in Algorithm 3) and of $d_u^{(j-1)}$ in LazyFastGreedy (Line 6 in Algorithm 4) because some off-diagonals of $L^{(j)}$ will not be computed in (Lazy)FastGreedy.

Our first step for stabilizing (Lazy)FastGreedy is to rewrite the right-hand side of (17) using L and C as follows.

Lemma 1.3 *Let $j \in [k]$, $v \in T^{(j)}$, $T' := T^{(j)} \setminus \{v\}$, and $J := [j]$. Then, we have*

$$d_v^{(j)} = \sum_{u \in T'} \langle C_{uJ}, C_{vJ} \rangle - \sum_{u \in T'} L_{uv}. \tag{18}$$

Proof By (17) and Lemma 1.1, we have

$$\sum_{u \in T'} \langle C_{uJ}, C_{vJ} \rangle - \sum_{u \in T'} L_{uv}. \tag{19}$$

as desired. □

The summation in the right-hand side of (18) avoids subtraction of like-sign numbers thanks to Proposition 1.4. Hence, Lemma 1.3 can be used to stabilize FastGreedy because C_{ul} have been settled for all $u \in T^{(j)}$ and $l \in [j-1]$ at Line 9 in Algorithm 3. However, FastGreedy stabilized in this way is no longer faster than the RCMC method. Furthermore, LazyFastGreedy does not compute some entries of C due to lazy updates, hindering us from applying Lemma 1.3. Calculating the omitted entries, of course, would ruin the benefits derived from the lazy heuristic.

Algorithm 5 Stable computation of $d_u^{(j-1)}$

Input: $j \in [k]$, $u \in T^{(j-1)}$
Output: $d_u^{(j-1)}$

```
function StablyComputeDiagonal(j, u)
    for l = 1, ..., j − 1 do
        C^{[j,u]}_{⋆l} := (L^{[j,u]}_{⋆s^{(l)}} − ⟨C^{[j,u]}_{⋆J}, C_{s^{(l)}J}⟩)/C_{s^{(l)}l} with J := [l − 1]
    return ⟨C^{[j,u]}_{⋆J}, C_{uJ}⟩ − L^{[j,u]}_{⋆u} with J := [j − 1]
```

Our idea to alleviate this dilemma is to introduce a "compressed" graph Laplacian matrix defined below. Let $j \in [k], u \in T, S' = S^{(j-1)} \cup \{u\}$, and $T' = V \setminus S' = T^{(j-1)} \setminus \{u\}$. Recall that $\mathbf{1}$ denotes the all-one vector of an appropriate dimension. The *compression* of L with respect to u at the jth iteration is a $(j+1) \times (j+1)$ matrix $L^{[j,u]}$ is defined by

$$L^{[j,u]} = \begin{array}{c} \\ S' \\ \star \end{array} \begin{array}{c} \begin{array}{cc} S' & \star \end{array} \\ \begin{pmatrix} L_{S'S'} & L_{S'T'}\mathbf{1} \\ \mathbf{1}^\top L_{T'S'} & \mathbf{1}^\top L_{T'T'}\mathbf{1} \end{pmatrix} \end{array},$$

where the rows and columns of $L^{[j,u]}$ are indexed by a set $S' \cup \{\star\}$ with a new element $\star$. Namely, $L^{[j,u]}$ is a matrix obtained by aggregating rows and columns belonging to T' into a single row and column, respectively, indexed by $\star$. It is easily checked that $L^{[j,u]}$ is again a graph Laplacian matrix. We denote by $C^{[j,u]}$ the Cholesky factor of $L^{[j,u]}$ with the pivoting order $s^{(1)}, \ldots, s^{(j-1)}, u$. The following easy observation on $C^{[j,u]}$ plays an important role.

Lemma 1.4 *For $j \in [k]$, $u \in T^{(j-1)}$, and $l \in [j-1]$, we have $C^{[j,u]}_{\star l} = \mathbf{1}^\top C_{T'l}$, where $T' = T^{(j-1)} \setminus \{u\}$.*

Proof Immediately follows from Line 8 in Algorithm 2. □

By Lemmas 1.3 and 1.4, $d_u^{(j-1)}$ can be expressed by

$$d_u^{(j-1)} = \langle C^{[j,u]}_{\star J}, C_{uJ} \rangle - L^{[j,u]}_{\star u} \tag{20}$$

with $J = [j-1]$. Furthermore, as with Algorithm 3, for $l \in [j-1]$, the $(\star, l)$ entry $C^{[j,u]}_{\star l}$ can be computed from $C^{[j,u]}_{\star 1}, \ldots, C^{[j,u]}_{\star, l-1}$ as

$$C^{[j,u]}_{\star l} = (L^{[j,u]}_{\star s^{(l)}} - \langle C^{[j,u]}_{\star J}, C^{[j,u]}_{s^{(l)}J} \rangle)/C^{[j,u]}_{s^{(l)}l} = (L^{[j,u]}_{\star s^{(l)}} - \langle C^{[j,u]}_{\star J}, C_{s^{(l)}J} \rangle)/C_{s^{(l)}l} \tag{21}$$

with $J = [l-1]$.

Based on (20) and (21), we describe in Algorithm 5 a procedure, StablyComputeDiagonal(j, u), to compute $d_u^{(j-1)}$ via $C^{[j,u]}$. Since $C^{[j,u]}$ is the Cholesky factor of a graph Laplacian matrix $L^{[j,u]}$, we have $C^{[j,u]}_{\star l} \leq 0$ for $l \in [j-1]$ by Proposition 1.4, and thus Algorithm 5 avoids subtractions of like-sign numbers.

Algorithm 4 with Line 6 being replaced by StablyComputeDiagonal(j, u) is refereed to as StableLazyFastGreedy.

4.2 Efficient Implementation by Segment Trees

Naively implementing Algorithm 5 results in the running time of StablyComputeDiagonal$(j, \cdot)$ being $\Theta(j^2 + \mathrm{nnz}(L_{S'T'}))$. This is because the inner product calculation at Line 3 takes $\Theta(j^2)$ time in total, and computing $L^{[j,u]}_{\star v} = \mathbf{1}^\top L_{T'v}$ for all $v \in S'$ requires $\Theta(\sum_{v \in S'} \mathrm{nnz}(L_{vT'})) = \Theta(\mathrm{nnz}(L_{S'T'}))$ time. The $\Theta(\mathrm{nnz}(L_{S'T'}))$ term becomes a bottleneck if L is dense or j is small. To reduce this complexity while maintaining numerical stability, we employ a data structure known as a *segment tree* or *range tree* (Chazelle 1988). For ease of exposition, we assume $V = [n]$ in this section.

We explain segment trees. Let $(X, \oplus)$ be a semigroup; that is, $\oplus : X \times X \to X$ is a binary operation on a set X satisfying the associative low $a \oplus (b \oplus c) = (a \oplus b) \oplus c$ for all $a, b, c \in X$. As a computational model, assume that every element in X can be stored in a unit cell of memory, and the operation $\oplus$ can be performed in constant time. A segment tree $\mathcal{T}$ over X is a data structure that manages an array $(a_1, \dots, a_n)$ of n elements $a_1, \dots, a_n \in X$ and supports the following operations:

- $\mathcal{T}$.INIT$(a_1, \dots, a_n)$: initialize a new segment tree $\mathcal{T}$ by $a_1, \dots, a_n \in X$,
- $\mathcal{T}$.SUM(i, j): compute $a_i \oplus a_{i+1} \oplus \cdots \oplus a_j$ $(1 \le i \le j \le n)$,
- $\mathcal{T}$.UPDATE(i, a): update a_i to a $(i \in [n], a \in X)$.

$\mathcal{T}$.INIT$(a_1, \dots, a_n)$ takes $\Theta(n)$ time and the other two operations can be done in $\mathrm{O}(\log n)$ time.

We employ segment trees in StableLazyFastGreedy by regarding $\mathbb{R}_{\ge 0}$ as a semigroup $(\mathbb{R}_{\ge 0}, +)$. We maintain n segment trees $\mathcal{T}_v$ $(v \in V)$, each of which is associated with a column v of L. At every beginning of **while** loop in Algorithm 4, we ensure that

$$\mathcal{T}_v.\mathrm{SUM}(1, n) = \begin{cases} \sum_{u \in T^{(j-1)}} L_{uv} & (v \in S^{(j-1)}), \\ \sum_{u \in T^{(b_v)} \setminus \{v\}} L_{uv} & (v \in T^{(j-1)}). \end{cases} \tag{22}$$

First, each $\mathcal{T}_v$ is initialized by the vth column of L with the diagonal entry L_{vv} replaced with 0, i.e.,

$$\mathcal{T}_v.\mathrm{INIT}(L_{1v}, \dots, L_{v-1,v}, 0, L_{v+1,v}, \dots, L_{nv}), \tag{23}$$

which clearly satisfies (22) as $b_v = 0$ and $T^{(0)} = V$. When b_u is updated from b_u^{old} to $j-1$ at Line 7, we execute $\mathcal{T}_u$.UPDATE$(s^{(l)}, 0)$ for $l = b_u^{\mathrm{old}} + 1, \dots, j-1$. In addition, when $s \in V$ is removed from $T^{(j-1)}$ at Line 12, we execute $\mathcal{T}_v$.UPDATE$(s, 0)$ for all $v \in S^{(j-1)}$. These two updates keep (22) to hold.

By the invariant condition (22), for any $u \in T^{(j-1)}$ with $b_u = j-1$ and $v \in S^{(j-1)} \cup \{u\}$, we have

$$L_{\star v}^{[j,u]} = \sum_{w \in T^{(j-1)} \setminus \{u\}} L_{wv} = \mathcal{T}_v.\textsc{sum}(1, n) - L_{uv} = \mathcal{T}_v.\textsc{sum}(1, u-1) + \mathcal{T}_v.\textsc{sum}(u+1, n). \tag{24}$$

Hence, we can replace every computation of $L_{\star v}^{[j,u]}$ at Algorithms 3 and 4 in Algorithm 5 with the twice calls of $\mathcal{T}_v.\textsc{sum}$, which cost $\mathrm{O}(\log n)$ time. Furthermore, this computation avoids catastrophic cancellations since the segment tree performs only the operation of the semigroup $(\mathbb{R}_{\geq 0}, +)$.

For further speeding up on sparse matrices, we can restrict the entries of $\mathcal{T}_v$ to only the non-zeros in the vth column L_{Vv} of L to improve n to $\mathrm{nnz}(L_{Vv}) = \mathrm{nnz}(K_{Vv})$ in the query complexity.

4.3 Compelxity Analysis for StableLazyFastGreedy

We estimate the complexity of StableLazyFastGreedy. The value M_{offdiag}, which is the total dimension of vectors for which the inner products are computed at Line 5 in Algorithm 4, is still given by (15) in the same way as LazyFastGreedy. In addition to M_{offdiag}, the runtime of StableLazyFastGreedy also depends on the total dimension, denoted by M_{diag}, of vectors for which the inner products are computed at Algorithms 3 and 4 in Algorithm 5. Recall from Sect. 3.2 that c_j denotes the count of **while** iterations with j, which is the same as the number of calls of StablyComputeDiagonal$(j, \cdot)$. Using c_j, we can represent M_{diag} as

$$M_{\text{diag}} = \sum_{j=1}^{k+1} c_j \left(\sum_{l=1}^{j-1} (l-1) + j - 1 \right) = \frac{1}{2} \sum_{j=1}^{k+1} j(j-1) c_j. \tag{25}$$

By Lemma 1.2, we have

$$\frac{k(k+1)(k+2)}{6} \leq M_{\text{diag}} \leq \frac{k(k+1)(k+2)(4n-3k-1)}{24} \tag{26}$$

and the lower and upper bounds are attained if the lazy heuristics work the best and worst, respectively.

The running time of StableLazyFastGreedy can be described by using M_{diag} and M_{offdiag} as follows.

Theorem 1.6 StableLazyFastGreedy *runs in* $\Theta(M_{\text{offdiag}} + M_{\text{diag}} + \mathrm{nnz}(K)) + O(\sum_{j=1}^{k+1} jc_j \log n)\mathrm{O}(\sum_{j=1})^{k+1} jc_j \log n$ *time. If the lazy heuristic works the best* *and worst, the running time becomes* $\Theta(k^3 + \mathrm{nnz}(K)) + \mathrm{O}(k^2 \log n)$ *and* $\Theta(k^3 n + \mathrm{nnz}(K)) + \mathrm{O}(k^2 n \log n)$*, respectively.*

Proof The most costly parts of StableLazyFastGreedy are Line 5 in Algorithm 4 and Line 3 (and 4) in Algorithm 5, which run in $\Theta(M_{\mathrm{offdiag}})$ and $\Theta(M_{\mathrm{diag}})$ time, respectively. We estimate complexities on queries for segment trees. The initialization $\mathcal{T}_v$.Init for all $v \in V$ takes $\Theta(\sum_{v \in V} \mathrm{nnz}(K_{Vv})) = \Theta(\mathrm{nnz}(K))$ time. The sum query $\mathcal{T}_v$.Sum takes $\mathrm{O}(j \log n)$ time in every invocations of StablyComputeDiagonal$(j, \cdot)$, hence $\mathrm{O}(\sum_{j=1}^{k+1} jc_j \log n)$ time in total. The update query $\mathcal{T}_v$.Update is called k times for $s \in S^{(k)}$ and b_v times for $v \in T^{(k)}$, meaning that the total computational time taken by the update queries is $\mathrm{O}((k^2 + \sum_{v \in V} b_v) \log n))$ and is $\mathrm{O}(\sum_{j=1}^{k+1} jc_j \log n)$ by Lemma 1.2. The number of heap queries at Line 3 is less than that of segment trees because every execution of Line 3 is followed by the call of StablyComputeDiagonal at Line 6. the best and worst time complexities are obtained from (16) and (26), and by substituting $c_j = 1$ and $c_j = n - j + 1$, respectively, for $j \in [k+1]$. □

5 Relaxing StableLazyFastGreedy

5.1 Acceleration by Relaxing Stabilization

If the lazy heuristic is effective, StableLazyFastGreedy runs in $\Theta(k^3)$ time by Theorem 1.6, assuming $k = \Omega(\log n)$. This performance is comparable to that of LazyFastGreedy. However, if the lazy heuristic performs poorly, the running time increases significantly to $\Theta(k^3 n)$. The bottleneck is the inner product calculation at Line 3 in Algorithm 5. To speed up the computation, we introduce a further heuristic based on the following error analysis on catastrophic cancellations.

Suppose that a computer stores an approximation $\hat{a} \in \mathbb{R}$ of a non-zero real number $a \in \mathbb{R}$, where the error arises from previous computations using finite-precision arithmetic. The *relative error* of $\hat{a}$ with respect to a is defined as $\mathrm{err}_a(\hat{a}) := |a - \hat{a}\|/|a\|$. It is easily checked that for two positive reals a and b with $a > b$ and their approximations $\hat{a}$ and $\hat{b}$, the relative error of $\hat{x} := \hat{a} - \hat{b}$ with respect to $x := a - b$ is bounded as

$$\mathrm{err}_x(\hat{x}) \le \frac{a+b}{a-b} \max\{\mathrm{err}_a(\hat{a}), \mathrm{err}_b(\hat{b})\} = \frac{1+r}{1-r} \max\{\mathrm{err}_a(\hat{a}), \mathrm{err}_b(\hat{b})\}, \tag{27}$$

where $r := b/a$ (Higham 2002, Sect. 1.7). Therefore, if $r \le \varepsilon/(2+\varepsilon)$ for some error tolerance parameter $\varepsilon \ge 0$, then $\mathrm{err}_x(\hat{x})$ is bounded by $(1+\varepsilon)\max\{\mathrm{err}_a(\hat{a}), \mathrm{err}_b(\hat{b})\}$. This implies that if an increase in relative error by a factor of at most $1 + 10^{-6}$ is acceptable, for example, we can subtract two numbers whose ratio is at most $10^{-6}/(2 + 10^{-6}) \approx 5 \times 10^{-7}$. Although this requirement may seem hardly attainable, it is often met in rate constant matrices of chemical reaction networks. However, since algorithms only have the approximations $\hat{a}$ and $\hat{b}$, but their true values, the ratio b/a is not directly computable. The following theorem shows that the same error bound holds even if $\hat{b}/\hat{a}$ is used instead of b/a under the mild assumption that $1 + \varepsilon \leq \max\{\mathrm{err}_a(\hat{a}), \mathrm{err}_b(\hat{b})\}^{-1}$.

Theorem 1.7 *Let $a, \hat{a}, b, \hat{b} \in \mathbb{R}$ be non-zero reals of the same sign with $|a| > |b|$, $|\hat{a}| > |\hat{b}|$, $x := a - b$, $\hat{x} := \hat{a} - \hat{b}$, and $e := \max\{\mathrm{err}_a(\hat{a}), \mathrm{err}_b(\hat{b})\}$. For $\varepsilon \in [0, e^{-1} - 1]$, if $\hat{b}/\hat{a} \leq \varepsilon/(2 + \varepsilon)$, then $\mathrm{err}_x(\hat{x}) \leq (1 + \varepsilon)e$ holds.*

Proof We can assume $e \leq 1$ because the claim is vacuously true for $e > 1$ as $[0, e^{-1} - 1]$ is empty. We further assume that a, b, $\hat{a}$, and $\hat{b}$ are positive without loss of generality. Let δ_a and δ_b be the values such that $\hat{a} = (1 + \delta_a)a$ and $\hat{b} = (1 + \delta_b)b$, respectively, and $r := \hat{b}/\hat{a}$. Note that $e = \max\{|\delta_a\|, |\delta_b\|\}$ holds by $\hat{a}, \hat{b} > 0$ and $r < 1$ follows from $\hat{a} > \hat{b}$. The relative error of $\hat{x}$ with respect to x is written as

$$\mathrm{err}_x(\hat{x}) = \frac{|x - \hat{x}||}{x} = \frac{|a\delta_a + b\delta_b\|}{a - b} = \frac{|\frac{\hat{a}\delta_a}{1+\delta_a} + \frac{\hat{b}\delta_b}{1+\delta_b}\|}{\frac{\hat{a}}{1+\delta_a} - \frac{\hat{b}}{1+\delta_b}} = \frac{|\frac{\delta_a}{1+\delta_a} + \frac{\delta_b}{1+\delta_b}r\|}{\frac{1}{1+\delta_a} - \frac{1}{1+\delta_b}r} = |f_r(\delta_a, \delta_b)\|, \tag{28}$$

where $f_r(p, q)$ is a function defined by

$$f_r(p, q) := \frac{\frac{p}{1+p} + \frac{q}{1+q}r}{\frac{1}{1+p} - \frac{1}{1+q}r} \quad \left(-e \leq p, q \leq +e,\ \frac{1}{1+p} > \frac{1}{1+q}r\right). \tag{29}$$

Note that the domain of f_r can be restricted to (p, q) with $1/(1 + p) > r/(1 + q)$ because $1/(1 + \delta_a) > r/(1 + \delta_b)$ is implied by $a > b$.

In the following, we show that $f_r(p, q)$ is monotone non-decreasing with respect to p and q for fixed q and p, respectively. In fact, since $(p, q) = (+e, +e)$ and $(-e, -e)$ are in the domain of $f_r(p, q)$, the monotonicity gives lower and upper bounds on $f_r(\delta_a, \delta_b)$ as follows:

$$-(1 + \varepsilon)e \leq -\frac{1 + r}{1 - r}e = f_r(-e, -e) \leq f_r(\delta_a, \delta_b) \leq f_r(+e, +e) = +\frac{1 + r}{1 - r}e \leq +(1 + \varepsilon)e, \tag{30}$$

where the first inequality is obtained from $r \leq \varepsilon/(2 + \varepsilon)$, showing the claim of the theorem.

Reformulating the expression, we have

$$f_r(p, q) = \frac{1 - \frac{1-q}{1+q}r}{\frac{1}{1+p} - \frac{1}{1+q}r} - 1 = \frac{r - \frac{1-p}{1+p}}{\frac{1}{1+p} - \frac{1}{1+q}r} + 1. \tag{31}$$

Therefore, $f_r(p, q)$ is monotone non-decreasing with respect to p and q if and only if $1 - \frac{1-q}{1+q}r \geq 0$ and $r - \frac{1-p}{1+p} \leq 0$, respectively. These inequalities follow from

$$r \leq \frac{\varepsilon}{2+\varepsilon} \leq \frac{1-e}{1+e} \leq \min\left\{\frac{1-p}{1+p}, \frac{1+q}{1-q}\right\}, \tag{32}$$

where the second and third inequalities are from the facts that a function $u \mapsto (1-u)/(1+u)$ is monotone decreasing and $(1+\varepsilon)^{-1} \geq e \geq +p, -q$. Hence, the monotonicity of $f_r(p,q)$ has been proved. □

We incorporate Theorem 1.7 into Line 3 in Algorithm 5 using the following relationship:

$$C_{\star l}^{[j,u]} = C_{\star l}^{[j-1,s^{(j-1)}]} - C_{ul}, \tag{33}$$

which comes from $C_{\star l}^{[j,u]} = \sum_{v \in T^{(j-1)} \setminus \{u\}} C_{vl}$ and $C_{\star l}^{[j-1,s^{(j-1)}]} = \sum_{v \in T^{(j-1)}} C_{vl}$. Note that $C_{\star l}^{[j-1,s^{(j-1)}]}$ is readily available since it is computed in StablyComputeDiagonal$(j-1, s^{(j-1)})$, which is previously invoked. Thus, according to Theorem 1.7, we can obtain $C_{\star l}^{[j,u]}$ using (33) when $C_{ul}/C_{\star l}^{[j-1,s^{(j-1)}]} \leq \varepsilon/(2+\varepsilon)$ with a prescribed error tolerance ε, avoiding the computation of the inner product. We term this heuristic the *relaxing heuristic* and the improved algorithm as RelaxedStableLazyFastGreedy.

5.2 Complexity Analysis for RelaxedStableLazyFastGreedy

We finally analyze the running time of RelaxedStableLazyFastGreedy. Recall that M_{offdiag} and M_{diag} denote the total dimensions of vectors for which the inner products are computed at Line 5 in Algorithm 4 and Line 3, 4 in Algorithm 5, respectively. Since the difference between RelaxedStableLazyFastGreedy and StableLazyFastGreedy is only Line 3 in Algorithm 5, the running time estimated in Theorem 1.6 is still valid for RelaxedStableLazyFastGreedy; that is, RelaxedStableLazyFastGreedy runs in $\Theta(M_{\text{offdiag}} + M_{\text{diag}} + \text{nnz}(K)) + O(\sum_{j=1}^{k+1} jc_j \log n)$ time. Here, the equality (15) for M_{offdiag} holds in both StableLazyFastGreedy and RelaxedStableLazyFastGreedy, whereas the relaxing heuristics improves the equality (25) for M_{diag} into the following inequality.

Lemma 1.5 *In* RelaxedStableLazyFastGreedy, *M_{diag} satisfies*

$$\sum_{j=1}^{k+1}(j-1)c_j \leq M_{\text{diag}} \leq \frac{1}{2}\sum_{j=1}^{k+1} j(j-1)c_j \tag{34}$$

and the lower and upper bounds are attained if the relaxing heuristics work the best and worst, respectively. In particular, M_{diag} is equal to

- $k(k+1)/2 = \Theta(k^2)$ *if both the lazy and relaxing heuristics work the best,*

- $k(k+1)(3n-2k-1)/6 = \Theta(k^2 n)$ *if the lazy and relaxing heuristics work the worst and best, respectively,*
- $k(k+1)(k+2)/6 = \Theta(k^3)$ *if the lazy and relaxing heuristics work the best and worst, respectively, and*
- $k(k+1)(k+2)(4n-3k-1)/24 = \Theta(k^3 n)$ *if both the lazy and relaxing heuristics work the worst.*

Proof If the relaxing heuristic works perfectly, Line 3 in Algorithm 5 does not compute an inner product and Line 4 contributes to M_{diag} by $j-1$, resulting in $M_{\text{diag}} = \sum_{j=1}^{k+1}(j-1)c_j$. If the relaxing heuristic does not work at all, RELAXEDSTABLELAZYFASTGREEDY is the same as STABLELAZYFASTGREEDY and (25) holds. The latter part of the claim is obtained by substituting the bounds on c_j given in Lemma 1.2. □

Lemma 1.5 leads us to the following time complexity on RELAXEDSTABLELAZYFASTGREEDY.

Theorem 1.8 *The running time of* RelaxedStableLazyFastGreedy *is* $\Theta(M_{\text{offdiag}} + M_{\text{diag}} + \text{nnz}(K)) + \text{O}(\sum_{j=1}^{k+1} jc_j \log n)$*, which boils down to*

- $\Theta(k^3 + \text{nnz}(K)) + \text{O}(k^2 \log n)$ *if the lazy heuristics works the best,*
- $\Theta(k^2 n + \text{nnz}(K)) + \text{O}(k^2 n \log n)$ *if the lazy and relaxing heuristics work the worst and best, respectively, and*
- $\Theta(k^3 n + \text{nnz}(K)) + \text{O}(k^2 n \log n)$ *if both the lazy and relaxing heuristics work the worst.*

Even if the lazy heuristics works poorly, the relaxing heuristics can improve the dominating term in the complexity from $\Theta(k^3 n)$ up to $\Theta(k^2 n)$.

6 Numerical Experiments

In this section, we show experimental results on the running time and numerical stability of the five algorithms: GREEDY, FASTGREEDY, LAZYFASTGREEDY, STABLELAZYFASTGREEDY, and RELAXEDSTABLELAZYFASTGREEDY (abbreviated as RSLFG). All the algorithms were implemented using C++11 with Eigen 3.4.0[3] for matrix computations and compiled by GCC 10.2.0. Experiments were conducted on a computer with Intel® Xeon® Gold 5222 CPU (3.8 GHz, 4 Cores) and 755 GB RAM. Only double-precision arithmetic has been employed.

[3] https://eigen.tuxfamily.org/ (accessed Aug 21, 2024).

6.1 Data

We employ six publicly available chemical reaction path networks from Searching Chemical Action and Network (SCAN) platform (Kuwahara et al. 2023): DFG2 (fluoroglycine synthesis), WL2 (Wohler's urea synthesis), ALD2 (base-catalyzed aldol reaction), Strecker (strecker reaction), HDF2 (cobalt-catalyzed hydroformylation), and Passerini (Passerini reaction). Detailed steps for converting a chemical reaction path network into a rate constant matrix can be found, for instance, in (Sumiya et al. 2017). Characteristic values of the rate constant matrices are summarized in Table 3.

Table 3 Summary of dataset

Data	n	nnz(K)	$\max_{i\neq j:\, K_{ij}\neq 0} \lvert K_{ij}\rvert$	$\min_{i\neq j:\, K_{ij}\neq 0} \lvert K_{ij}\rvert$
DFG2	1,765	7,980	3.8×10^{14}	9.4×10^{-229}
WL2	1,776	8,742	1.2×10^{20}	1.7×10^{-149}
ALD2	6,199	20,848	3.1×10^{15}	2.9×10^{-182}
Strecker	9,203	27,092	4.2×10^{17}	1.8×10^{-134}
HDF2	10,810	29,802	9.7×10^{27}	3.1×10^{-210}
Passerini	12,215	30,746	1.2×10^{126}	1.2×10^{-144}

6.2 Results

Figure 1 shows the running times of the greedy parts (Step 1) of the algorithms, where RelaxedStableLazyFastGreedy was executed with four different error tolerance parameters $\varepsilon \in \{10^0, 10^{-16}, 10^{-32}, 10^{-48}\}$. While LazyFastGreedy ran fastest on all data, it selected incorrect $s^{(j)}$ at some j due to catastrophic cancellations. RelaxedStableLazyFastGreedy with $\varepsilon = 10^0$ was the fastest among the algorithms that produced the correct solution, whereas that with other values of ε ran in comparable times. This means that the choice of ε does not affect both the numerical stability and the running time significantly. Considering the fact that even additions of like-sign numbers may increase the error by the factor of the machine epsilon, which is roughly 10^{-16} for double-precision numbers, we recommend employing $\varepsilon = 10^{-16}$.

Figure 2 shows the running times of the methods including both Steps 1 and 2 with opt = `full`. To measure the running time, we recorded the time τ_j taken to compute $q^{(j)}$ by (3) for each j and regarded $\sum_{j=1}^{k} \tau_j$ as the running time of Step 2 until the kth iteration. We observe that Step 1 is actually a bottleneck of the original RCMC method, especially for small k, and our fast greedy methods can accelerate the entire procedure even with the Full option. Note that the running time with opt = `last` is almost the same as drawn in Fig. 1 as τ_k is negligibly small compared to the running time of Step 1.

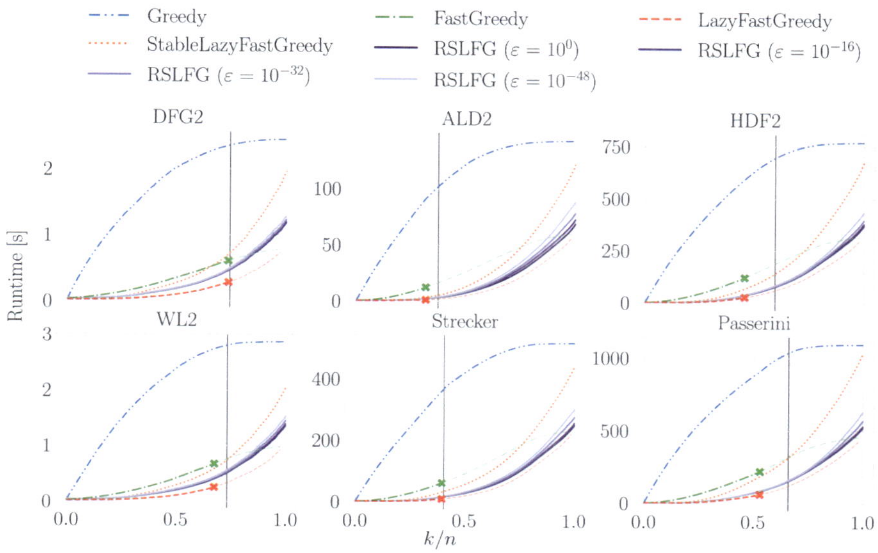

Fig. 1 The iteration counts k over the matrices size n versus the running times of Step 1. The cross marks with the horizontal coordinate k indicate the smallest k where the algorithm selected an incorrect element as $s^{(k)}$ due to catastrophic cancellation. The vertical black lines indicate the value of k corresponding to $t_{\max} = 86,400$ sec

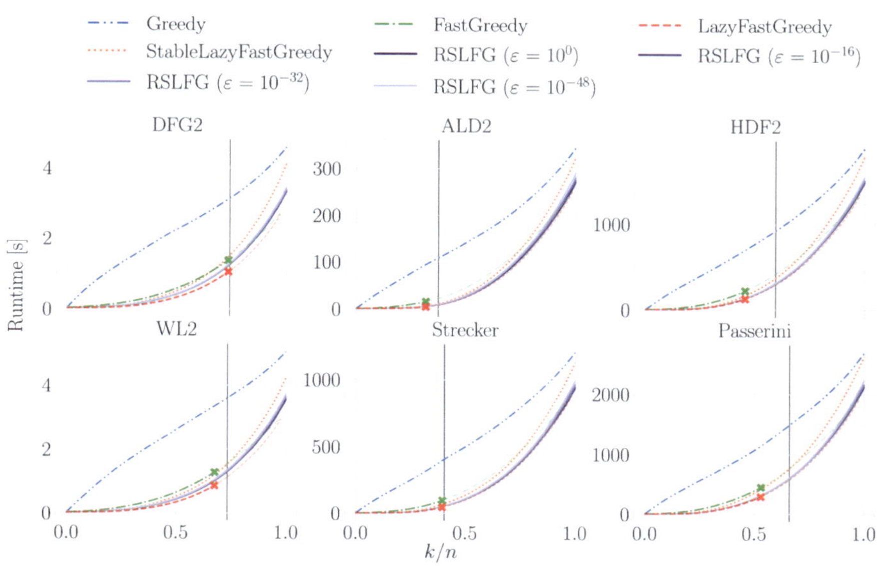

Fig. 2 The iteration counts k over the matrices size n versus the total running times of the methods (Steps 1 and 2). The cross marks and vertical lines indicate the same as Fig. 1

Table 4 describes the running times of Step 1 (GREEDY, STABLELAZYFASTGREEDY, and RELAXEDSTABLELAZYFASTGREEDY) and Step 2 at k corresponding to $t_{\max} = 86,400$ sec. We can observe for any data that the running time of GREEDY is greater than that of Step 2, whereas RELAXEDSTABLELAZYFASTGREEDY runs faster than Step 2. This means that our proposed method sufficiently speeds up Step 1, and the bottleneck of the RCMC method is now moved to Step 2.

Table 4 Running times (seconds) at k corresponding to $t_{\max} = 86,400$ sec. (R)SLFG stands for (RELAXED)STABLELAZYFASTGREEDY. The error tolerance ε in RSLFG is set to 10^{-16}

Data	k	Step 1			Step 2
		GREEDY	SLFG	RSLFG	
DFG2	1311	2.35	0.73	0.48	0.78
WL2	1304	2.81	0.75	0.52	0.79
ALD2	2318	102.18	5.33	3.53	5.79
Strecker	3729	369.14	34.16	17.33	38.39
HDF2	6427	696.10	140.51	79.54	226.71
Passerini	8057	1038.79	314.30	154.81	450.37

To take a closer look at how well the heuristics work, we illustrate in Figs. 3 and 4 the values of M_{offdiag} and M_{diag}, respectively. Figure 3 indicate that M_{offdiag} attains almost the lower bounds in (16) at any k for all data, meaning that the lazy heuristics work quite well in practice. Similarly, the curves of RELAXEDSTABLELAZYFASTGREEDY lie lower than that of STABLELAZYFASTGREEDY in Fig. 4, hence the relaxing heuristics is also effective.

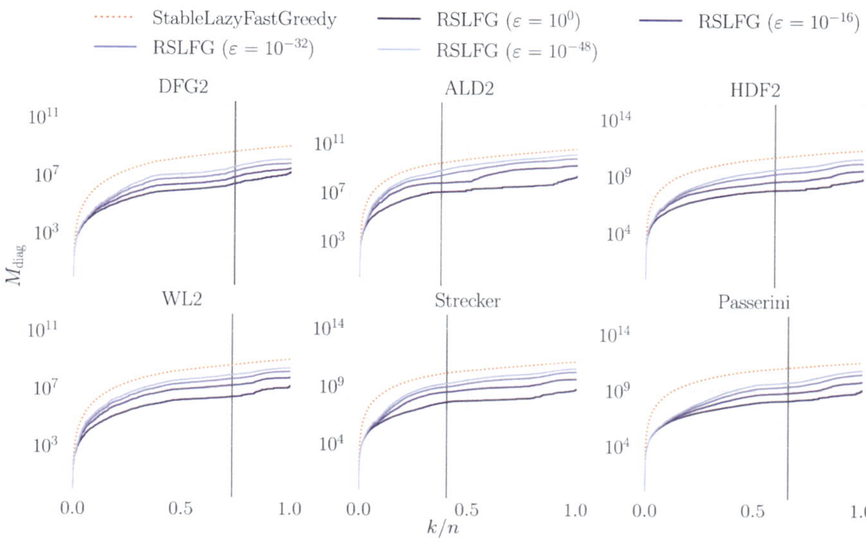

Fig. 3 The iteration counts k over the matrices size n versus M_{offdiag} (defined in Sect. 3.3). The gray areas indicate the ranges of possible values of M_{offdiag} for every k given by (16). The cross marks and vertical lines indicate the same as Fig. 1

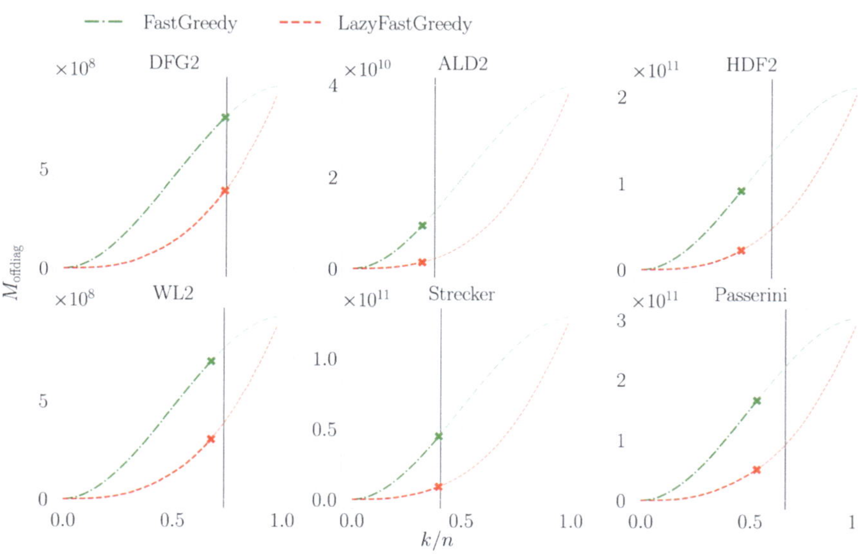

Fig. 4 The iteration counts k over the matrices size n versus M_{diag} (defined in Sect. 4.3). The gray areas indicate the ranges of possible values of M_{offdiag} for every k given by Lemma 1.5. The vertical lines indicate the same as Fig. 1

7 Conclusion

In this paper, we have proposed a fast and numerically stable implementation of the greedy part (Step 1) of the RCMC method for chemical kinetics simulation. Our technique is to modify the lazy greedy for the DPP MAP inference to avoid subtractions of like-sign numbers, leveraging properties of rate constant matrices. For faster computation, we employ segment trees and partially allow like-sign subtractions if they do not cause catastrophic cancellations. Using real instances from chemical reactions, we have confirmed that the presented algorithm runs numerically stably and much faster than the original method. Accelerating Step 2, a new bottleneck, is left for future investigation.

Acknowledgements The authors thank Satoshi Maeda and Yu Harabuchi for their helpful comments, Yutaro Yamaguchi for providing information on the reference (Chazelle 1988) of segment trees, and anonymous referees for careful reading. This work was supported by JST ERATO Grant Number JPMJER1903, JST CREST Grant Number JPMJCR24Q2, JST Forest Grant Number JPMJFR232L, and JSPS KAKENHI Grant Number JP22K17853.

References

Anderson, W.J.: Continuous-Time Markov Chains. Springer Series in Statistics. Springer, New York (1991). https://doi.org/10.1007/978-1-4612-3038-0

Chazelle, B.: A functional approach to data structures and its use in multidimensional searching. SIAM J. Comput. **17**(3), 427–462 (1988). https://doi.org/10.1137/0217026

Chen, L., Zhang, G., Zhou, E.: Fast greedy map inference for determinantal point process to improve recommendation diversity. In: Advances in Neural Information Processing Systems 31 (NeurIPS '18), pp. 5627–5638 (2018). https://doi.org/10.5555/3327345.3327465

Doolittle, M.H.: Method employed in the solution of normal equations and the adjustment of a triangulation. US Coast Geod. Surv. Rep. **1878**, 115–120 (1878)

Fan, K.: An inequality for subadditive functions on a distributive lattice, with application to determinantal inequalities. Linear Algebra Appl. **1**(1), 33–38 (1968)

Gauss CF (1809) Theoria motus corporum coelestum. Werke

Grcar, J.F.: How ordinary elimination became Gaussian elimination. Hist. Math. **38**(2), 163–218 (2011). https://doi.org/10.1016/j.hm.2010.06.003. https://www.sciencedirect.com/science/article/pii/S0315086010000376

Harabuchi, Y., Yokoyama, T., Matsuoka, W., Oki, T., Iwata, S., Maeda, S.: Differentiating the yield of chemical reactions using parameters in first-order kinetic equations to identify elementary steps that control the reactivity from complicated reaction path networks. J. Phys. Chem. A **128**(14), 2883–2890 (2024). https://doi.org/10.1021/acs.jpca.4c00204. https://pubs.acs.org/doi/full/10.1021/acs.jpca.4c00204

Hemmi, S., Oki, T., Sakaue, S., Fujii, K., Iwata, S.: Lazy and fast greedy map inference for determinantal point process. In: Advances in Neural Information Processing Systems 35 (NeurIPS '22), pp. 2276–2789 (2022)

Higham, N.J.: Accuracy and stability of numerical algorithms. SIAM (2002). https://doi.org/10.1137/1.9780898718027

Iwata, S., Oki, T., Sakaue, S.: Rate constant matrix contraction method for stiff master equations with detailed balance. SIAM J. Sci. Comput. (2023). arXiv:2312.05470

Ko, C.W., Lee, J., Queyranne, M.: An exact algorithm for maximum entropy sampling. Oper. Res. **43**(4), 684–691 (1995). http://www.jstor.org/stable/171694

Kulesza, A., Taskar, B.: Determinantal point processes for machine learning. Found. Trends® Mach. Learn. **5**(2–3), 123–286 (2012)

Kuwahara, M., Harabuchi, Y., Maeda, S., Fujima, J., Takahashi, K.: Searching chemical action and network (SCAN): an interactive chemical reaction path network platform. Digit. Discov. **2**, 1104–1111 (2023). https://doi.org/10.1039/D3DD00026E

Macchi, O.: The coincidence approach to stochastic point processes. Adv. Appl. Probab. **7**(1), 83–122 (1975). https://doi.org/10.2307/1425855

Maeda, S., Taketsugu, T., Morokuma, K.: Exploring transition state structures for intramolecular pathways by the artificial force induced reaction method. J. Comput. Chem. **35**(2), 166–173 (2014). https://doi.org/10.1002/jcc.23481

Minoux, M.: Accelerated greedy algorithms for maximizing submodular set functions. In: Stoer, J. (ed.) Optimization Techniques, pp. 234–243. Springer, Berlin, Heidelberg (1978)

Nakamura, T., Sakaue, S., Fujii, K., Harabuchi, Y., Maeda, S., Iwata, S.: Selecting molecules with diverse structures and properties by maximizing submodular functions of descriptors learned with graph neural networks. Sci. Rep. **12**(1), 1124 (2022)

Nemhauser, G.L., Wolsey, L.A.: Best algorithms for approximating the maximum of a submodular set function. Math. Oper. Res. **3**(3), 177–188 (1978)

Nemhauser, G.L., Wolsey, L.A., Fisher, M.L.: An analysis of approximations for maximizing submodular set functions-i. Math. Program. **14**(1), 265–294 (1978)

Ohno, K., Maeda, S.: A scaled hypersphere search method for the topography of reaction pathways on the potential energy surface. Chem. Phys. Lett. **384**(4), 277–282 (2004). https://doi.org/10.1016/j.cplett.2003.12.030

Sumiya, Y., Maeda, S.: Rate constant matrix contraction method for systematic analysis of reaction path networks. Chem. Lett. **49**(5), 553–564 (2020). https://doi.org/10.1246/cl.200092. http://www.journal.csj.jp/doi/10.1246/cl.200092

Sumiya, Y., Nagahata, Y., Komatsuzaki, T., Taketsugu, T., Maeda, S.: Kinetic analysis for the multistep profiles of organic reactions: significance of the conformational entropy on the rate constants of the claisen rearrangement. J. Phys. Chem. A **119**(48), 11641–11649 (2015). https://doi.org/10.1021/acs.jpca.5b09447

Sumiya, Y., Taketsugu, T., Maeda, S.: Full rate constant matrix contraction method for obtaining branching ratio of unimolecular decomposition. J. Comput. Chem. **38**(2), 101–109 (2017). https://doi.org/10.1002/jcc.24526

Open Access This chapter is licensed under the terms of the Creative Commons Attribution 4.0 International License (http://creativecommons.org/licenses/by/4.0/), which permits use, sharing, adaptation, distribution and reproduction in any medium or format, as long as you give appropriate credit to the original author(s) and the source, provide a link to the Creative Commons license and indicate if changes were made.

The images or other third party material in this chapter are included in the chapter's Creative Commons license, unless indicated otherwise in a credit line to the material. If material is not included in the chapter's Creative Commons license and your intended use is not permitted by statutory regulation or exceeds the permitted use, you will need to obtain permission directly from the copyright holder.

Studying Morphological Variation: Exploring the Shape Space in Evolutionary Anthropology

Shira Faigenbaum-Golovin and Ingrid Daubechies

Abstract We present results of a long-term team collaboration of mathematicians and biologists. We focus on building a mathematical framework for the shape space constituted by a collection of homologous bones or teeth from many species. The biological application is to quantitative morphological understanding of the evolutionary history of primates in particular, and mammals more generally. Similar to the practice of biologists, we leverage the power of the whole collection for results that are more robust than can be obtained by only pairwise comparisons, using tools from differential geometry and machine learning. This paper concentrates on the mathematical framework. We review methods for comparing anatomical surfaces, discuss the problem of registration and alignment, and address the computation of different distances. Next, we cover broader questions related to cross-dataset landmark selection, shape segmentation, and shape classification analysis. This paper summarizes the work of many team members other than the authors; in this paper that unites (for the first time) all their results in one joint context, space restrictions prevent a full description of the mathematical details, which are thoroughly covered in the original articles. Although our application is to the study of anatomical surfaces, we believe our approach has much wider applicability.

1 Introduction

Machine learning methods provide tools to tackle a wide array of questions raised by large and complex data sets, with a core focus on examining and using the similarities and the differences among samples within a dataset. In some applications, observed scattered high-dimensional data 'points' (each of which can be quite complex and have significant internal structure–e.g. images, curves, surfaces) are sufficiently similar to their neighbors that it is reasonable to view them as lying on an unknown manifold that one wants to understand better.

S. Faigenbaum-Golovin (✉) · I. Daubechies
Duke University, Durham, England
e-mail: shira.golovin@math.duke.edu

© The Author(s) 2026

S. Oishi et al. (eds.), *Recent Developments in Industrial and Applied Mathematics*, ICIAM2023 Springer Series 1, https://doi.org/10.1007/978-981-95-1446-5_3

In our particular application, we typically consider a collection of anatomical shapes, usually with some rough correspondence between the shapes, and we aim to study the underlying shape space. In the current paper, we survey how applying tools from the field of differential geometry and machine learning aids in studying anatomical surfaces, in the framework of Evolutionary Anthropology. Comparing such surfaces is vital for the understanding of the evolutionary history of primates; quantitative comparisons help in answering questions of interest in evolutionary anthropology Morphology and relative sizes of the anterior dentition have long played a role in the interpretation of systematics and evolution of early primates Simons (1961), Rose and Bown (1984) and Rasmussen et al. (1995). Understanding the relationship between skeletal morphology, function, and ecology in this group is critical for understanding the evolutionary origins of the order to which humans belong. It is generally thought that particular tooth forms are suited to processing essential food items of respectively particular forms and material properties, and that mammalian teeth/dentitions are often unique taxonomic identifiers for more details see Boyer (2008). Examples of the different teeth of a fruit eater, a leaf eater, a predator (of arthropods), and an omnivore are shown in Fig. 1. All the teeth used in this study can be downloaded from MorphoSource Morphosource (2024). Although the surfaces illustrated in Fig. 1 differ considerably, tooth surfaces in a much wider collection acquired from many species show more gradual transitions, in particular between species closely related to each other; the variations are sufficiently incremental that the surfaces can reasonably be viewed as samples lying on a manifold.

In many cases, only a surface cast of the original tooth or bone is available in our collection; for this reason, we work with surfaces rather than volumetric data. The surfaces are scanned as scattered data and later represented as triangular meshes. Compiling the geometric information in these triangular meshes is a challenge, due

(a) (b) (c) (d)

Fig. 1 Comparison of dental morphology based on the diet. **a** Fruit eater; **b** Leaf eater; **c** Predator; **d** Omnivore (*Source* images wikimedia commons, meshes taken from morphosource (AMNH:Mammals:M-80072, AMNH:Mammals:M-100832, AMNH:Mammals:M-196480, AMNH:Mammals:M-174537) Morphosource (2024) to the nuanced variations in anatomical shapes. Our team follows a workflow that consists of several steps. First, pre-processing steps are performed, starting from denoising and cleaning the triangulated arrays followed by a coarse alignment of these surfaces Puente (2013) as well as a first estimation of a robust and biologically meaningful distance for each pair of surfaces. Next, the variation between the shapes, using the power of *collection* of surfaces is addressed. A (up-to -date only until the time of its writing) survey of several automated geometric morphometric methods (some also discussed below) can be found in Gao et al. (2018).

Although surface analysis, such as comparison and registration, is extensively addressed in the computer graphics or Computer Aided Geometric Design (CAGD) communities, the algorithms developed there cannot simply be ported to our application. On the one hand, the features of interest in anatomical shapes and their variations are typically too subtle to be considered useful features for computer graphics methods; on the other hand, it turns out that leveraging manifold analysis to study a whole collection of surfaces leads to much better results. In addition, the research questions stemming from evolutionary biology introduce a unique perspective.

In this paper, we survey the research on anatomical surfaces carried out by a team of biologists and mathematicians over more than 15 years, involving a total of more than 20 researchers (at different times) focusing on the mathematical aspects of the work. We cover the methods that were developed for the comparison of the surfaces, discuss the problem of registration and alignment, and review the calculation of the different distances. Next, we cover broader questions related to cross-dataset landmark selection, shape segmentation, and shape classification analysis. We conclude this paper with our view of the future of the shape space realm. These steps are illustrated in Fig. 2. Our main analysis focuses on disc-type surfaces, such as the part of a tooth above the gum); most of our methods can be (and some have been) applied to sphere-type surfaces (which have no boundary). So far, we have not extended the work to higher genus surfaces (which have holes).

The remainder of the paper is organized as follows. In Sect. 2 we review the data acquisition process, from CT scan to clean surfaces. Next, in Sect. 3 we review the methods that result in robust distance metrics between the shapes, starting with

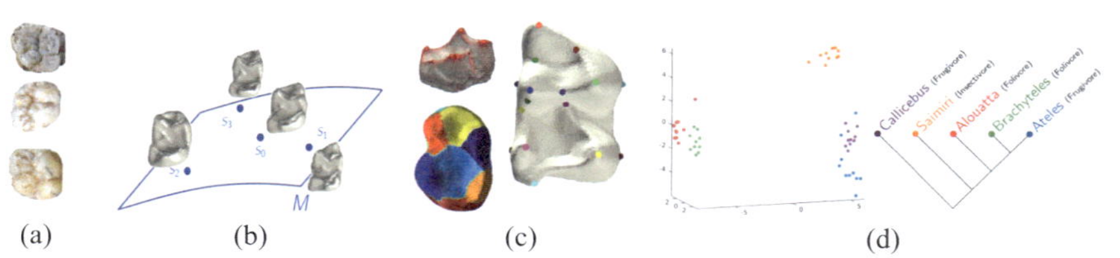

Fig. 2 Working flow in a shape space. **a** Data acquisition; **b** Alignment, registration, and distance calculation on the manifold of shapes; **c** Automatic Shape analysis (e.g., from top left clockwise, curvature estimation (ariaDNE Shan et al. (2019)), landmark sampling, consistent segmentation of shapes in the collection; **d** Phylogeny of different species groups. Images adopted from Shan et al. (2019) and Gao (2015)

alignment, landmarking, and registration. Then in Sect. 3.2 we detail several methods for distance calculation (among them is the Conformal Wasserstein local distance, continuous Procrustes distance, as well as the Horizontal-based diffusion distance). Quality control aspects are provided with the distance accuracy validation in Sect. 3.3. Next, we dive into the manifold learning analysis in Sect. 4, where we briefly survey the classical tool for studying manifold geometry, Diffusion Map in Sect. 4.2. Next, the theoretical and practical aspects of decomposing manifolds by exploring fibers and the base manifold are explained. Later, we provide details about the preeminent approach of horizontal random walks and diffusion processes on fibre bundles. In Sect. 5 we are finally ready to shed light on morphology via statistical analysis and machine learning methods. In Table 1 links to all the code discussed in this paper appear, while the available datasets used for the study are listed in Table 2.

List of abbreviations used in this survey

MSE—Mean Square Error
MST—Minimum Spanning Tree
PCA—Principal Component Analysis
Auto3DGM—Automated 3D Geometric Morphometrics
MSPE—Maximum Mean Squared Prediction Error
DM—Diffusion Map
SGE—Semi-group Error
cP—Continuous Procrustes
cPDist/CPD—Continuous Procrustes distance
cWn—Conformal Wasserstein Neighborhood Dissimilarity Distance
HDM—Horizontal Diffusion Maps
HBD—Hypoelliptic Diffusion Distance
HBDM—Hypoelliptic Base Diffusion Maps
HBDD—Hypoelliptic Base Diffusion Distance
DNE—Dirichlet Normal Energy
ariaDNE—a Robustly Implemented Algorithm for Dirichlet Energy of the Normal
EOP—Eyes on the Prize.

2 Data Acquisition and Pre-Processing

We shall visit two types of example surfaces: disk-type surfaces obtained by scanning with high resolution the portion above the gumline of second mandibular molars of prosimian primates and nonprimate close relatives, and sphere-type surfaces obtained by scanning talus bones of primates and close relatives. Digitized surfaces were obtained from high-resolution X-ray computed tomography (μCT) scans, using the A Nikon XT H 225 ST μCT machine at Duke University's Shared Materials Instrumentation Facility (SMIF). For greater detail on the protocol followed, see e.g. the

Table 1 Studying morphological variation: available code

Method	Short description	Read more	Link to the code
cPD	Find the best correspondence of a pair of surfaces via optimization process	Sect. 3.2, Al-Aifari et al. (2013)	https://github.com/trgao10/cPdist
Auto3dgm	Automated 3D Geometric Morphometric Aligment	Sect. 3, Boyer et al. (2015)	https://github.com/ToothAndClaw/Auto3dgm_Python
Gaussian Process Landmarking	Landmarking via Gaussian Kriging method, and later registration via landmark matching	Sect. 3.1, Gao et al. (2019a)	https://github.com/shaharkov/GPLmkBDMatch
HDM	Horizontal random walk on the fibre bundle	Sect. 4.3, Gao (2021)	https://github.com/trgao10/HDM
ariaDNE	Robust curvature estimation	Sect. 5, Shan et al. (2019)	https://sshanshans.github.io/articles/ariadne.html
Hecate	Consistent segmentation of biological surface regions by spectral clustering from a hypoelliptic diffusion map, and Continues Procrustes distance, and HDM	Sect. 4.4, Fulwood et al. (2021)	https://github.com/JulieWinchester/hecate
Eyes on the Prize (EOP) Registration—SAMS	Matlab package to automatically process, register, and analyze collections of anatomical surfaces at the resolution higher	Sect. 3.1, Ravier (2018a, b)	https://github.com/RRavier/SAMS
SlicerMorph	Platform to retrieve, visualize, and analyze 3D morphology	Rolfe et al. (2021)	https://slicermorph.github.io/#six & https://toothandclaw.github.io/installations/

description in Yapuncich et al. (2019).[1] Figure 3 illustrates the steps to acquire a surface mesh by scanning a (small) real object using an X-ray μCT system.

[1] CT scan images courtesy of The Duke Lemur Center that provided access to these data, originally appearing in Yapuncich et al. (2019), the collection of which was funded by NSF BCS 1540421 to Gabriel S. Yapuncich and Doug M. Boyer. The files were downloaded from www.MorphoSource.org, Duke University.

Table 2 Studying morphological variation: 3D specimen data

Data set name	Short description	Ref.	Link
MorphoSource	3D meshes, and CT-images of biological skeletal material	Boyer et al. (2016)	https://www.morphosource.org/users/sign_in?locale=en and visualization: https://gaotingran.com/research/auto3dgm.html
Second mandibular surfaces	116 s mandibular molars of prosimian primates and non-primate, as well as observer-determined landmark	Boyer et al. (2011)	https://www.wisdom.weizmann.ac.il/~ylipman/CPsurfcomp/

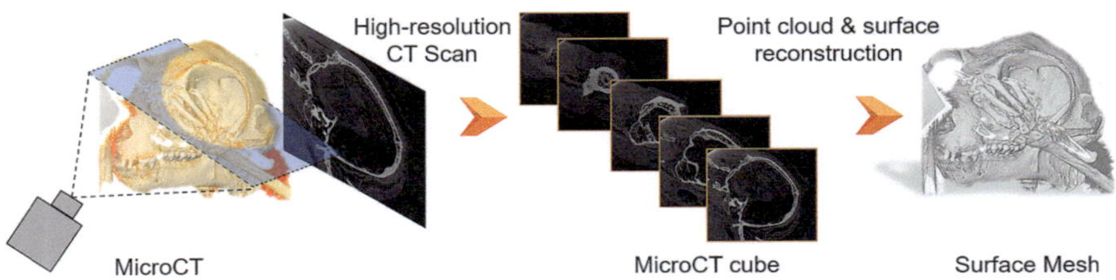

Fig. 3 Illustration of the scanning process of a skull: from CT scan to MicroCT cube, to point cloud and surfaces reconstructed and cleaned[1]

The MicroCT cubes were used to create a 3D digital point cloud collection using Avizo or dragonfly software Avizo (2024) and dragonfly (2024), after which a mesh was created using via standard triangulation algorithms. The procedure of converting the CT scan images to point clouds and meshes is typically prone to many errors, leading to inconsistencies in the meshes; among such artifacts are incorrect identification of nearest neighbors of certain points, reversed neighboring connectivity, resulting in failure to generate a smooth surface (with some faces flipped), spikes, and holes. These are then manually identified, and off-the-shelf software GeoMagic is used to address these issues. All scans are publicly available on MorphoSource Boyer et al. (2016), an online repository specifically designed to archive 3D data.

3 From Scattered Surfaces to Metric Shape Space

Once the shapes are scanned, the point clouds extracted, and surfaces are created and cleaned, we are ready to dive into the shape analysis and the comparison question.

The main challenges for calculating the similarity between shapes are alignment, registration, and distance calculation. Alignment and registration are tightly connected but represent different tasks. While *alignment* moves one object into alignment with another object so that they both share the same coordinate system, *registration* is the mapping from one surface to the other that connects areas with corresponding function and geometry.

Calculating the distance between two registered objects typically amounts to integrating (or summing), over one of the objects, the distance of each point to its corresponding point on the other object. On the other hand, the optimal correspondence map from one object to the other is usually found by optimization, minimizing the value of the distance between the objects. Pairwise registration and distance computation thus go hand-in-hand. In this section, we review several algorithms used by our team for these tasks. Often our analysis workflow uses several different approaches, with simple algorithms at first, and more sophisticated ones at a later stage.

3.1 Alignment, Landmarking, and Registration

Alignment: The goal of shape alignment is to establish a meaningful correspondence between corresponding points, features, or structures on different shapes, ensuring that they are spatially aligned or registered with each other. For computational purposes, the meshes are first down-sampled prior to analysis. It is useful to note that finding a joint coordinate system is easier for similar shapes than for different shapes. This idea is incorporated into the Automated 3D Geometric Morphometrics (Auto3DGM) tool Auto3dgm (2024), one of the earliest tools developed by our team. It first uses PCA to compute the three principal axes for each shape. For each pair of shapes, an alignment is carried out based on best matching their principal axis, after which the associated Procrustes distance between the shapes is calculated. (This distance is defined later in (5)). These are only intermediate steps–the alignment and distance obtained in this way are typically not very good when the two shapes are too dissimilar. However, the approximate distances can be used to determine a minimum spanning tree (MST) for the collection; between two elements that are close to each other, the alignment is typically quite good. By aligning shapes with each other sequentially, following the MST, even very dissimilar bones can be aligned correctly, using intermediate bones in the process, as illustrated in Fig. 4. More details can be found in Boyer et al. (2015)).

Registration and Landmarking: Registration is the mapping of every piece of one shape to another while taking into account the corresponding features. Shape matching is an area of active research in computer graphics Deng et al. (2022) and Van Kaick et al. (2011) and also the prior art appearing in Ravier (2018a). In order for the registration to be useful for our biological collaborators, it must (a) be consistent with the field knowledge about the anatomically equivalent (biologically homologous parts); (b) take into account the variation present in other species of the sample; and (c) correspond to a *smooth* map between shapes: the ability to compute high-quality

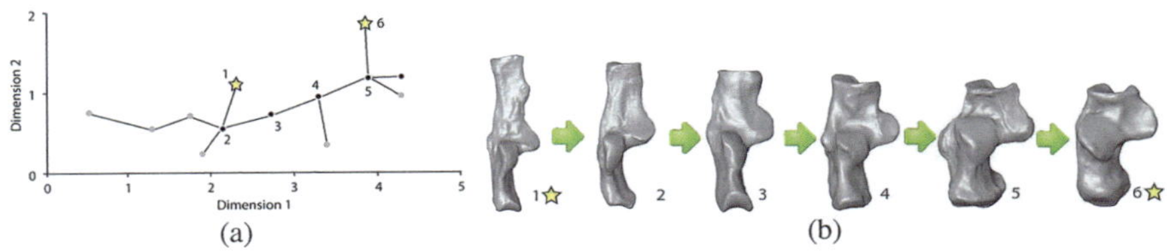

Fig. 4 Alignment of two different shapes labeled as 1 and 6, by composing a series of incremental different bones (right) found via the MST (left) Boyer et al. (2015)

diffeomorphisms between shapes would allow the use of a wider range of statistical techniques in order to compare shapes in a collection.

One plausible solution is to choose landmarks that capture some geometric features of the surfaces. 'Landmarking by experts' is a manual process (now typically carried out on a computer screen) where an equal number of geometrically or semantically meaningful feature points, the *landmarks*, are marked on a surface. The selected landmarks are certified by domain experts to be in consistent one-to-one correspondences. The methodology of landmark selection emphasizes a comprehensive and balanced decision between sharp geometric features and points taken farthest away from other points under certain metrics (for a discussion about the popular methods see Watanabe (2018) and Wärmländer et al. (2019)). While experts can manually suggest the important feature points, and put into correspondence areas of high and low curvature, they have a harder time prescribing the correspondence map in intermediate regions, between these feature points. In rare cases, experts can also disagree on landmark correspondences. The imposition that the same exact number of landmarks be picked on all the surfaces in the collection can also be a limiting aspect. Therefore, we considered several automatic procedures for locating the important points, described below. In our workflow at present, we typically start with a simple pairwise approach the results of which provide the point of departure for more sophisticated methods, used further in the workflow.

Continuous Procrustes Landmarking: In the discrete implementation of the continuous Procrustes distance (defined below in Eq. (5)), one starts by distributing equal numbers of approximately evenly spaced points on two surfaces; an iterative procedure then determines the correspondence between the two surfaces that minimizes the Procrustes distance between the two point sets. This provides us with an out-of-the-box landmarking. However, since this landmarking is performed pairwise, the propagation of landmarks between shapes is not always consistent, and can change if a morphologically close shape is added: in the left panel of Fig. 5 landmark propagation from shape A to B changes if we first propagate from A to C and then from C to B.

Matching Pursuit Algorithm Landmarking: An alternative method for landmark selection was proposed in Gao (2015), using a Matching Pursuit algorithm. It exploits the similarity (or rather dissimilarity) between surfaces, and among different portions of the surfaces, to create a large matrix, $L^{H}_{\alpha,*}$, the graph hypoelliptic

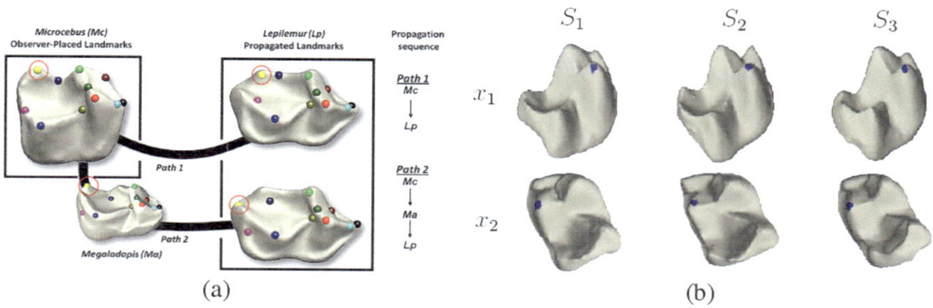

Fig. 5 **a** Landmark propagation via continuous Procrustes distance. Propagating from Microcebus to Lepilemur is not equivalent if their ancestor Megaladopis is also used Boyer et al. (2011); **b** Landmarking via Matching Pursuit algorithm. Points that are in correspondence with each other using an eigenvector (each row is a correspondence to a different eigenvector) Gao (2015)

Laplacian, generalising the graph Laplacian, on the union of all the sample points of all the surfaces. An exact correspondence map between the surfaces (normalized to all have area 1) would lead to a sparse eigenvector of that matrix. This suggested the following optimization approach to find a vector x that would (approximately) encode corresponding landmarks:

$$min_{||x||_2=1} \left[||L^H_{\alpha,*}x||_2^2 + \lambda ||x||_1 \right] \tag{1}$$

More details can be found in in Gao (2015). The algorithm produces corresponding points on the entire collection of surfaces, even though this is not built into the functional *a priori*. This method was the first, developed by our team, that leverages the power of considering a *collection* of surfaces. See Fig. 5b for an illustration.

Gaussian Process Landmarking: A different approach to the 'automatic' identification of salient points, which can be interpreted as landmarks, was introduced in Gao et al. (2019b, b). It is essentially a Kriging method, based on successively selecting points of maximum mean squared prediction error (MSPE) with respect to the Gaussian process on the anatomical surface, with a variance-covariance structure determined by the heat kernel of the surface, viewed as a Riemannian manifold; the landmarks are then selected successively, each time picking a new landmark as the point on the surface with the largest variance conditioned on all the previously selected landmarks. Figure 6 compares two methods for landmark selection: produced by geodesic farthest point sampling (A) Moenning and Dodgson (2003), and by Gaussian processing landmarking (B) Gao et al. (2019a). As can be seen, The Gaussian sampling (B) prioritizes choosing semantically meaningful features for geometric morphometrics and agrees better with points selected by biological experts.

After landmarks are identified on two surfaces S_1 and S_2, the best registration via landmark matching is determined. Among all the candidates obtained by Gaussian process landmarking on each of the surfaces, a maximal subset of geometrically

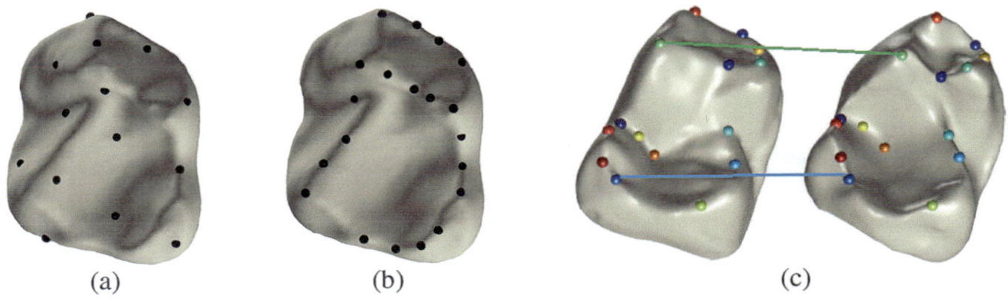

Fig. 6 Shape registration via automatic Landmarking Gao et al. (2019a). **a** and **b** demonstrate different landmarking methods: geodesic farthest point sampling in (**a**), by Gaussian process landmarking in (**b**). **c** After Gaussian process landmarking on each of the two surfaces, bounded distortion matching identifies corresponding pairs of landmarks (spheres with matching colors); note that each surface may have many landmarks (not shown) that do not correspond to a corresponding landmark on the other surface

consistent correspondences is determined; next interpolation is used found to obtain a diffeomorphism between the surfaces S_1 and S_2 that connects the landmarks in this subset; this results in two fully registered surfaces. See Fig. 6c for an illustration of landmark matching. MATLAB code for the surface registration algorithm is publicly available at GPLmkBDMatch (2024).

Consistent registration methods: As pointed out above, in the discussion of Auto3DGM Alignment, propagating landmarks via best correspondences that are determined by pairwise comparisons of surfaces (whether done via the Continuous Procrustes approach, identifying best possible rigid rotations and pairwise mapping, or via e.g. Gaussian landmarking), following paths on the corresponding distance graph for a whole collection by determining the Minimum Spanning Tree, can lead to inconsistencies (see Fig. 5a, as well as the discussion below about the accuracy of the distance induced by the Continuous Procrustes process (Sect. 3.3). To avoid this sensitivity to the selection of surfaces included in the study, we have also introduced alternative methods that use diffusion networks to propagate information, or to define a consistent reparametrization of the full collection. We shall come back to these methods below, after discussing different distances defined on collections of surfaces.

3.2 *Distances in the Shape Space*

We turn now to the question of comparing the shapes qualitatively, studying them statistically, and testing different hypotheses of evolution. Quantifying and comparing complex shapes is a key component of evolutionary morphology. The standard approach used by biological morphologists computes the *Procrustes distance* between surfaces, based on the preliminary identification of homologous features (corresponding points or *landmarks* on each of the surfaces in the collection to be studied, based on the domain knowledge of the landmarking expert of the functionality of different geometric features of the bones or teeth). The collection of the landmark points on each surface can then serve as a proxy for its surface, and distances between shapes are computed based on the following definition.

Definition 1.1 (*Procrustes distance*) The *Procrustes distance* between two surfaces S_1, S_2 is computed using the following procedure:

1. Specify two sets of (operationally) homologous landmarks $\left\{x_\ell^{(1)} \mid 1 \le \ell \le L\right\}$ and $\left\{x_\ell^{(2)} \mid 1 \le \ell \le L\right\}$ on S_1 and S_2, respectively.
2. Compute the distance between S_1 and S_2 by minimizing the energy functional

$$d_{\mathrm{cP}}\left(S_1, S_2\right) = \inf_{T \in \mathbb{E}(3)} \left(\frac{1}{L} \sum_{\ell=1}^{L} \left\| T\left(x_\ell^{(1)}\right) - x_\ell^{(2)} \right\|^2 \right)^{\frac{1}{2}} \tag{2}$$

   where $\mathbb{E}\left(3\right)$ is the group of rigid motions in $\mathbb{R}^3$.

The main disadvantage of this method is that it requires corresponding landmarks as input; pre-determining these can be a time-consuming process.

One of the first tasks our team was asked to tackle, in our collaboration with biological morphologists, was to define distances (and algorithms to compute them) that would not require preliminary expert-defined landmarking, but that would nevertheless be as useful for quantification of similarities and differences in their biological studies as their usual Procrustes distances. In our first approach, we used mass transportation techniques to define the *Conformal Wasserstein neighborhood distance*. In a first step, the disk-type surfaces are mapped conformally to a disk; the conformal factor for each is then interpreted as a "landscape" on the disk to which we apply a mass transportation principle. (A similar approach is also possible for sphere-type surfaces, mapping them conformally to a sphere instead of a disk.) To define a mass transportation distance between the two landscapes, we need to introduce a (dis)similarity metric distance between points v, w in the disk adapted to our problem: to make sure that the resulting distance between the surfaces is independent of the (non-unique) conformal flattenings that were performed first, we define this metric distance in a conformally invariant way. The distance between points z (in the landscape determined by μ) and w (in the landscape for ν) is set to be the L^1–distance between the landscapes around the points, restricted to a small local hyperbolic neighborhood of each, after moving them via a Möbius transformation (i.e. a conformal map on the disk) to the center of the disk. This pointwise distance is then used to define a Wasserstein distance Rubner et al. (2000) between the two configurations. Lipman and Daubechies (2011). We have thus:

Definition 1.2 (*Conformal Wasserstein neighborhood distance (cWn)*) Given two disk-type surfaces S_1, S_2 we map each conformally to the hyperbolic disk $D = \{z||z| < 1\}$, with corresponding conformal factors μ, ν. For a fixed choice of $R <$

< 1, we define the *conformal neighborhood dissimilarity distance* $\mathbf{d}^R_{\mu,\nu}(v, w)$ as the minimum over $\theta \in [0, 2\pi]$ of the L^1-distance (with respect to the hyperbolic surface area of the disk D) between the restrictions to the disk $N(0, R) = \{z\,;\ |z| < R\}$ of the functions $\mu \circ \tau_1{}^{-1}$ and $\nu \circ \tau_2{}^{-1} \circ R_\theta$, where τ_1, τ_2 are, respectively, Möbius transformations that map v and w to the center 0 of D, and R_θ is the rotation around 0 by an angle θ. Equivalently, $\mathbf{d}^R_{\mu,\nu}(v, w)$ is defined as:

$$\mathbf{d}^R_{\mu,\nu}(v, w) = \inf_{m \in \mathcal{M}_{\mathcal{D}}, m(v)=w} \left[\int_{N(v,R)} |\mu(z) - (m * \nu)(z)\,|\, d_{vol_H}(z) \right], \tag{3}$$

where $\mathcal{M}_{\mathcal{D}}$ is disk-preserving Mobius group, $N(v, R) = \tau_1{}^{-1}\,(N(0, R))$, and d_{vol_H} is the volume form for the hyperbolic disk.

The distance between the two surfaces S_1 and S_2 is then defined by the standard Kantorovich approach to mass transportation distances:

$$\mathcal{D}^R_{\text{cWn}}(S_1, S_2) = \inf_{\pi \in \Pi(\mu,\nu)} \int_{\mathbf{D}\times\mathbf{D}} \mathbf{d}^R_{\mu,\nu}(v, w)\, d\pi(v, w)\,, \tag{4}$$

where $\Pi(\mu, \nu)$ is the set of all probability measures on $\mathbf{D}\times\mathbf{D}$ of which the marginals coincide with respectively μ and ν.

An attractive feature of this distance is that it determines a (dis)similarity between points on S_1 and S_2 by comparing their respective surroundings on each surface, which is quite local; on the other hand, it also takes into account the global setting, by integrating over the whole surface.

In practice, the cWn distance, with its multiple steps each involving optimization (including one over the 5-dimensional set of Möbius transformations), was fairly heavy computationally. We therefore also introduced another approach, the *Continuous Procrustes distance* (cP), Al-Aifari et al. (2013), inspired by the Procrustes distance but landmark-free.

When landmarks are given on a collection of surfaces, one could conceivably interpolate between them and thus use this to define correspondence maps defined on the full surfaces (and not just on the landmark points), or, in a computational setting, on a much larger number of discrete points than only the landmarks. In the absence of landmark points between which to interpolate, a reasonable approach is to pick a large number N of "reasonably well distributed" points on each surface and view bijections between the $N-$tuples on different surfaces as discrete approximations to possible correspondence maps. If "reasonably well distributed" means that the Voronoi region around a random sample point should have an area of (approximately) $1/N$ (assuming the total area of each surface has been normalized to 1), then these bijections are discrete stand-ins for area-preserving maps. This motivated the following definition of the *Continuous Procrustes distance*:

Definition 1.3 (*Continuous Procrustes distance between surfaces (cP)*) Given two surfaces S_1, and S_2, we denote by $\mathcal{A}(S_1, S_2)$ the set of all area-preserving Al-Aifari et al. (2013) diffeomorphisms, and let $\mathbb{E}_3$ be the Euclidean group on $\mathbb{R}^3$. Then the Continuous Procrustes distance between S_1 and S_2 is given as

$$\mathbf{D}_{cP}(S_1, S_2) = \inf_{C \in \mathcal{A}(S_1, S_2)} \left[\inf_{R \in \mathbb{E}_3} \left(\int_{S_1} ||R(x) - C(x)||^2 \, d\mathrm{vol}_{S_1}(x) \right)^{1/2} \right]. \quad (5)$$

Note that the cWn-distance was *intrinsic*: it used only information "visible" from within each surface, such as geodesic distances between pairs of points; consequently it would not be able to distinguish a surface from any of its isometric embeddings in 3-D. (For instance, it would not distinguish a Swiss roll 2D-surface from a rectangle.) The cP uses *extrinsic* information about the surfaces; its computation also provides, as a side result, a mapping between the surfaces, which is the continuous Procrustes registration described in the previous section.

Comparing the results of the cWn and cP distances with the Procrustes distances derived from expert-placed landmarks (which to our biological collaborators served as "ground truth") showed that the cP distance performs better Boyer et al. (2011); we have therefore used it more systematically, at least in its numerical implementation. Moreover, we proved in Al-Aifari et al. (2013) that when the cP distance between two surfaces is small, the corresponding minimizing area-preserving map is also close to the minimizing conformal map; this has the important consequence that in a numerical implementation, the (practically infeasible) search over area-preserving maps can be replaced by a (much more manageable) search over conformal maps, with a guarantee on the error on the estimate for the cP distance made by substituting this easier optimization.

The cP correspondence maps are determined pairwise, and turn out (unsurprisingly) to be non-transitive, which is a disadvantage for collections of bone or tooth shapes from species that are sufficiently closely related that biologists expect transitive correspondence maps between all of them. Uniqueness of correspondence could be restored by considering only those maps associated with a Minimizing Spanning Tree in the distance graph, but this then suffers from the sensitivity to inserting extra specimens in the collection pointed out before. Finally, as reported in Boyer et al. (2012) there exist anomalies in some of the pairwise correspondence maps between specimens. Thus, it is not clear how the resulting cP correspondence maps could be incorporated in a more traditional geometric morphometric workflow.

In a later subsection, we introduce a new mechanism that uses tools from differential geometry, that uses the cP correspondence maps (or other correspondence maps, if available) as input to a reparametrization that leads to higher-quality, transitive correspondence maps and an improved distance graph. In order to define this new distance, we first need to introduce some extra mathematical machinery. We will discuss Diffusion maps, the corresponding classical Diffusion Distances, and finally Horizontal-based diffusion in full length in Sect. 4.3. However, for the sake of completeness in our list of distances, we provide here place-holding definitions for Diffusion distance and Horizontal-based diffusion distance and refer the reader to additional details in Sects. 4.2 and 4.3.

Definition 1.4 (*Diffusion Distance (DD)*) Given a collection of objects, endowed with a distance metric. Let $(W_t)_{t>0}$ be the diffusion matrices constructed by means of these distances, as defined in (11), and let $\psi_1, \ldots, \psi_L$ be the first few dominant non-trivial eigenvectors of W_τ, for one appropriately chosen $\tau > 0$ (see below), with corresponding eigenvalues $\lambda_1, \ldots, \lambda_L$. Then we define a parametrization of the objects by means of the values of these eigenvectors (i.e. object k is parametrized by the values $\psi_\ell(k)$, with ℓ ranging over $1, \ldots, L$) and for each $\gamma > 0$ we define a γ-dependent diffusion map D_γ by setting

$$D_\gamma(k, k')^2 = \sum_{\ell=1}^{L} \lambda_\ell^\gamma \, |\psi_\ell(k) - \psi_\ell(k')|^2 . \tag{6}$$

The Diffusion Distance can be (and has been) used for much more general collections than the bones and teeth we consider here; in our case, we shall apply the Diffusion distance approach to approximative cP distances between bone or teeth surfaces that we compute from "well-distributed" samples on those surfaces. We shall see in Sect. 4.2 that this leads to results that are much more meaningful for our biological collaborators than the cP distances themselves. Taking into account not only the (approximative) cP distances but also the approximative correspondence maps will lead to even better distances, via the Horizontal Base Diffusion Distance.

Definition 1.5 (*Horizontal Base Diffusion Distance (HBDD)*) Given a collection of surfaces $(S_k)_{k=1,\ldots,K}$, endowed with a distance between the surfaces and pairwise correspondence maps between the surfaces. Let $(\mathbf{W}_t)_{t>0}$ be the bundle diffusion matrices constructed by means of these distances and correspondence maps, as defined in (11), and let $\Psi_1, \ldots, \Psi_L$ be the first few dominant non-trivial eigenvectors of $\mathbf{W}_\tau$, for one appropriately chosen $\tau > 0$. with corresponding eigenvalues $\lambda_1, \ldots, \lambda_L$. Each of these eigenvectors has an entry for each surface k and for each point $x \in S_k$. For each surface k, we consider the matrix G_k^γ of the restrictions to S_k of the L eigenvectors, weighted by the γ-th power of their corresponding eigenvalues. Then the Horizontal-based diffusion distance (HBDM) between surfaces k and k' is given as a matrix-distance between G_k^γ and $G_{k'}^\gamma$,

$$d_{\mathrm{HBDD},\gamma}\,(S_k, S_{k'})) = \left\| G_k^\gamma - G_{k'}^\gamma \right\| . \tag{7}$$

3.3 *Quantifying Errors of Shape Distances*

One of the pillar-stones of good research is the ability to measure the error of the proposed method, providing insight into the reliability and accuracy of measurements and algorithms. Error evaluation helps not only to identify weaknesses in our methods but also allows us to refine and improve them for future use. Ultimately, by rigorously evaluating errors, we can increase confidence in our findings and advance knowledge and innovation across various domains. Although we shall not go into detail of the error analysis for many of the results reported here, we do want to list a few of the measurement tools we use.

In the paper Gao et al. (2018) three methods are proposed to evaluate the cP registration. First, as initially proposed in Boyer et al. (2011) the Mean Square Error (MSE) can be calculated as the squared Euclidean distances between the expert-determined landmarks on a surface and those propagated, via the cP correspondence map, of expert-determined landmarks on other surfaces. Larger MSEs indicate greater deviations between the computed correspondence map and the (unknown) "true" biological correspondence map. Second, the correspondence can be evaluated by comparing the surface characteristics, e.g. the curvature. Last Gao et al. (2018) considered the skewness of the distribution of candidate maps, and assessed the distinctiveness of the optimal map. Although alignment errors occur with undesirable frequency (27.6%), most of them are relatively small and represent some landmark distortion at either the anterior or posterior end of the tooth; the most egregious errors arose when landmarks were propagated between very dissimilar teeth. As will be shown in the next subsections, such alignment errors between pairs of surfaces with a large cP distance have very little influence on the next-level maps constructed by applying (horizontal) diffusion to the surface collection, considered as a fibre bundle.

4 Manifold Learning

Given a data collection in which the samples are observed or assumed to vary gradually as some parameters governing the samples are varied, it is natural to view the collection as having a manifold structure. The manifold assumption provides us with the tools to study the geometry of the Riemannian manifold. Dimension reduction is a prominent approach to dealing with high-dimensional data (PCA Pearson (1901), Multidimensional Scaling Cox and Cox (2000), Linear Discriminant Analysis Fisher (1936), Locality Preserving Projections He and Niyogi (2004), Locally Linear Embedding Roweis and Saul (2000), ISOMAP Tenenbaum et al. (2000), to mention just a few); other methods reconstruct and study the manifold geometry in high dimension (for a detailed survey about such methods please see Faigenbaum-Golovin and Levin (2023)). We focus here on Diffusion Maps Coifman et al. (2005), which play a big role in our geometric analysis. In this section, we review the mathematical apparatus that is used in the next section. Our main vehicle for studying the geometry of the base manifold is the Diffusion Maps on Fibre Bundles Gao (2021), and utilize the Procrustes distance between surfaces defined in Sect. 3.2 as "input".

4.1 Diffusion Map

The Diffusion Map (DM) Coifman et al. (2005) is a frequently used non-linear method to perform dimensional reduction. It proposes a probabilistic interpretation for graph-Laplacian-based dimensionality reduction algorithms. Under the assumption that the discrete graph is appropriately sampled from a smooth manifold, it assigns transition probabilities from a vertex to each of its neighbors (vertices connected to it) according to the edge weights, thus defining a graph random walk, the continuous limit of which is a diffusion process over the underlying manifold. The eigenvalues and eigenvectors of the graph Laplacian, which converge to those of the manifold Laplacian under mild assumptions Belkin and Niyogi (2008, 2006), then reveal intrinsic information about the smooth manifold. The associated operators form a semi-group, as discussed below.

Definition 1.6 (*Diffusion Map*) Let $\{x_i\}$ be a data set of points in $\mathbb{R}^n$ sampled from a Riemannian manifold M whose intrinsic dimension is low. We define the diffusion operator on the discrete data samples as $D_t^{-1} W_t$ where $W_{t;i,j} = \exp(||x_i - x_j||^2/4t)$, and $D_t = [D_{t;i,i}]$ is the diagonal matrix with entries defined by $D_{t;i,i} = \sum_j W_{t;i,j}$ (non-diagonal entries of D_t are 0).

Explicitly the diffusion map is given as

$$H_{t;i,j} = N_{t;i} \exp\left(\frac{||x_i - x_j||^2}{4t}\right), \tag{8}$$

where $N_{t;i}$ is a normalizing factor ensuring that $\sum_j H_{t;i,j} = 1$ for all i.

We omit here a detailed discussion of the theory underlying the use of the diffusion map; see e.g. Coifman et al. (2005); Singer and Coifman (2008). When the x_i correspond to a uniform sampling of points on a compact Riemannian manifold $\mathcal{M}$, the eigenvalues and eigenvectors of the matrix H approximate, in the limit as the sampling density goes to ∞, and for sufficiently small t, the eigenvalues and eigenvectors of the (heat) diffusion on $\mathcal{M}$ induced by the Laplace-Beltrami operator on $\mathcal{M}$. The argument uses that the heat diffusion kernel has the same form on all Euclidean spaces (only its normalization factor, reflects the dimension). For x_i and x_j sufficiently close together on $\mathcal{M}$, the difference between the distance $\|x_i - x_j\|$ in the ambient high-dimensional Euclidean space and the distance to the origin of the pull-back to the tangent space at x_i of the point x_j becomes negligible, so that the kernel in (8) is appropriate in this limit. The main advantage of the diffusion map is that it allows to perform a local, non-linear dimensional reduction, and enables dealing with manifolds with various geometries. In practice, one carries out a spectral analysis of H, and one parametrizes each $N-$dimensional data point x_i by the $k-$tuple $\left(\lambda_\ell^\tau\, u_\ell(i)\right)_{\ell=1,\ldots,k}$, where the u_ℓ and λ_ℓ are the eigenvectors and eigenvalues of H, k is a (smallish) integer, and τ is set to a value dependent on the problem.

Diffusion Semi-Group Property

Tuning the diffusion time is crucial, and is tightly connected to the question of studying local and global geometry. Picking a too large diffusion time can result in misreading local manifold geometry, since it may give too much weight to large distances, the estimates of which are expected to be much less trustworthy. On the other hand, if the diffusion time is too small, then there is no or barely any diffusion, so that the analysis fails to capture the geometric properties of the underlying manifold. In order to choose the optimal diffusion time, the paper Shan and Daubechies (2022); Shan (2019) proposed to use the extent to which the candidate-diffusion operator exhibits Semi-group behavior in order to calibrate the t. Let us first define the Semi-group Property as

Definition 1.7 (*Semi-group Property*) A family of operators $\{T^t\}$ for $0 \leq t < \infty$, in which T^0 is the identity operation, is said to have semi-group property if $T^{t_1+t_2} = T^{t_1}T^{t_2}$.

The family of heat diffusion operators on a manifold forms a strongly continuous semigroup. The discrete operators H_t can be reasonably expected to be good discretizations of a true manifold diffusion operator only if they approximately obey the Semi-Group Property. This motivates the following definition:

Definition 1.8 (*Semi-group Error*) The (t-dependent) Semi-group Error of the operators $(H_t)_{t>0}$ is defined as

$$SGE(t) = ||(H_t^2 - H_{2t}|| \tag{9}$$

If $(H_t)_{t>0}$ were a true semi-group, then $SGE(t)$ would be zero for all t. As shown in Shan and Daubechies (2022) and Shan (2019), in practice $SGE(t)$ typically decreases as t increases (starting from t close to zero), then reaches a minimum, after which it starts increasing again. Picking the diffusion time t around the value for which this minimum is reached is a heuristic way of "tuning" the diffusion time that produces good results in applications, as illustrated in Shan and Daubechies (2022) and Shan (2019). In what follows, we shall assume that the value t picked for a diffusion operator H_t has been "tuned" in this way.

4.2 *Diffusion Distance*

Although the operator H_t is not symmetric (because of its non-symmetric normalization), one can determine its eigenvectors and eigenvalues by simply considering the symmetric operator $D_t^{1/2} H_t D_t - 1/2$; this symmetric operator has an orthonormal basis of eigenvectors, applying $D_t - 1/2$ to each of these eigenvectors turns it into an eigenvector of H_t, with the same eigenvalue. Typically only the first non-trivial dominant eigenvectors of H_t are "trustworthy": the distance matrix estimate from

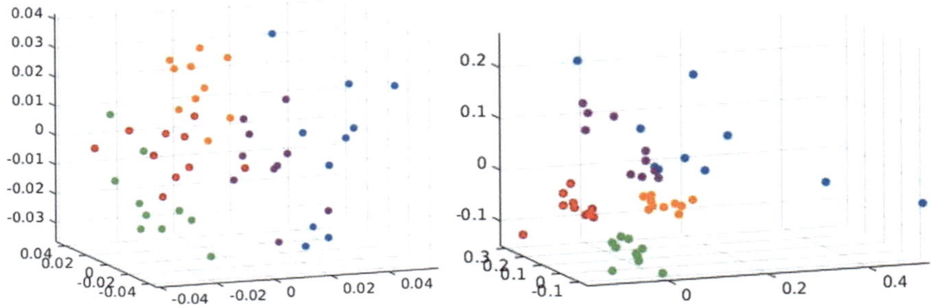

Fig. 7 Comparing the cP distances (left) and DD distances (right) for a dataset of second molars from 50 species belonging to 5 genera of monkeys; in each case a 3D MDS representation of the distances are shown, with one colored dot for each molar, where species belonging to the same genus are given the same color

which we start is typically only an approximate estimate, and their inaccuracy affects the higher-order eigenvectors and eigenvalues more than the first few.

Note that we use only the non-trivial eigenvectors: because of the normalization of H_t, its very first eigenvector is the constant-1 vector, with eigenvalue 1; it is non-informative and typically not considered. Given those eigenvectors and eigenvalues, the diffusion distance is later computed as indicated in (6). The non-linear embedding identifies a new domain that represents meaningful geometric descriptions of data sets, by generating efficient representations of complex geometric structures.

This can be seen in Fig. 7 which illustrates the power of this diffusion approach. Its left panel shows a multi-dimensional scaling (MDS) representation in 3D of the cP distances for a data set of 50 s molars of different monkey species, belonging to 5 genera (10 species per genus). The right panel of the figure uses those *same* cP distances as input, but analyzes them via the manifold diffusion processing described above (and in greater detail in Gao (2015)); the figure shows the MDS representation in 3D of the corresponding Diffusion Distances computed from this input. It is clear that this extra step in the analysis provides a much better separation of the genera.

4.3 Manifolds in which Each Sample has Structure: Fiber Bundles

The previous subsection illustrated that diffusion distances provide a powerful tool to analyze (dis)similarities of teeth and bones in a biological morphology framework. In using the cP distances as input for a manifold diffusion analysis, we took into account only part of the information produced by the continuous Procrustes approach: as indicated by the formula (5), the cP distance between surfaces S_1 and S_2 is the minimal value obtained in an optimization procedure that also determines a corresponding minimizing correspondence map between S_1 and S_2. In the Horizontal Base Diffusion Distance (HBDD) construction, we use not only the cP distances but also the correspondence amps as input; we do this by considering the unknown manifold describing (or parametrizing, or underlying) the collection of surfaces as a *base manifold*, and each of the surfaces themselves as the *fibers* in a *fiber bundle* framework. In the next few paragraphs, we briefly review this framework, in which we can naturally define a diffusion analysis that lends itself to our purposes.

Definition 1.9 (*Fibre Bundle*) Assume that M and F are orientable Riemannian manifolds. A *Fibre bundle* with *base space* M and *fibre space* F is a quadruple (E, M, F, π) where E (called the *total space*) is a differentiable Riemannian manifold and $\pi : E \to M$ is a surjective smooth map from E to M that satisfies the local triviality condition, meaning that for every $p \in E$, $\exists$ an open neighborhood $U \subset M$ of $\pi(x)$ and a diffeomorphism

$$\phi : \pi^{-1}(U) \to U \times F \tag{10}$$

such that $\pi|_U = \pi_1 \circ \phi$, where $\pi|_U$ is the restriction of π to U and π_1 (from $U \times F$ to U) is the projection to the first components (i.e. it maps (u, f) to u).

For any point $p \in M$, $\pi^{-1}(\{p\})$ is diffeomorphic to F. We then denote $F_p = \pi^{-1}(\{p\})$ as the *fibre* over p. The local trivialization allows a way to describe a specific fibre in a space with a simpler topology.

In our case, we also assume that the bundle comes equipped with correspondence maps between fibres, $P_{yx} : F_x \to F_y$. This function maps points on F_x to points on F_y in a smooth, bijective way. It associates the two fibres and identifies their points to each other. (In Shan (2019) these are interpreted as *parallel transport* in a probabilistic setting.)

An illustration of these concepts appears in the toy example illustrated in the left panel of Fig. 8. Here the two-dimensional manifold $E = \{(x, y, z) \in \mathbb{R}^3;\ x^2 + y^2 = 1\}$ is a cylinder. Then one can view the circle S_1 (the cross-section of the cylinder) can be viewed as the base manifold, and $\mathbb{R}$ is the fibre. The coordinates x, y, z of $p \in E$ can be viewed as a combination of $s := (x, y) \in S_1$, $z \in \mathbb{R}$. Now, the projection map $\pi : E \to S_1$ is defined as $\pi(x, y, z) = (x, y) = s$ and its inverse $\pi(x)^{-1} \cong \mathbb{R}_s$ i.e. the entire line (the fibre over $s \in S_1$). (Note: in this case, E can be identified simply with the product space of M and F; in cases like these, the bundle is said to be *trivial*. For general fibre bundles this need not be the case–for instance, one can add a slight twist to this toy construction and obtain a Moebius-like total space E that is no longer diffeomorphic to $M \times F$.)

In our setting, the total space E consists of the collection of all the points in the union of a collection of (tooth) surfaces that we assume can be parametrized smoothly by a base manifold M that we want to study. As long as we look at species that are closely related to each other, we believe that we can bring them back, following a reasonably short path in a phylogenetic tree, to a common ancestor with reasonably similar teeth. In that case, the mapping from one surface S_i to another S_j is the same

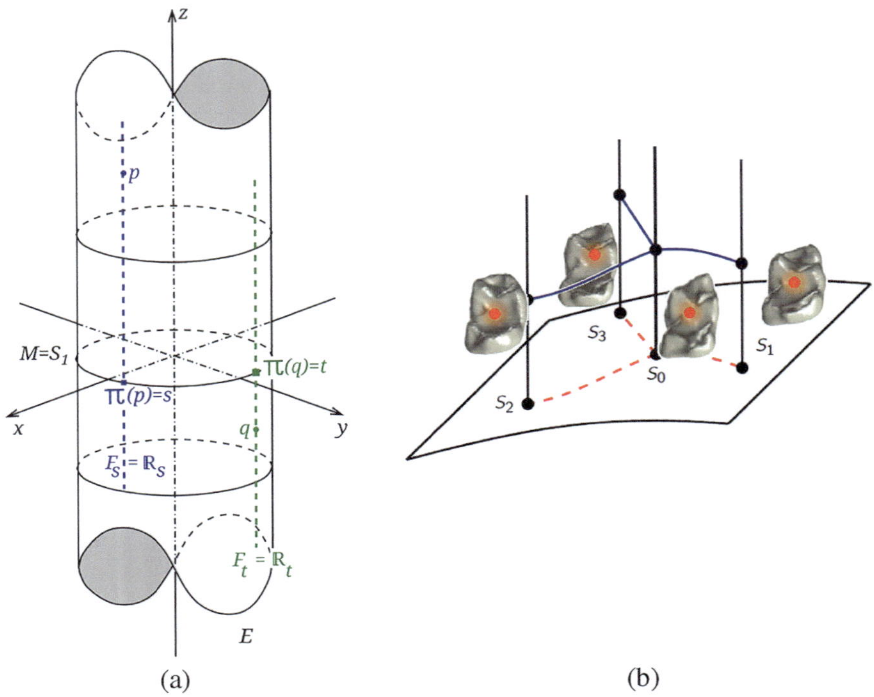

Fig. 8 Left: Illustration of the decomposition of the cylinder manifold into the base manifold S_1, and the vertical fibers, where the total space can be represented as a Cartesian product of the base manifold, and the fiber; Right: Fiber bundle of teeth. Each anatomical surface corresponds to a point on the base manifold. Each surface constitutes a fibre

as the mapping from S_i to an ancestral surface S_A, and back from S_A to S_j. This would then mean that the mappings are all consistent, over the whole collection–that mapping from S_i to S_j and then from S_j to S_k produces the same result as directly from S_i to S_k. (When we will start looking at surfaces for more distantly related species, this will no longer be the case, and we expect our fibre bundles to become non-trivial.) This is illustrated in the right panel of Fig. 8.

Horizontal random walks and diffusion processes on fibre bundles

We shall use the fibre bundle model to define diffusion operators on the total space E, and hope to learn from its spectral analysis more than from simple diffusion on the base manifold M, using the cP distances as input.

To this end, we have to define distances on E, or, equivalently, define a diffusion kernel on $E \times E$. In practice, we have on each of the surfaces S_k, with k ranging from 1 to K, a collection of N_k sample points $\left(x_{i,k}\right)_{i=1,\dots,N_k}$, where the N_k for different k could be (but need not be) equal to each other; for simplicity, we shall assume they are equal in this exposition. We also have (from the computational optimization inspired by the theoretical definition of cP distance), for each pair of surfaces

k, ℓ, correspondence maps $P_{k,\ell}$ that map the collection $\left(x_{i,k}\right)_{i=1,\ldots,N}$ bijectively to $\left(x_{j,\ell}\right)_{j=1,\ldots,N}$ (we are assuming $N_k = N_\ell = N$ here); we shall consider these $P_{k,\ell}$ to be (noisy) approximations for the (unknown) correspondence maps between the continuous surfaces. "Random Walking" on the total surface will consist of walking from surface to surface and, on each surface, from point to point. We don't have an a priori way to "synchronize" the walking in the fibre direction with that along the base direction, so we shall use two different time constants, and explore later what it means to vary one vs. the other. Taking all this into account, we define the (discrete and approximate) diffusion kernel on E as

$$\mathbf{W}_{\tau_1,\tau_2;j,\ell,i,k} = N_{j,\ell} \exp\left(-\frac{||P_{k\ell}(x_{\ell,j}) - x_{k,i}||^2 + ||x_{\ell,j} - P_{\ell k}(x_{k,i})||^2}{4\tau_1}\right) \exp\left(-\frac{D_{k,\ell}^2}{2\tau_2}\right) \quad (11)$$

where $D_{k,\ell}$ denotes the cP distance between S_ℓ and S_k and $P_{k\ell}$ is the map from $\left(x_{j',\ell}\right)_{j'=1,\ldots,N}$ to $\left(x_{i',k}\right)_{i'=1,\ldots,N}$ described above; the normalization $N_{j,\ell}$ ensures that the sum over all i and k of the $\mathbf{W}_{\tau_1,\tau_2;j,\ell,i,k}$ equals 1. The diffusion parameters τ_1 and τ_2 control the amount of information that is taken into account from the base manifold versus from the fibers.

Thus the HDM matrix is a block matrix in which the (ℓ, k)-th block captures the comparison of surface ℓ and surface k while relying on the correspondence map between them. The (j, i) entry in the (ℓ, k)-th block of the HDM matrix stands for the edge weight between $x_{\ell,j}$, and $x_{k,i}$. Choosing τ_1 and τ_2 can be tricky; increasing one while keeping the other fixed corresponds to diffusing more rapidly in one of the two directions (base manifold or fibre) than the other. We'll assume that both values are chosen in a regime where the semigroup error is small (see above); within that region, some latitude is possible. Depending on what we want to explore, we may adjust the ratio τ_1/τ_2; we shall come back to this below.

Spectral Distances and Embeddings

The eigenvectors Ψ_r of $\mathbf{W}$ have $N_1 + N_2 + \cdots + N_K$ entries (or KN if all the N_k are equal to N). As expected from a diffusion kernel, due to the normalization of the rows of H, the largest eigenvalue is a trivial eigenvector, with all entries equal to 1, and eigenvalue 1. We shall systematically disregard this (uninformative) eigenvector Ψ_0, and consider only Ψ_ℓ with $\ell \geq 1$. We can consider the eigenvectors as concatenations of K segments of lengths N_k, each indexed according to surface S_k, with k ranging from 1 to K; let's denote these segments by $\psi_{\ell,k}$.(Equivalently, for $m = 1, \ldots, N_k$, $\left(\psi_{\ell,k}\right)_m = (\Psi_\ell)_{N_1+\cdots+N_{k-1}+m}$.) Because the data from which the spectral analysis is constructed were noisy (the points picked on the different S_k are not in strict correspondence to each other, since they were picked randomly on each surface, independently for each surface, leading to errors not only in the correspondence maps but also in the approximate cP distances between the surfaces), the eigenvectors are noisy as well; this expresses itself all the more as the corresponding eigenvalue decreases. In practice, we therefore use only the L dominant eigenvectors and eigenvalues, i.e. we assume the ordering is such that the λ_ℓ decrease as ℓ increases, and we use only the Ψ_ℓ and λ_ℓ, for $\ell \leq L$; the value we pick for L can vary depending on the application. Below we shall see two applications, one with $L = 3$ and another with $L = 100$; in all cases L is significantly smaller than any of the N_k.

One way to "distill" the geometric content from the Ψ_ℓ, $\ell = 1, \ldots L$, as it pertains to each of the surfaces S_k, is to consider, for each k, the $R \times R$ Gramm matrix G_k defined by

$$(G_k)_{\ell,\ell'} = \langle \psi_{\ell,k}, \psi_{\ell',k} \rangle = \sum_{m=1}^{N_k} (\Psi_\ell)_{N_1+\cdots+N_{k-1}+m} \overline{(\Psi_{\ell'})_{N_1+\cdots+N_{k-1}+m}} \;; \quad (12)$$

we also consider the weighted Gramm matrices G_k^t defined by

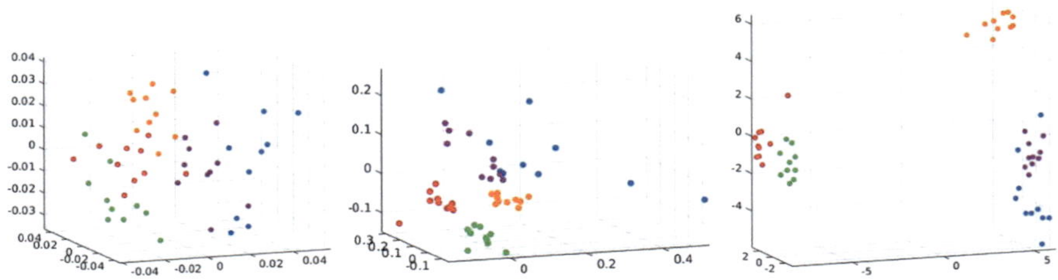

Fig. 9 Geometric Similarity of Anatomical Surfaces using different distance calculations. Left: Embeddings into $\mathbb{R}^3$ using *Multi-dimensional Scaling* of CPD (5); Middle: DM calculated after the diffusion map embedding (8); Right: and HBDD

$$\left(G_k^{\gamma}\right)_{\ell,\ell'} = \lambda_\ell^{\gamma/2}\lambda_{\ell'}^{\gamma/2}\left(G_k\right)_{\ell,\ell'} \ . \tag{13}$$

In Gao (2021), the *HBDD distance* $d_{\text{HBDD},\gamma}(S_k, S_{k'})$ is then defined as a matrix distance between the two $R \times R$ matrices G_k^{γ} and $G_{k'}^{\gamma}$. (Several possible candidates can be considered for this matrix distance; if one uses the Hilbert-Schmidt norm, then the HBDD distance can be easily reformulated in terms of the original $\psi_{\ell,k}$ vectors, which is then closer to the embeddings considered in Coifman and Lafon (2006), extended to this fiber bundle setting.)

To illustrate the improvement of the HBDD distance over the Diffusion Distance, we revisit the comparison provided in Fig. 7, adding now a third panel on the right in which we show the MDS embedding of the HBDD distances for the same data set. It is apparent that the surfaces (2nd molars of monkeys in this case) for the five different genuses are clustered much better with HBDD than with the Diffusion distance. In addition, the HBDD reflects the dietary groups within the data set: the folivores Alouatta (red) and Brachyteles (green) are adjacent to each other in the rightmost panel of Fig. 9, and so are the frugivores Ateles (blue) and Callicebus (purple); the insectivore Saimiri (orange) is far from the other herbivorous groups.

4.4 *Improved Spatial Correspondence*

The spectral analysis of the Bundle Diffusion provides a different parametrization for the surfaces: for each $k = 1, \ldots, K$, we associate to each point $p_m \in S_k$, with $m = 1, \ldots, N_k$ the L-tuple $\left(\psi_{\ell,k}\right)_m = (\Psi_\ell)_{N_1+\cdots+N_{k-1}+m}$, with ℓ ranging from 1 to some predetermined L. This defines a mapping from the entire collection of surfaces into $\mathbb{R}^L$; we call this the Horizontal Diffusion Map (HDM). See Fig. 11 for an illustration of embedding two surfaces (outmost left and right) into the diffusion space to recover the fiber manifold (in the middle). The main advantage of HDM embedding is that not only does it uncover a new domain that captures meaningful geometric characteristics of datasets, but also preserves the intrinsic geometry of each surface S_k.

In this section, we review a few different applications of the HDM, related to improving spatial correspondences between the S_k.

Slice and Dice: Segmenting Surfaces to Match Evolutionary Equivalents

Once we have embedded a collection of surfaces into $\mathbb{R}^L$, we can use standard clustering algorithms to determine the best way to separate the collection into some predetermined number M of clusters. Figure 10 shows the result on 10 different primate teeth (part of a larger collection of 32 that were analyzed with Bundle Diffusion); in this case L was picked to be 100, and $M = 12$. To visualize the clustering done in HDM space in terms of the surfaces themselves, one of 12 colors was assigned

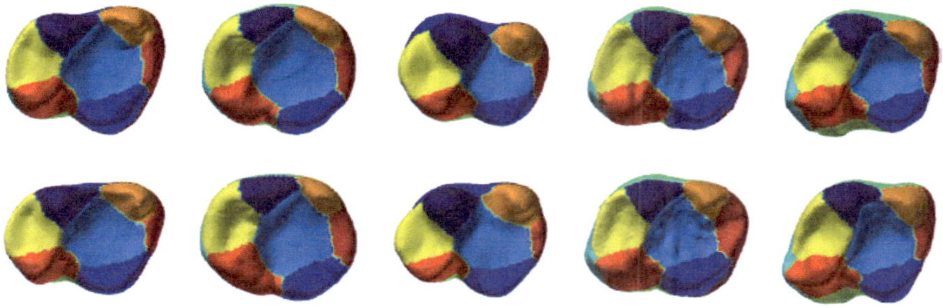

Fig. 10 Globally consistent segmentation by spectral clustering for HDM. From the left to the right row: Alouatta, Ateles, Brachyteles, Callicebus, Saimiri, image source Gao (2021). In each column are different teeth examples from the same species

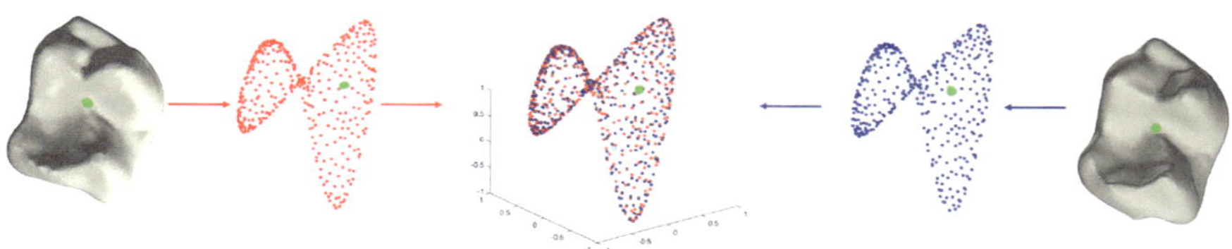

Fig. 11 Fiber bundle diffusion map calculated based on a collection of teeth. Its eigenvectors are later used to embed the surface coordinates into 3-dimensional space (in the middle). The different colors (red, and blue) represent points from the two different teeth projected to the diffusion space. As can be seen, the projected surfaces closely recover the fiber manifold

to each of the 12 clusters, and points on the surfaces in the collection were given the color of the cluster to which their HDM embedding belonged.

Although there is no a priori constraint imposing it, each cluster color is represented on each of the S_k, and the zones that they delimit on the different teeth appear to be in biological homologous correspondence. Note that the same-color zones on different teeth do not have the same area–even though the analysis started from correspondence maps (the cP maps) that were (approximately) area-preserving. Our biological colleagues are very excited by this potential of the HDM method to segment the original meshes in a consistent way across the entire shape space, similar to what is done in classical anatomical studies, where features are commonly not addressed as points (or landmarks), but rather as sub-regions of the anatomy, and they are starting to use this as a tool in their own statistical analyses.

Improved Spatial Registration

A different but likewise very important application uses a very low value of L, more precisely $L = 3$. Upon examination of the resulting embedding into $\mathbb{R}^3$, one finds that all the S_k are mapped to essentially the same 2-dimensional surface (up to noise effects), as illustrated in Fig. 11. We call this surface the *template surface*.

Of course, the images on the template surface of the randomly picked points $p_{k,m}$ on each S_k are typically all different. By using the template surface, it is now possible, however, to compute for each point $p_{k,m}$ on S_k a corresponding interpolation of the points $p_{k',m'}$ on another S_k' that is much better correspondence with $p_{k,m}$ than any of the individual $p_{k',m'}$. To do this, one considers the image $P_{k,m}$ on the template space, under the embedding, of $p_{k,m}$, and determines weights for the nearby images $P_{k',m'}$ that would make their weighted average coincide with $P_{k,m}$. Taking the same weighted average of the $p_{k',m'}$ on the original $S_{k'}$ defines then a point on $S_{k'}$ that is a better candidate for the point on $S_{k'}$ "truly" corresponding to $p_{k,m}$. One can then use these improved correspondence maps to repeat the entire procedure of constructing the Bundle Diffusion, identifying its Eigenvectors, determining a template surface, and going through another step of improving the correspondence maps. Numerical experiments on toy examples suggest that it may be useful to repeat this iterative procedure a few times; we don't know whether or under what conditions it would converge.

Collection-Based Gaussian Process Landmarking

In Sect. 3.1 we described Gaussian-Process Landmarking, a technique that identifies "geometrical landmarks" on individual surfaces by selecting successively those points that contribute the most new information about the surface, given the points chosen earlier. Once the surfaces S_k in a collection have been jointly registered via the HDM (as described above), one can use a similar approach for the whole collection. In this case one defines a Gaussian kernel on the template surface using the covariance between points over *all* the surfaces S_k simultaneously, and then selects successive landmark points on the template surface, each time picking a new landmark as the point with the largest variance conditioned on all the previously selected landmarks. This automatically defines corresponding landmarks on all the S_k via the HDM map. This is described in more detail in Shan and Daubechies (2025).

5 Other Differential Geometry Tools

The preceding sections have concentrated on tools from differential geometry that we brought to bear on the analysis *collections* of surfaces. However, we have also developed tools to reliably compute geometric properties for individual surfaces; in fact, the very first collaboration of our team combining applied mathematicians and biological morphologists Bunn et al. (2011), Winchester et al. (2014), Winchester (2016) and Pampush et al. (2016) concerned the use of effective discrete algorithms to compute the Dirichlet Normal Energy (DNE) of a surface and its use on molar shapes to distinguish dietary preferences of different mammal species. The DNE is a measure of curvature, computed as the integral, over the surface (normalized to have area 1), of the change in direction of the normal vector. Compared with some other methods used in morphological studies, the DNE had the advantage of being landmark-free and independent of the surface's initial position, orientation, and scale, resulting in a more robust metric, less sensitive to noise than some other tools. Nevertheless, our first implementation for the calculation of the DNE was strongly tied to the mesh representing the surface; once it was applied to a wider range of mesh preparation protocols, it turned out to be overly sensitive to variation in some of the mesh preparation parameters, especially for the boundary triangles.

In Shan et al. (2019) a more robust implementation of the same mathematical concept was introduced, called *ariaDNE* (for **a** **r***obustly* **i***mplemented* **a***lgorithm* for **D***irichlet* **N***ormal* **E***nergy*). To compute the DNE numerically, ariaDNE uses a local PCA to determine the tangent plane as well as the curvature (both approximately)–at sample points P where the surface does not have overly sharp ridges or cusps, the first two principal components of the PCA of the sample points in a neighborhood of P are tangent to the surface, the third principal component approximates the normal direction, and the smallest principal value corresponds to the curvature. Near a ridge or cusp, it may be necessary to adjust the procedure, since the normal direction may then no longer be the one with the smallest PCA eigenvalue; in this case, it is however easy to determine which of the three eigenvectors is normal to the surface at P, and select the corresponding eigenvalue. For numerical stability, ariaDNE uses weighted PCA (with a Gaussian kernel centered on P) rather then local PCA.

Figure 12 shows DNE and ariaDNE values on an example tooth and various meshes representing the same surface and different amounts of noise: (from left to right) 2k triangles, 20k triangles, a

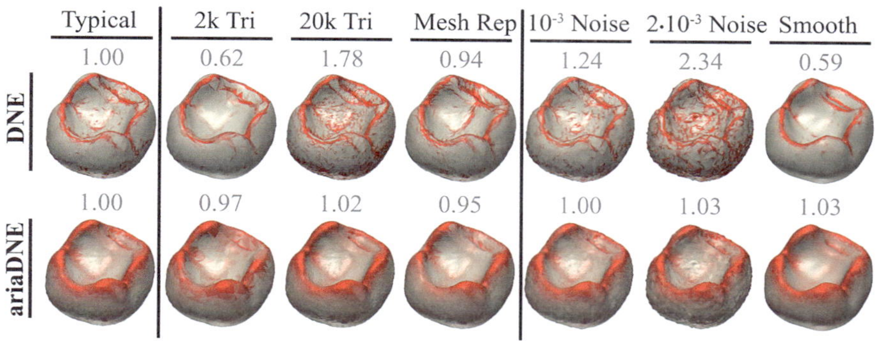

Fig. 12 Comparing effects of triangle count, mesh representation, noise, and smoothing on DNE (top) and ariaDNE (bottom). The distribution of curvature as measured by each method overlaid in shades of red on a grey 3D rendering of the surface is presented Shan et al. (2019)

different mesh representation, 0.001 noise magnitude, 0.002 noise, and smoothing (commonly used to eliminate noise produced during mesh preparation). The red surface shading indicates the value of curvature as measured by each method. As can be seen, by the max curvatures value heading the surfaces, the ariaDNE is remarkably stable to noise and different mesh configurations.

6 Summary and Future Research Directions

This paper summarizes methods that were developed, based on differential geometry and machine learning, to analyze collections of anatomical surfaces within the framework of evolutionary anthropology. The surfaces, representing teeth or bones, are scanned and processed as triangular meshes; they pose challenges, due to nuanced anatomical variations, that are quite different from the classical problems in Computer Graphics and CAGD. To address these challenges, tools are required that differ significantly from classical surface analysis methods. Through a comprehensive survey spanning more than 15 years of research, the paper covers methods for surface comparison, registration, alignment, and distance calculation. It also addresses broader topics such as landmark selection, shape segmentation, and classification analysis, and showcases the advantages of manifold analysis in addressing evolutionary biology inquiries. While focusing primarily on disc-type surfaces resembling teeth, the methods discussed are adaptable (and in several instances, have been adapted) to sphere-type surfaces like bones, suggesting broader applicability and potential avenues for future research. We are currently working also on combining some of the approaches outlined above, e.g. in joint Gaussian Landmarking for collection of surfaces and iterative increasingly accurate registration (both mentioned above), but also incorporating not only spatial coordinates but also other features (such as the scalar DNE field) in computing generalized continuous Procrustes distances to be used in subsequent diffusion-based learning. We also hope to learn how to coarse-grain the analysis so as to extend our analysis to species that are more distantly related. Finally, we are convinced our bundle-diffusion approach is applicable to many other manifold learning settings beyond anatomical surfaces, in which the data "points" have internal structure and the (dis)similarity between them relies on comparing these internal structures.

The tools outlined in this paper are summarized in the following table (Table 1), providing references to both the original paper and a link to the corresponding code repository.

While most of the 3D surfaces in this paper appear in the Morphosource (2024), there are other datasets that we thought are worth mentioning.

Acknowledgements The authors gratefully acknowledge support from the Simons Foundation Math + X grant 400837. Shira is grateful to the Eric and Wendy Schmidt Fund for Strategic Innovation, the Zuckerman-CHE STEM Program.

References

Al-Aifari, R., Daubechies, I., Lipman, Y.: Continuous procrustes distance between two surfaces. Commun. Pure Appl. Math. **66**, 934–964 (2013)

Auto3dgm. https://github.com/ToothAndClaw/Auto3dgm_Python. Accessed 27 Mar 2024

Avizo. www.thermofisher.com. Accessed 27 Mar 2024

Belkin, M., Niyogi, P.: Convergence of Laplacian eigenmaps. Adv. Neural Inf. Process. Syst. **19**, (2006)

Belkin, M., Niyogi, P.: Towards a theoretical foundation for Laplacian-based manifold methods. J. Comput. Syst. Sci. **74**, 1289–1308 (2008)

Boyer, D., Costeur, L., Lipman, Y.: Earliest record of Platychoerops (primates, Plesiadapidae), a new species from Mouras quarry, Mont de Berru, France. Am. J. Phys. Anthropol. **149**, 329–346 (2012)

Boyer, D., Lipman, Y., St. Clair, E., Puente, J., Patel, B., Funkhouser, T., Jernvall, J., Daubechies, I.: Algorithms to automatically quantify the geometric similarity of anatomical surfaces. Proc. Natl. Acad. Sci. **108**, 18221–18226 (2011)

Boyer, D.: Relief index of second mandibular molars is a correlate of diet among prosimian primates and other euarchontan mammals. J. Hum. Evol. **55**, 1118–1137 (2008)

Boyer, D., Puente, J., Gladman, J., Glynn, C., Mukherjee, S., Yapuncich, G., Daubechies, I.: A new fully automated approach for aligning and comparing shapes. Anat. Rec. **298**, 249–276 (2015)

Boyer, D., Gunnell, G., Kaufman, S., McGeary, T.: Morphosource: archiving and sharing 3-D digital specimen data. Paleontol. Soc. Pap. **22**, 157–181 (2016)

Bunn, J., Boyer, D., Lipman, Y., St. Clair, E., Jernvall, J., Daubechies, I.: Comparing Dirichlet normal surface energy of tooth crowns, a new technique of molar shape quantification for dietary inference, with previous methods in isolation and in combination. Am. J. Phys. Anthropol. **145**, 247–261 (2011)

Coifman, R., Lafon, S.: Diffusion maps. Appl. Comput. Harmon. Anal. **21**, 5–30 (2006)

Coifman, R., Lafon, S., Lee, A., Maggioni, M., Nadler, B., Warner, F., Zucker, S.: Geometric diffusions as a tool for harmonic analysis and structure definition of data: diffusion maps. Proc. Natl. Acad. Sci. **102**, 7426–7431 (2005)

Cox, T., Cox, M.: Multidimensional scaling. Chapman (2000)

Deng, B., Yao, Y., Dyke, R., Zhang, J.: A survey of non-rigid 3D registration. Comput. Graph. Forum **41**, 559–589 (2022)

Dragonfly. https://www.theobjects.com/dragonfly. Accessed 27 Mar 2024

Faigenbaum-Golovin, S., Levin, D.: Manifold reconstruction and denoising from scattered data in high dimension. J. Comput. Appl. Math. **421**, 114818 (2023)

Fisher, R.: The use of multiple measurements in taxonomic problems. Ann. Eugen. **7**, 179–188 (1936)

Fulwood, E., Shan, S., Winchester, J., Gao, T., Kirveslahti, H., Daubechies, I., Boyer, D.: Reconstructing dietary ecology of extinct strepsirrhines (Primates, Mammalia) with new approaches for characterizing and analyzing tooth shape. Paleobiology **47**, 612–631 (2021)

Gao, T.: Hypoelliptic Diffusion Maps and their Applications in Automated Geometric Morphometrics. Duke University (2015)

Gao, T.: The diffusion geometry of fibre bundles: Horizontal diffusion maps. Appl. Comput. Harmon. Anal. **50**, 147–215 (2021)

Gao, T., Yapuncich, G., Daubechies, I., Mukherjee, S., Boyer, D.: Development and assessment of fully automated and globally transitive geometric morphometric methods, with application to a biological comparative dataset with high interspecific variation. Anat. Rec. **301**, 636–658 (2018)

Gao, T., Kovalsky, S., Boyer, D., Daubechies, I.: Gaussian process landmarking for three-dimensional geometric morphometrics. SIAM J. Math. Data Sci. **1**, 237–267 (2019a)

Gao, T., Kovalsky, S., Daubechies, I.: Gaussian process landmarking on manifolds. SIAM J. Math. Data Sci. **1**, 208–236 (2019b)

GPLmkBDMatch. https://github.com/shaharkov/GPLmkBDMatch. Accessed 27 Mar 2024

Greenwade, G.: The comprehensive TEX archive network (CTAN). TUGBoat **14**, 342–351 (1993)

He, X., Niyogi, P.: Locality preserving projections. *Advances In Neural Information Processing Systems*. pp. 153–160 (2004)

Herrera, J., Dávalos, L.: Phylogeny and divergence times of lemurs inferred with recent and ancient fossils in the tree. Syst. Biol. **65**, 772–791 (2016)

Intanagonwiwat, C., Govindan, R., Estrin, D., Heidemann, J., Silva, F.: Directed diffusion for wireless sensor networking. IEEE/ACM Trans. Netw. **11**, 2–16 (2003)

Kaick, O., Zhang, H., Hamarneh, G., Cohen-Or, D.: A survey on shape correspondence. Comput. Graph. Forum **30**, 1681–1707 (2011)

Lipman, Y., Daubechies, I.: Conformal Wasserstein distances: comparing surfaces in polynomial time. Adv. Math. **227**, 1047–1077 (2011)

Moenning, C., Dodgson, N.: Fast marching farthest point sampling. University of Cambridge, Computer Laboratory (2003)

Morphosource. https://www.morphosource.org/users/sign_in?locale=en. Accessed 27 Mar 2024

Nagai, Y., Ohtake, Y., Suzuki, H.: Tomographic surface reconstruction from point cloud. Comput. & Graph. **46**, 55–63 (2015)

Pampush, J., Winchester, J., Morse, P., Vining, A., Boyer, D., Kay, R.: Introducing molaR: a new R package for quantitative topographic analysis of teeth (and other topographic surfaces). J. Mamm. Evol. **23**, 397–412 (2016)

Pearson, K.: LIII. On lines and planes of closest fit to systems of points in space. Lond. Edinb. Dublin Philos. Mag. J. Sci. **2**, 559–572 (1901)

Puente, J.: Distances and Algorithms to Compare Sets of Shapes for Automated Biological Morphometrics. Princeton University, Princeton, NJ (2013)

Rasmussen, D., Shekelle, M., Walsh, S., Riney, B.: The dentition of Dyseolemur, and comments on the use of the anterior teeth in primate systematics. J. Hum. Evol. **29**, 301–320 (1995)

Ravier, R.: Algorithms with Applications to Anthropology. Duke University (2018a)

Ravier, R.: Eyes on the prize: improved biological surface registration via forward propagation (2018b). arXiv:1812.10592

Rolfe, S., Pieper, S., Porto, A., Diamond, K., Winchester, J., Shan, S., Kirveslahti, H., Boyer, D., Summers, A., Maga, A.: SlicerMorph: an open and extensible platform to retrieve, visualize and analyse 3D morphology. Methods Ecol. Evol. **12**, 1816–1825 (2021)

Rose, K., Bown, T.: Gradual phyletic evolution at the generic level in early Eocene omomyid primates. Nature **309**, 250–252 (1984)

Roweis, S., Saul, L.: Nonlinear dimensionality reduction by locally linear embedding. Science **290**, 2323–2326 (2000)

Rubner, Y., Tomasi, C., Guibas, L.: The earth mover's distance as a metric for image retrieval. Int. J. Comput. Vision **40**, 99–121 (2000)

Shan, S., Daubechies, I.: Diffusion maps: using the semigroup property for parameter tuning. In: Theoretical Physics, Wavelets, Analysis, Genomics: An Indisciplinary Tribute To Alex Grossmann, pp. 409–424 (2022)

Shan, S., Daubechies, I.: Gaussian process joint landmarking for collections of surfaces (forthcoming)

Shan, S.: Probabilistic Models on Fibre Bundles. Duke University (2019)

Shan, S., Kovalsky, S., Winchester, J., Boyer, D., Daubechies, I.: ariaDNE: a robustly implemented algorithm for Dirichlet energy of the normal. Methods Ecol. Evol. **10**, 541–552 (2019)
Simons, E.: The dentition of Ourayia: its bearing on relationships of omomyid prosimians (1961)
Singer, A., Coifman, R.: Non-linear independent component analysis with diffusion maps. Appl. Comput. Harmon. Anal. **25**, (2008)
Sober, B., Levin, D.: Manifold approximation by moving least-squares projection (MMLS). Constr. Approx. **52**, 433–478 (2020)
Steenrod, N.: The Topology of Fiber Bundles. Volume 14 of Princeton Mathematical Series, vol. 16 (1951)
Tenenbaum, J., De Silva, V., Langford, J.: A global geometric framework for nonlinear dimensionality reduction. Science **290**, 2319–2323 (2000)
Turner, K., Mukherjee, S., Boyer, D.: Persistent homology transform for modeling shapes and surfaces. Inf. Inference J. IMA **3**, 310–344 (2014)
Wang, B., Sudijono, T., Kirveslahti, H., Gao, T., Boyer, D., Mukherjee, S., Crawford, L.: A statistical pipeline for identifying physical features that differentiate classes of 3D shapes. Ann. Appl. Stat. **15**, 638–661 (2021)
Wärmländer, S., Garvin, H., Guyomarc'h, P., Petaros, A., Sholts, S.: Landmark typology in applied morphometrics studies: what's the point? Anat. Rec. **302**, 1144–1153 (2019)
Watanabe, A.: How many landmarks are enough to characterize shape and size variation? PLoS One **13**, e0198341 (2018)
Winchester, J., Boyer, D., St. Clair, E., Gosselin-Ildari, A., Cooke, S., Ledogar, J.: Dental topography of platyrrhines and prosimians: convergence and contrasts. Am. J. Phys. Anthropol. **153**, 29–44 (2014)
Winchester, J.: MorphoTester: an open source application for morphological topographic analysis. PLoS One **11**, e0147649 (2016)
Yapuncich, G., Kemp, A., Griffith, D., Gladman, J., Ehmke, E., Boyer, D.: A digital collection of rare and endangered lemurs and other primates from the Duke Lemur Center. PLoS One **14**, e0219411 (2019)

Open Access This chapter is licensed under the terms of the Creative Commons Attribution 4.0 International License (http://creativecommons.org/licenses/by/4.0/), which permits use, sharing, adaptation, distribution and reproduction in any medium or format, as long as you give appropriate credit to the original author(s) and the source, provide a link to the Creative Commons license and indicate if changes were made.

The images or other third party material in this chapter are included in the chapter's Creative Commons license, unless indicated otherwise in a credit line to the material. If material is not included in the chapter's Creative Commons license and your intended use is not permitted by statutory regulation or exceeds the permitted use, you will need to obtain permission directly from the copyright holder.

Existence of Minimizers in the Variational Approach to Brittle Fracture

Antonin Chambolle

Abstract We review the variational approach to crack growth of Francfort and Marigo, based on incremental *free discontinuity problems* in linearized elasticity, and the successive techniques used to establish the existence of minimizers for such problems. In particular, we mention the crucial Korn-Poincaré inequality for functions with small jump set (proved with S. Conti and G. Francfort) which is at the root of many subsequent important results. In the second half of the article we give, in the much simpler scalar case, an overview of the proofs of a Poincaré inequality with small jump set.

1 Introduction: The Variational Approach to Brittle Fracture

The classical theory of crack growth in linear-elastic brittle materials assumes that the bulk (elastic) energy dissipated to increase a crack by a certain amount is proportional to the length (in 2D) or surface (in 3D) of the new crack. This was formalized by Griffith (1920, 1924) in the early XXth century. In a modern formulation of that theory, the following situation is considered:

- One is given a reference configuration $\Omega \subset \mathbb{R}^2$ of a linear-elastic material;
- On some part of the boundary $\partial^D\Omega \subset \partial\Omega$, a boundary load U^0, possibly depending on time, is prescribed;
- One is also given a crack path $\Gamma(\ell)$, $\ell \geq 0$, which is a curve parameterized by its length.

A. Chambolle (✉)
CEREMADE, CNRS, Université Paris-Dauphine PSL, Paris, France
e-mail: antonin.chambolle@ceremade.dauphine.fr

Mokaplan, INRIA, Paris, France

© The Author(s) 2026

S. Oishi et al. (eds.), *Recent Developments in Industrial and Applied Mathematics*, ICIAM2023 Springer Series 1, https://doi.org/10.1007/978-981-95-1446-5_4

For any $\ell \geq 0$, one can compute the energy given the crack $\Gamma([0,\ell])$ of length ℓ:

$$\mathcal{W}(\ell) = \min\left\{\frac{1}{2}\int_{\Omega\setminus\Gamma([0,\ell])} \mathbb{C}e(u) : e(u)\,dx \;:\; u = U^0 \text{ on } \partial^D\Omega\right\}. \tag{1}$$

Here, one assumes $u \in H^1_{\text{loc}}(\Omega \setminus \Gamma([0,\ell]); \mathbb{R}^2)$, $e(u) = (Du + Du^T)/2$ is its symmetrized gradient, and $\mathbb{C}$ is a symmetric tensor defining a coercive quadratic form of $e(u)$ (yet which does not control the skew-symmetric part of the gradient). Denoting σ the internal stress, the relationship $\sigma = \mathbb{C}e(u)$ is the *Hooke's law*. The modeling behind this form is very classical (Ciarlet 1988, Sect. 6.2): if the configuration Ω is deformed by some smooth diffeomorphism into $\Phi(\Omega)$, then typically a first order, frame invariant energy will depend only on the singular values of $D\Phi$, hence be a unitary invariant function of $(D\Phi)^T D\Phi$. Hence, if the deformation is given by an infinitesimal displacement of the form $\Phi(x) = x + \varepsilon u(x)$, using that $(D\Phi)^T D\Phi = I + \varepsilon(Du^T + Du) + \varepsilon^2 Du^T Du$, at the first order the elastic energy is a function of $e(u)$. The specific form (1) is rigorously derived by Γ-convergence arguments in Maso et al. (2002). A typical Hooke tensor $\mathbb{C}$, for an isotropic linear-elastic material, is defined by

$$\mathbb{C}e(u) = \lambda \operatorname{Tr} e(u) I + 2\mu e(u)$$

with λ, μ the *Lamé constants*, see Ciarlet (1988, Sect. 3.8).

It is clear that if the crack length increases (freezing the boundary load at the current time), then the value of $\mathcal{W}(\ell)$ will decrease since the minimization in (1) is performed on a larger set. Then, a fundamental quantity is the *energy release rate* $-\mathcal{W}'(\ell) \geq 0$, which expresses the amount of elastic energy which is released by the growth of the crack of one unit of length. The idea behind Griffith's theory is that the amount of energy dissipated to increase the crack by a given length should be proportional, with some factor $\gamma > 0$ (the *toughness*), to this additional length, so that the crack will actually grow as soon as this release rate reaches this threshold.

At time $t \geq 0$, assuming the crack has reached a length $\ell = \ell(t)$ (and is thus given by $\Gamma([0,\ell(t)])$), Griffith's idea postulates therefore that at equilibrium:

- either $-\mathcal{W}'(\ell(t)) < \gamma$ and ℓ remains constant;
- or $-\mathcal{W}'(\ell(t)) = \gamma$ and ℓ may increase.

This is summarized by the following rules:

$$\begin{cases} -\mathcal{W}'(\ell) \leq \gamma; \\ (\gamma + \mathcal{W}'(\ell))\dot{\ell} = 0. \end{cases} \tag{2}$$

In general, to consider a situation where a crack should actually grow, one may consider an increasing boundary load. In practice, in this short note, we will consider as in Francfort and Marigo (1998) a load of the form $t \mapsto tU_0$ on $\partial^D\Omega$. More general assumptions require that the boundary load be absolutely continuous (for instance, Dal Maso and Toader (2002b) and subsequent works).

This theory of Griffith is widely considered as valid (in a very simplified framework) by the mechanics community, has a good predicting power, and is the basis for many finite elements (mostly "XFEM" elements, which incorporate discontinuities) implementations in the mechanical engineering community (Moës et al. 1999; Abuaisha et al. 2015; Podgórski and Gontarz 2021), yet it has some limitation:

1. crack initiation is theoretically impossible;
2. the crack path should be known in advance;
3. it is hard to extend properly to the more realistic 3D setting.

For the first point, to be more precise, it is known that in a smooth enough domain, if the minimizer of (1) for $\ell = 0$ has no strong singularity (that is, a point x such that $\int_{B_r(x)} |e(u)|^2 dx \sim r$ as $r \to 0$), then the derivative $\mathcal{W}'(0) = 0$. In particular, in the proposed setting where the load is of the form tU^0 on $\partial^D \Omega$, the solution will be at each "time" tu^0, where u^0 solves (1) for $\ell = 0$, and the energy release rate at any possible "crack tip" (for no crack, though) will remain 0 whatever t and the load. This is of course absurd for a brittle material. On the other hand, if $\ell(0) > 0$, then it is known (Grisvard 1985; Costabel and Dauge 2002; Babadjian et al. 2015) that at the tip $x = \Gamma(\ell(0))$, one will have $\int_{B_r(x)} |e(u^0)|^2 dx \sim r$ as $r \to 0$, and in particular $\mathcal{W}'(\ell(0)) < 0$. Then for the load $t \mapsto tU^0$, the energy release rate will grow quadratically with t and certainly reach γ at some point, where the crack will start to grow (continuously).

For the second point: there have been many attempts to understand the possible direction of an increase of a crack at a tip, when the load increases. Typically, one should evaluate the energy release rate as a function of the direction of the crack starting from the tip. Then, one selects a rule to choose the direction. This is done at the cost of complicated calculation (see for instance the non exhaustive list (Leblond 1989; Amestoy and Leblond 1992; Argatov and Nazarov 2002; Destuynder and Djaoua 1981)), and the rule seems not entirely clear (Amestoy and Leblond 1992): whether one should favor stability (and hence ask that the new crack grows in a direction where the crack opening is in "mode I", that is, without shearing further the tip and provoking immediately a new turn), or an energetic criterion (and hence ask that the crack grows in the first direction where the release rate reaches the toughness). Both criteria seem logical yet they also seem incompatible, and one way to resolve this paradox could be to admit that, as well as for crack initiation, in case the crack has to kink and change direction it can only do so in a "brutal" way, by increasing of a finite (nonzero) amount of additional length (Chambolle et al. 2009, 2010). Of course, we reach here a situation where the model is too simplified to describe what really should happen at small scales for a material in such a physical situation. In addition, these models which are purely quasi-static cannot capture the dynamics of brutal changes in the material configuration. For extensions to 3D linearized elasticity, the mathematical analysis is much harder and more technical, see for instance Sack (1946), Leblond (1999), Leblond et al. (1999), Lazarus and Leblond (1998a, b).

In the paper (Francfort and Marigo 1998), the authors considered Griffith's theory with a new point of view. Instead of fixing the crack path and viewing $\ell(t)$ as the unknown, they proposed to view also the path as a global variable. This was inspired by previous models of damage (Francfort and Marigo 1993) and the emerging interest for *free-discontinuity problems* (De Giorgi and Ambrosio 1988; Ambrosio et al. 2000), and in particular the study of the *Mumford-Shah functional* (Mumford and Shah 1989) for image segmentation. These new models led to the introduction of a whole lot of analytic tools (based on geometric measure theory and BV spaces (Federer 1969; Evans and Gariepy 1992; Ziemer 1989; Ambrosio et al. 2000)) and it was early proposed to use them in the modeling of fracture (Francfort and Marigo 1999; Ambrosio and Braides 1995). In Francfort and Marigo (1998, 1999), the authors pushed the idea a bit further by introducing an incremental model for computing crack evolutions under a growing load, and a limiting "time"[1]-continuous model which generalizes Griffith's evolutions. The setting is as before (in what follows, Ω may be a subset of $\mathbb{R}^d$ for any $d \geq 1$, the dimensions $d = 2, 3$ being the physical dimensions), except that now $\Gamma \subset \Omega \cup \partial^D\Omega$ is part of the unknown. In the incremental model, one will start with $t_0 = 0$, Γ^0 possibly empty, and find for any $k \geq 1$, given $t_k \geq t_{k-1}$, a new configuration (u^k, Γ^k) which solves:

$$\min\left\{\frac{1}{2}\int_{\Omega\setminus\Gamma} \mathbb{C}e(u) : e(u)\,dx + \gamma\mathcal{H}^{d-1}(\Gamma)\ :\ u = t_k U^0 \text{ on } \partial^D\Omega\setminus\Gamma,\ \Gamma \supseteq \Gamma^{k-1}\right\}. \tag{3}$$

Here, $\mathcal{H}^{d-1}$ is the $(d-1)$-dimensional Hausdorff measure (imposing thus that Γ be an at most $(d-1)$-dimensional set, and in particular with vanishing Lebesgue measure), see for instance Federer (1969), Evans and Gariepy (1992), Ambrosio et al. (2000). The last constraint accounts for the non-reversibility of the crack, which may only grow. We will denote by $\mathcal{E}(u, \Gamma)$ the energy appearing in (3).

A time-continuous evolution is obtained in the limit $\sup_k(t_{k+1} - t_k) \to 0$, and one hopes to converge to a configuration $u(t), \Gamma(t)$ which satisfies

- *(irreversibility)* $\Gamma^0 \subseteq \Gamma(t) \subseteq \Gamma(s)$ for any $0 \leq t < s$,
- *(unilateral minimality)* for each $t > 0$, $u(t) = tU^0$ on $\partial^D\Omega \setminus \Gamma(t)$ and $(u(t), \Gamma(t))$ is a *unilateral minimizer*:

$$\mathcal{E}(u(t), \Gamma(t)) \leq \mathcal{E}(v, K)$$

  for each $K \supseteq \bigcup_{s<t}\Gamma(s)$ and any $v \in H^1_{\mathrm{loc}}(\Omega \setminus K; \mathbb{R}^d)$ with $v = tU^0$ on $\partial^D\Omega \setminus K$.
- *(continuity and energy conservation)* $t \mapsto \mathcal{E}(u(t), \Gamma(t))$ is absolutely continuous and one has $d\mathcal{E}(u(t), \Gamma(t))/dt = \int_{\partial^D\Omega\setminus\Gamma(t)} \mathbb{C}e(u(t)) : (U^0 \odot \nu_\Omega)d\mathcal{H}^{d-1}$ for a.e. $t \geq 0$.

(Here, for $a, b \in \mathbb{R}^d$, $a \odot b = \frac{1}{2}(a_i b_j + a_j b_i)_{i,j=1}^d$.) The last condition is written in a weaker sense in the original work (Francfort and Marigo 1998), and is derived in

[1] In the quasistatic models described in this note, the time is just a parameter accounting for non-reversibility but should not be considered as a physical time.

this form first in Dal Maso and Toader (2002b) in dimension 2, for a simplified scalar setting. The linearized-elasticity counterpart of the results of Dal Maso and Toader (2002b) can be found in Chambolle (2003). A fundamental result which links the variational theory to the classical one is Francfort and Marigo (1998, Proposition 4.8). This result shows that if $\Gamma(t)$ is growing along a rectifiable curve and satisfies the properties above, then it is a classical evolution in the sense of Griffith's theory.

The great improvement of this approach upon the classical one is that it solves all three issues (1)–(2)–(3) mentioned before, that is, initiation (yet, most probably brutal, that is, not satisfying the continuity axiom above at least at $t = 0$ (Chambolle et al. 2008, 2009)) becomes possible (and will always occur), the crack path is determined by the process, and the formulation is the same in two or three dimensions (or any dimension). This comes at several costs. One important issue, for instance, is the fact that the *global* minimization introduced in the model (or in the axiom of unilateral minimality) is nonphysical (in particular taking into account internal or boundary forces would be impossible, since one may reach arbitrarily low energies by breaking the material into small pieces sent to infinity, corresponding to an infinite work of the forces). Some progress towards considering local minimization was developed in Dal Maso and Toader (2002a) and Larsen (2010, 2013). An analysis of the differences between local and global minimizations is proposed in Negri and Ortner (2008). The authors provide an example where the variational solution is different from the classical one.

Yet, the fact that evolutions computed with this approach are compatible with Griffith's theory, and even more the link with the phase-field approach independently developed in the physics community (Karma et al. 2001; Hakim and Karma 2009; Spatschek and Karma 2011) (this link is rigorously established through Γ-convergence arguments for the Ambrosio-Tortorelli approximation of (3), see Ambrosio and Tortorelli (1990), Bourdin et al. (2000), Chambolle (2004), Chambolle (2005), Iurlano (2014), Chambolle and Crismale (2019a) and Freddi and Royer-Carfagni (2010)), have turned this theory more and more into a "standard tool" for fracture modeling and computation (see for instance Miehe et al. (2010), Schreiber et al. (2021), Kristensen et al. (2021)). The quality of the crack growth computed by this approach and their fidelity to the physical experiments are sometimes remarkable: for instance, Fig. 1, computed by Bourdin et al. (2014), shows numerical simulations in accordance with the experiments from Shao et al. (2010) (a hot ceramic sample is brutally cooled and exhibits oscillating crack patterns starting from the stressed bottom boundary). See also Bourdin et al. (2014, Fig. 3).

On the other hand, this theory has raised numerous mathematical difficulties. While existence of minimizers for the Mumford-Shah functional was proved only a few years after its introduction in the late 80 s (Mumford and Shah 1989; Dal Maso et al. 1992; De Giorgi et al. 1989), it was not until about 2010 that Gianni Dal Maso introduced in Dal Maso (2013) an appropriate energy space for Griffith's problem (3), the space "$GSBD_2(\Omega)$". Before this, although many results were already available in simplified settings (with additional bounds in the space SBD (Ambrosio et al. 1997a; Bellettini et al. 1998), or in dimension 2 with a connectedness assumption on the crack (Dal Maso and Toader 2002b; Chambolle 2003)), it was difficult to tackle the

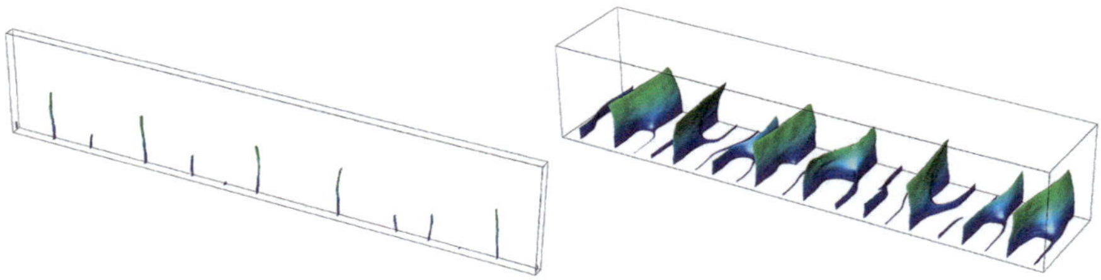

Fig. 1 Numerical simulation: an almost 2D hot ceramic plate (left) and a thicker plate (right) plunged into cold water at the bottom (experimental setup described in Bourdin et al. (2014), simulation by Blaise Bourdin)

problem in its full generality and reproduce the programme developed for Mumford-Shah. We explain this in this note: in the next Section, we show how existence for the so-called "weak" formulation has been established in a series of papers after 2016. Then in Sect. 2.2 we sketch an explanation of how the weak solutions are shown to be also strong minimizers of (3). Some open issues are raised in Sect. 3. In Sect. 4, we describe in the scalar setting some of the techniques for proving the rigidity estimate of Chambolle et al. (2016), stated below. This result is the starting point for most of the recent progress in this field.

2 Existence for Griffith's Variational Problem

2.1 Poincaré-Korn Inequality with Small Jump Set and Compactness

There are two main issues for showing existence of the minimizers of (3). First, one needs to get a control on u (or on a minimizing sequence). This is not natural as a displacement u with a very large value in a tiny ball B with $\partial B \subset K$ may have a very low energy. Then, one has to show that an almost minimizer (or the limit of a minimizing sequence) has a crack set K which can be properly described as a closed $(d-1)$-dimensional set with bounded surface, and does not spread out everywhere in the domain. In the scalar setting, the first issue corresponds to the compactness and lower-semicontinuity result shown in Ambrosio (1989) for the Mumford-Shah energy, the second to the density estimates proved in De Giorgi et al. (1989) and which allow to show that weak minimizers are strong solutions of the Mumford-Shah problem (see also Dal Maso et al. (1992) and Maddalena and Solimini (2001) for an alternative approach). In the setting of linearized elasticity, the situation is far more complex. The main issue for showing compactness is the lack of control of the maximum of $|u|$, indeed, if one adds an artificial box constraint on the displacement, then the compactness and lower-semicontinuity result proved in Bellettini et al. (1998) is sufficient to show existence of weak minimizers to (3), in the framework of the space $SBD(\Omega)$. This latter space is a subset of $BD(\Omega)$, the space of displacements $u \in L^1(\Omega; \mathbb{R}^d)$ such that the distributional symmetrized gradient Eu is a bounded Radon measure (Suquet 1978; Temam and Strang 1980). Yet, without such a box constraint, there is no result which ensures that a minimizing sequence should remain in that space.

A first progress was accomplished in Chambolle et al. (2016) where the following *Poincaré-Korn inequality for functions with small jump set* was proved. We quote (Chambolle et al. 2016, Proposition 3.1).

Proposition 1 *Let $Q = (-r, r)^d$, $Q' = (-r/2, r/2)^d$, $u \in SBD(Q)$ and $p \in [1, \infty)$. We have the following:*

(i) There exist a set $\omega \subset Q'$ and an affine function $a : \mathbb{R}^d \to \mathbb{R}^d$ with $e(a) = 0$ such that

$$|\omega| \leq cr\mathcal{H}^{d-1}(J_u)$$

and

$$\int_{Q'\setminus\omega} |u - a|^{dp/(d-1)} dx \leq cr^{d(p-1)/(d-1)} \left(\int_Q |e(u)|^p \right)^{d/(d-1)}.$$

(ii) If, additionally, $p > 1$, then there is $q > 0$ (depending on p and n) such that, for a given mollifier $\varphi_r \in C_c^\infty(B_{r/4})$, $\varphi_r(x) = r^{-d}\varphi_1(x/r)$, the function $v = u\chi_{Q'\setminus\omega} + a\chi_\omega$ obeys

$$\int_{Q''} |e(v * \varphi_r) - e(u) * \varphi_r|^p dx \leq c \left(\frac{\mathcal{H}^{d-1}(J_u)}{r^{d-1}} \right)^q \int_Q |e(u)|^p dx,$$

where $Q'' = (-r/4, r/4)^d$.

The constant in (i) depends only on p and d, the one in (ii) also on φ_1.

In this statement, $SBD(Q)$ stands for the space of "special functions with bounded deformation". These functions u are such that their symmetrized distributional gradient $Eu = \frac{1}{2}(Du + Du^T)$ is a bounded Radon measure decomposed as

$$Eu = e(u)dx + (u^+ - u^-) \odot \nu_u \mathcal{H}^{d-1} \llcorner J_u.$$

In this expression, $e(u)$ is the absolutely continuous part of Eu with respect to the Lebesgue measure dx, and J_u is the set of jump points x, that is, where the blow-ups $y \mapsto u(x + ry)$ converge, as $r \to 0$ and in $L^1(B_1; \mathbb{R}^d)$, to a function taking two values $u^\pm$ on both sides of a hyperplane $\nu_u^\perp$. This intrinsic jump set is known to be a $(d-1)$-countably rectifiable set, that is, it may be covered by countably many C^1 hypersurfaces up to a $\mathcal{H}^{d-1}$-negligible set with normal vector ν_u $\mathcal{H}^{d-1}$-almost everywhere. We refer to Ambrosio et al. (1997a), Bellettini et al. (1998) and Del Nin (2021) for details.

We mention that roughly at the same time, a series of similar Poincaré-Korn-type inequalities were proved by Manuel Friedrich. In a first paper (Friedrich 2017), the result above (point (i)) was proved in dimension 2 with different techniques, leading in particular (for $p = 2$) to a control on the *perimeter* of the exceptional set ω, and an actual *Korn* inequality, that is, a control on $|\nabla u|$, outside of ω. This was remarkable as in particular, the techniques of Chambolle et al. (2016), described in Sect. 4, certainly cannot give any control on the perimeter of ω. A following paper (Friedrich 2018) was stating a much more precise inequality in dimension 2 (splitting Ω into a partition of sets where u was controlled up to a different rigid motion in each set), and showing, in arbitrary dimensions, a variant of our main result (Chambolle et al. 2016, Theorem 1.1), valid for $p = 2$ only but also with an estimate of the perimeter of the exceptional set, and a more precise control of u (in particular, $u\chi_{\omega^c}$ is shown to be in BV): see Friedrich (2018, Theorem 2.10). Similar results were proved afterwards with the techniques of Chambolle et al. (2016), first in dimension 2 (Conti et al. 2017), then in general (Cagnetti et al. 2022). In both latter references, the inequality is for any $p > 1$, the perimeter of ω is controlled by the size of the jump, and it is shown that out of ω, u coincides with a function $v \in W^{1,p}(Q; \mathbb{R}^d)$ so that in particular also ∇u is controlled.

While the proofs in Chambolle et al. (2016) are written in the so-called space $SBD(Q)$ (as that work started actually before (Dal Maso 2013) was announced), the proof applies to the more general space $GSBD(Q)$ introduced in Dal Maso (2013) as the correct energy space for (3). Indeed, the definition of this space is based only on 1D slicing properties of the functions, upon which the proof of Proposition 1 mostly relies: see Sect. 4 where we explain the main mechanisms in the simpler scalar case.

Proposition 1 turned out to be the starting point for many developments: first, after some time, we realized with Vito Crismale that this was sufficient to obtain compactness for minimizing sequences for (3). The idea, developed in Chambolle and Crismale (2021), is quite simple. One introduces a scale $\delta > 0$, covers the domain with cubes $Q_{z,\delta} = \delta z + [0, \delta)^d$, $z \in \mathbb{Z}^d$ (considering only the cubes inside Ω), and for each function of the minimizing sequence $(u_n, \Gamma_n)_n$ one introduces $\omega_n^{z,\delta}$ and $a_n^{z,\delta}$ obtained by invoking Proposition 1 in each cube $Q_{z,\delta}$ (or actually, in slightly larger cubes). We obtain piecewise affine functions $a_n = \sum_z a_n^{z,\delta} \chi_{Q_{z,\delta}}$ with

$$\int_{\Omega\setminus\omega_n} |u_n - a_n|^2 dx \lesssim \delta^2 \int_{\Omega} |e(u_n)|^2 \, dx,$$

where $\omega_n = \bigcup_z \omega_n^{z,\delta}$ (plus a small set near $\partial\Omega$) satisfies

$$|\omega_n| \lesssim \delta \mathcal{H}^{d-1}(\Gamma_n).$$

(In practice, one works with the "weak" formulation, where Γ_n is to be replaced with the intrinsic jump set J_{u_n} of u_n, which might not be closed.)

Thanks to the fact that $(a_n)_n$ is finite-dimensional, one can extract a subsequence which converges (up to pieces which may go to infinity), and as δ gets smaller, we end up with showing that (a subsequence of) $(u_n)_n$ is a Cauchy sequence (in some sense), converging either to infinity or to a limiting u.

The set (denoted by A in Chambolle and Crismale (2021)) where $|u_n| \to \infty$ is not really an issue, as replacing u with 0 in that set, one obtains (reproducing almost verbatim the arguments in Bellettini et al. (1998)) lower semicontinuity of the energy and that the limit is actually a minimizer of the weak formulation.

A natural question which emerges is "what happens in the set A"? This seems of little importance (as one eventually replaces u with 0 in A), yet in more general problems where the elastic energy is not homogeneous and may not be minimal for $e(u) = 0$ (consider for instance the case of a pre-stress as in Shao et al. (2010) and Bourdin et al. (2014), *cf* Fig. 1), then this trick will not work. This issue has been addressed in Chambolle and Crismale (2023, 2025). The main result in Chambolle and Crismale (2025), inspired from previous results in GBD (Almi and Tasso 2023) and $GSBV$ (Friedrich 2019), shows that if $(u_n)_n$ is a bounded sequence for the energy (3), then there exists a subsequence, a (Cacciopoli, that is with bounded perimeter) partition $(E_i)_i$ of Ω, and sequences of infinitesimal rigid motions $(a^i_n)_n$ (for each E_i), such that

$$u_n - \sum_i a^i_n \chi_{E_i} \to u \quad \text{a.e.}$$

and the energy of the limit u is bounded by the energies of $(u_n)_n$. See Chambolle and Crismale (2025) for details.

2.2 Existence of Strong Solutions

Once weak minimizers to (3) have been found, one should show that they are "strong", that is, their discontinuity set J_u (which, in general, might be not closed or might even be dense) is concentrated rather than spread out, so that its closure $\Gamma = \overline{J_u}$ will have the same surface (Hausdorff measure) and (u, Γ) is a minimizer of (3). This is where Proposition 1 (ii) played a major role.

It is well-known since this issue was solved in De Giorgi et al. (1989) for the Mumford-Shah functional, that this follows from a *lower density estimate* of the form:

There exists $r_0 > 0$ and $\beta > 0$ such that for any $x \in J_u$ and any ball $B_r(x) \subset \Omega$ with $r \leq r_0$, then

$$\mathcal{H}^{d-1}(J_u \cap B_r(x)) \geq \beta r^{d-1}.$$

Indeed, it follows easily that the same holds for $\overline{J_u}$, then that the latter set has finite surface (see for instance (Ambrosio et al. 2000, Theorem 2.56)), and that this surface is the same as the surface of J_u (at least inside Ω).

For (3), this was first shown in dimension 2 in Conti et al. (2016, 2019). Then, this was extended to arbitrary dimensions in Chambolle et al. (2019), while in the paper (Chambolle and Crismale 2019b), with Vito Crismale, we showed the full existence to minimizers of (3) by establishing the density estimate up to the Dirichlet boundary (when the crack is along this boundary), see Chambolle and Crismale (2019b, Corollary 5.8).

Proposition 1 (ii) is crucial for establishing the lower density in Chambolle et al. (2019) and Chambolle and Crismale (2019b). Indeed, it allows to build, in a ball B_r of radius r where $\mathcal{H}^{d-1}(J_u \cap B_r) \ll r^{d-1}$, a competitor v which coincides with u near ∂B_r but is smooth inside a smaller ball and has much smaller jump. This is the basic brick of a "Decay Lemma", following the main idea of De Giorgi et al. (1989): one shows that for $\tau < 1$, there exist r_0, ε, such that for a minimizer, if $\mathcal{H}^{d-1}(J_u \cap B_r(x)) < \varepsilon r^{d-1}$, $r \le r_0$ then the energy in the ball $B_{\tau r}(x)$ is controlled by τ^{d-1+s} times the energy in the ball $B_r(x)$, for some $s > 0$. This is shown by contradiction: assuming this is not true, one finds a sequence of (quasi) minimizers (u_r) in balls with vanishing radius r, with vanishing jump and for which the energy does not decay at the given rate. Scaling everything to the ball of radius 1 and using the previous construction, one can replace these functions by smooth competitors in most of the ball, showing that in the limit the energy should behave like the energy of smooth functions without jump, and decay like the volume of the ball (that is, τ^d), which gives a contradiction.

Once a Decay Lemma is proven, the conclusion follows: indeed, at rectifiability points x of J_u ($\mathcal{H}^{d-1}$-almost all points), one has $\lim_{r\to 0} \mathcal{H}^{d-1}(J_u \cap B_r(x))/r^{d-1} = \omega_{d-1}$, the volume of the $(d-1)$-ball. But if at larger scales $r \sim r_0$ one had $\mathcal{H}^{d-1}(J_u \cap B_r(x)) \le \varepsilon r$, the Decay Lemma would imply that the energy $\mathcal{E}(u, B_r(x)) = \frac{1}{2}\int_{B_r(x)} \mathbb{C}e(u) : e(u)dx + \gamma\mathcal{H}^{d-1}(J_u \cap B_r(x))$ decays too fast. It would follow that $\mathcal{E}(u, B_r(x))/r^{d-1} \to 0$ as $r \to 0$: a contradiction since this limit is at least $\gamma\omega_{d-1}$.

2.3 *Regularity of the Jump Set*

One natural question, once the existence of minimizers to (3) has been established, is the regularity one can hope to show for the crack set Γ. This is a tough question already in the Mumford-Shah case, the full "Mumford-Shah conjecture" (Mumford and Shah 1989) (the jump set in 2D is a finite union of smooth curves with possible endpoints, meeting only at triple point with three angles of 120°, and at the boundary with an angle of 90°) being still only proved for isolated connected components of Γ (Bonnet 1996; Bonnet and David 2001). In higher dimensions, (Ambrosio and Pallara 1997; Ambrosio et al. 1997b) have shown that Γ is regular up to a set of dimension smaller than $d - 1$ (see also Ambrosio et al. (2000)).

For the Griffith energy, very important progress has been accomplished in the past few years, towards showing similar results. First in Babadjian et al. (2022), it is proved (in 2D) that up to a $\mathcal{H}^1$-null set, any isolated connected component of Γ is $C^{1,\alpha}$ for some $\alpha > 0$. This is extended to arbitrary dimensions in Labourie and Lemenant (2023), yet under the assumption that locally, the jump set is separating a ball into two "big" pieces (and possibly smaller ones). Finally, the authors of Friedrich et al. (2023) have considerably improved in 2D the results of Babadjian et al. (2022), showing that K is in fact, almost everywhere locally a $C^{1,1/2}$ curve, up to a set of Hausdorff dimension strictly below 1, yet, as for the Mumford-Shah functional, it is not known whether the singular set is finite. In higher dimensions, removing the separating assumption is also still a challenge.

3 Extensions and Open Problems

Among the other difficult issues which remain open, the most important from the modeling point of view is the existence of generalized Griffith evolutions with continuously growing crack: to complete the theory introduced in Francfort and Marigo (1998), one would like to be able to send the "time-steps" $(t_{k+1} - t_k)$ to 0 in the most general setting. In the scalar setting, there are a few situations where this is understood, since the work (Dal Maso and Toader 2002b) (in 2D, with the assumption that the crack is connected), extended then in the strong setting to the general 2D case in Babadjian and Giacomini (2014). In the weak formulation, the general scalar case was first addressed in Francfort and Larsen (2003) where the celebrated "jump transfer lemma" was proven. This lemma shows that under some assumptions, a sequence of minimizers of Mumford-Shah type problems converges to a minimizer, and, unfortunately, no such results exists for the Griffith energy in the general setting. This lemma was proved in dimension 2 in Friedrich and Solombrino (2018), whose authors could therefore show existence of continuous evolutions in the weak setting. (Until then, the only clear situation was the extension of Dal Maso and Toader (2002b) to linearized elasticity in Chambolle (2003), assuming connectedness of the 1D crack.) Establishing the existence of continuous-time evolution in higher dimension, or simply of strong evolutions (with a closed crack Γ) in 2D as in Babadjian and Giacomini (2014), still seems very difficult.

For similar reasons, homogenization (most of the time discussed in the scalar setting (Maso and Zeppieri 2010; Zeppieri 2020) or with scalings which exclude (3) as the limit problem (Scardia 2010), see Cagnetti et al. (2019) for a general study) or linearization (of non-linear elasticity models with cracks) remain challenging. A generalization of Maso et al. (2002) in the Griffith setting has been proved in 2D by Manuel Friedrich (2020) and extended to incoporate a non-interpenetration condition in Almi et al. (2023). Important progress has been made very recently concerning the convergence and integral representation (Friedrich et al. 2023; Crismale et al. 2022) of Griffith-type functionals in arbitrary dimensions and one may expect more results in the next years.

Many other issues remain open in the framework of the phase-field approximations of Griffith's problem, based on Ambrosio and Tortorelli (1990) and widely adopted for simulations. Incorporating constraints, such as non-interpenetration (Amor et al. 2009; Lancioni and Royer-Carfagni 2009), is tricky to rigorously justify and well-understood only in 2D (Chambolle et al. 2018). One technical reason behind this issue is that in 2D, one can easily approximate a displacement with very small jump set by smooth functions without adding additional jump, as first observed in Conti et al. (2017) (as given a ball B_r such that $\mathcal{H}^1(B_r \cap J_u) \ll r$, there will be many circles ∂B_s, $s < r$, with $J_u \cap \partial B_s = \emptyset$ and one can select radii s where it is possible to wipe out the jump inside B_s without changing too much the energy), while this is in general impossible in higher dimensions without adding some spurious (small) crack set where the constraint may be impossible to check (as a very tiny jump set could cross all spheres ∂B_s, $s < r$, in the same setting as before). We observe that in this setting, existence of weak minimizers can be shown using the same techniques as described in this note, yet it remains unclear whether these are also strong minimizers.

4 Ideas of Proofs

4.1 Poincaré Inequality with a Small Jump Set (The Scalar Case)

We give here the main ideas of the proofs in Chambolle et al. (2016), in the (much) simpler case where the functions are real-valued with integrable gradient (and where many alternative results may be derived, such as De Giorgi et al. (1989, Theorem 3.1)). As for most proofs of Poincaré or Sobolev-type inequalities, the results rely on integrating along lines to get control on a value at some point with values at other points. The same proofs for vectorial fields, involving only the symmetrized gradients, are more involved, since such integrals can control only the difference of the components along the direction between the two points. Hence, each time such an argument is invoked below, it has to be replaced with a (quite) more complicated construction involving d different directions in order to gain control on all the components of the field. Throughout all this section, except otherwise stated, we work at scale 1 (and, even, in $(0,1)^d$), the corresponding statements in a cube $(-r,r)^d$ are recovered without effort by appropriately scaling the functions and sets.

Consider $Q = (0,1)^d$ and $u \in W^{1,p}(Q \setminus K)$ ($p \in [1,+\infty)$), for $K \subset Q$ a closed, rectifiable set. We assume $\mathcal{H}^{d-1}(K)$ is small. The setting could be as well $SBV(Q)$ or $GSBV(Q)$ (Ambrosio et al. 2000), with J_u replacing K, and it would not change anything to what follows. In a first step, we prove:

Proposition 2 *There exists C depending only on d and p such that, given u and K as above, there exists $c \in \mathbb{R}$ and $\omega \subset Q$ with:*

$$|\omega| \le C\mathcal{H}^{d-1}(K), \tag{4}$$

$$\int_{Q\setminus\omega} |u-c|^p dy \le C \int_Q |\nabla u|^p dx. \tag{5}$$

Proof We follow closely (Chambolle et al. 2016). Given $x, y \in Q$, we let $\nu(x, y) := \mathcal{H}^0(K \cap [x, y])$, and we observe that if $\nu(x, y) = 0$, then (for a.e. x, y, or assuming first u is smooth away from K and reasoning then by density):

$$u(y) = u(x) + \int_0^1 \nabla u(x + s(y - x)) \cdot (y - x) ds.$$

Then (extending all functions by 0 out of Q),

$$\begin{aligned}
&\int_Q \int_Q (1 - \nu(x, y))^+ |u(x) - u(y)|^p dy dx \\
&\quad \leq \int_Q dx \int_Q dy (1 - \nu(x, y))^+ \int_0^1 |\nabla u(x + s(y - x))|^p |y - x|^p ds \\
&\quad \leq \int_0^1 ds \int_Q dx \int_{[-1,1]^d} |\nabla u(x + s\xi)|^p |\xi|^p d\xi \\
&\quad = \int_0^1 ds \int_{[-1,1]^d} |\xi|^p d\xi \int_{Q - s\xi} |\nabla u(x + s\xi)|^p dx \leq 2^d \sqrt{d}^p \int_Q |\nabla u|^p dx,
\end{aligned}$$

while, similarly,

$$\int_Q \int_Q \nu(x, y) dx dy \leq \int_{[-1,1]^d} \int_Q \mathcal{H}^0((x + \mathbb{R}\xi) \cap K) dx d\xi \leq 2^d \mathcal{H}^{d-1}(K).$$

Denoting $\omega_x = \{y : \nu(x, y) \geq 1\}$, we deduce that for any $\eta > 0$,

$$\begin{aligned}
&\left| \left\{ x \in Q : |\omega_x| \geq \tfrac{2^d}{\eta} \mathcal{H}^{d-1}(K) \right\} \right| \leq \eta, \\
&\left| \left\{ x \in Q : \int_{Q \setminus \omega_x} |u - u(x)|^p dy \geq \tfrac{2^d \sqrt{d}^p}{\eta} \int_Q |\nabla u|^p dy \right\} \right| \leq \eta.
\end{aligned}$$

Hence, if $\eta < |Q|/2 = 1/2$, there is a set of positive measure of points x such that

$$|\omega_x| \leq \tfrac{2^d}{\eta} \mathcal{H}^{d-1}(K), \quad \int_{Q \setminus \omega_x} |u - u(x)|^p dy \leq \tfrac{2^d \sqrt{d}^p}{\eta} \int_Q |\nabla u|^p dy.$$

This shows the result. □

We now show how to improve the estimate (4) to the right dimensional scaling.

4.2 Improvement: Scaling of the Volume Estimate

Let now u, K and ω, c as in Proposition 2. Consider the set $\omega_1^* = \omega_1 \cup \tilde{\omega}_1$, where

$$\omega_1 = \{x \in Q : (x + \mathbb{R}e_1) \cap K \neq \emptyset\}, \quad \tilde{\omega}_1 = \left\{x \in Q : \mathcal{H}^1((x + \mathbb{R}e_1) \cap \omega) \geq \tfrac{1}{2}\right\}. \tag{6}$$

Here, $(e_1, \dots, e_d)$ is the canonical basis of $\mathbb{R}^d$. Clearly, $\omega_1^* = (0,1) \times \Pi_1(\omega_1^*)$ is invariant by translation (in Q) in the direction e_1, where Π_1 is the projection onto $e_1^\perp$. In addition, $|\omega_1| \le \mathcal{H}^{d-1}(K)$, while we have, letting $Q' = (0,1)^{d-1}$:

$$|\omega| = \int_{Q'} \int_0^1 \chi_\omega(x_1, x') dx_1 dx' = \int_{Q'} |\omega \cap ((0, x') + \mathbb{R}e_1)| dx' \ge \frac{|\tilde{\omega}_1|}{2},$$

so that (from (4)): $|\tilde{\omega}_1| \le 2C\mathcal{H}^{d-1}(K)$ and $|\omega_1^*| \le (2C+1)\mathcal{H}^{d-1}(K)$.

Now, if $x \in Q \setminus \omega_1^*$, for any t with $x + te_1 \in Q$,

$$u(x + te_1) = u(x) + \int_0^t \partial_1 u(x + se_1) ds,$$

and, averaging over $\{t \in \mathbb{R} : x + te_1 \in Q \setminus \omega\}$ we get:

$$u(x) = \fint_{(x+\mathbb{R}e_1)\cap Q\setminus\omega} u \, d\mathcal{H}^1 - \int_{-x_1}^{1-x_1} \frac{\chi_{Q\setminus\omega}(x+te_1)}{\mathcal{H}^1((x+\mathbb{R}e_1)\cap Q\setminus\omega)} \int_0^t \partial_1 u(x+se_1) ds dt.$$

It follows (for any $c \in \mathbb{R}$ but we consider here the constant from Proposition 2):

$$\begin{aligned} |u(x) - c| &\le \left| \fint_{(x+\mathbb{R}e_1)\cap Q\setminus\omega} u \, d\mathcal{H}^1 - c \right| + \int_{(x+\mathbb{R}e_1)\cap Q} |\partial_1 u| d\mathcal{H}^1 \\ &\le 2 \int_{(x+\mathbb{R}e_1)\cap Q\setminus\omega} |u - c| \, d\mathcal{H}^1 + \int_{(x+\mathbb{R}e_1)\cap Q} |\partial_1 u| d\mathcal{H}^1. \end{aligned} \tag{7}$$

In the last estimate we have used that $x \notin \tilde{\omega}_1$ so that $|(x+\mathbb{R}e_1)\cap Q \setminus \omega| \ge 1/2$. We deduce from Jensen's lemma (or Hölder's inequality):

$$|u(x) - c|^p \le 2^{p-1} \left(2^p \int_{(x+\mathbb{R}e_1)\cap Q\setminus\omega} |u - c|^p \, d\mathcal{H}^1 + \int_{(x+\mathbb{R}e_1)\cap Q} |\partial_1 u|^p d\mathcal{H}^1 \right). \tag{8}$$

Integrating (8) over $x \in Q \setminus \omega_1^*$, we get

$$\int_{Q\setminus\omega_1^*} |u - c|^p dx \le 2^{2p-1} \int_{Q\setminus\omega} |u - c|^p dx + 2^{p-1} \int_Q |\partial_1 u|^p d\mathcal{H}^1, \tag{9}$$

then thanks to (5) we get that for some constant, still denoted C (depending on d and p),

$$\int_{Q\setminus\omega_1^*} |u - c|^p dx \le C \int_Q |\nabla u|^p dx. \tag{10}$$

Letting $\omega^* = \bigcap_{i=1}^d \omega_i^*$ (where ω_i^* is defined as $\omega_1^* = \omega_1 \cup \tilde{\omega}_1$ in (6), but in the direction e_i instead of e_1), we find (changing again C, which is multiplied by d):

$$\int_{Q\setminus\omega^*} |u-c|^p dx \le C \int_Q |\nabla u|^p dx.$$

Using now that $|\Pi_i(\omega^*)| \le C\mathcal{H}^{d-1}(K)$ for $i = 1, \dots, d$, we deduce $|\omega^*| \le C\mathcal{H}^{d-1}(K)^{d/(d-1)}$ (From Loomis-Whitney's inequality (Loomis and Whitney 1949), see also Chambolle et al. (2016, Lemma 4.1), or follow the proof in Sect. 4.3 below, applied to the functions $f_i = \chi_{\omega_i^*}$, which are such that $\chi_{\omega^*} = \prod_{i=1}^d f_i$). We have shown:

Proposition 3 *In Proposition 2, one may replace* (4) *with the estimate*

$$|\omega| \le C\mathcal{H}^{d-1}(K)^{\frac{d}{d-1}}. \tag{11}$$

4.3 Higher Integrability

Now, we show how a better control on $u-c$ is obtained, using the standard approach to Sobolev's inequalities (see for instance (Evans and Gariepy 1992, Sect. 4.5)). We consider (8) for $i = 1, \dots, d$:

$$\chi_{Q\setminus\omega_i^*}|u(x)-c|^p \le 2^{p-1}\chi_{Q\setminus\omega_i^*} \int_{(x+\mathbb{R}e_i)\cap Q} \Big(2^p(1-\chi_\omega)|u-c|^p + |\partial_i u|^p\Big) d\mathcal{H}^1.$$

Denoting by $f_i(x)$ the right-hand side, we stress that it does not depend on x_i. We multiply these inequalities for $i = 1, \dots, d$ and obtain that for all $x \in Q$:

$$f(x) := \chi_{Q\setminus\omega^{**}}|u(x)-c|^{\frac{pd}{d-1}} \le \prod_{i=1}^d f_i(x)^{\frac{1}{d-1}},$$

where we have denoted $\omega^{**} = \bigcup_{i=1}^d \omega_i^*$. Then, we pursue as in Evans and Gariepy (1992, p. 139) or Loomis and Whitney (1949):

$$\begin{aligned}\int_0^1 f(x)dx_1 \le f_1(x)^{\frac{1}{d-1}} \int_0^1 \prod_{i=2}^d f_i(x)^{\frac{1}{d-1}} dx_1 &\le f_1(x)^{\frac{1}{d-1}} \left(\prod_{i=2}^d \int_0^1 f_i(x)dx_1\right)^{\frac{1}{d-1}} \\ &= \left(\prod_{i=1}^d \int_0^1 f_i(x)dx_1\right)^{\frac{1}{d-1}}\end{aligned}$$

(using Hölder's inequality);

$$\int_0^1\int_0^1 f(x)dx_1dx_2 \le \left(\int_0^1 f_2(x)dx_1\right)^{\frac{1}{d-1}}\int_0^1\left(\prod_{i\neq 2}\int_0^1 f_i(x)dx_1\right)^{\frac{1}{d-1}}dx_2 \le$$

$$\left(\int_0^1 f_2(x)dx_1\right)^{\frac{1}{d-1}}\left(\prod_{i\neq 2}\int_0^1\int_0^1 f_i(x)dx_1dx_2\right)^{\frac{1}{d-1}} = \left(\prod_{i=1}^d\int_0^1\int_0^1 f_i(x)dx_1dx_2\right)^{\frac{1}{d-1}},$$

and the same argument for $i = 3, \dots, d$. Eventually, we find

$$\int_Q f(x)dx \le \left(\prod_{i=1}^d\int_Q f_i(x)dx\right)^{\frac{1}{d-1}},$$

and it follows:

$$\int_{Q\setminus\omega^{**}} |u(x)-c|^{\frac{pd}{d-1}}dx \le 2^{\frac{(p-1)d}{d-1}}\prod_{i=1}^d\left(2^p\int_{Q\setminus\omega}|u-c|^pdx + \int_Q|\partial_i u|^pdx\right)^{\frac{1}{d-1}}$$
$$\le C\left(\int_Q|\nabla u|^pdx\right)^{\frac{d}{d-1}},$$

for some constant depending only on d, p. By construction, one has here $|\omega^{**}| \le C\mathcal{H}^{d-1}(K)$. By simple scaling arguments, we have eventually proven.

Proposition 4 *Let $Q = (-r,r)^d$, $K \subset Q$ with $\mathcal{H}^{d-1}(K) < +\infty$ and $u \in W^{1,p}_{\mathrm{loc}}(Q \setminus K)$ with $\int_{Q\setminus K}|\nabla u|^pdx < +\infty$, $p \in [1,+\infty)$. Then there exists $\omega \subset Q$ and $c \in \mathbb{R}$ such that*

$$|\omega| \le Cr\mathcal{H}^{d-1}(K)$$

and

$$\int_{Q\setminus\omega}|u-c|^{\frac{pd}{d-1}}dx \le Cr^{\frac{d(p-1)}{d-1}}\left(\int_Q|\nabla u|^pdx\right)^{\frac{d}{d-1}},$$

where the constant C depends only on d and p.

This is the scalar counterpart of Proposition 1, (i), except that in this setting it is easier to obtain the estimate in the whole cube Q and not a smaller one. Although we worked with compact sets K and $u \in W^{1,p}(Q \setminus K)$, the experts will notice that the result holds in $GSBV(Q)$ with the same proof and K replaced with J_u. Note that, similarly to the vectorial setting, one cannot easily obtain here the Sobolev exponent $pd/(d-p)$. In classical proofs of the Sobolev inequality, this is derived from the result in the case $p = 1$, applied to the function $|u|^q \in W^{1,1}_{\mathrm{loc}}(Q \setminus K)$ (for

$q = p(d-1)/(d-p)$, and assuming in a first stage that $|u|$ is bounded). Yet, doing so, one will obtain on the right-hand side an integral of $|u|^{q-1}|\nabla u|$ on the whole set Q, which cannot be compared to the left-hand side where $|u|^{qd/(d-1)}$ is integrated only on $Q \setminus \omega$. To get such a result, much more involved proofs have to be developed, still based on Proposition 1, yet using both (i) and (ii)), see Cagnetti et al. (2022) for details (in $GSBD$).

Here again the estimate on $|\omega|$ could be improved as in Proposition 3 at the expense of a more complicated construction, involving intersecting sets in various directions, see Chambolle et al. (2016, Sect. 4).

4.4 Proof of Proposition 1, (ii) in the Scalar Case

Here, there is a priori not much to simplify with respect to Chambolle et al. (2016). Yet we found some simplification (also possible for the vectorial case) which we detail below.

We consider as before $r = 1$, and as in the statement, a smooth mollifier φ with support in $B_{1/4}$. One will obtain a result which is valid in $Q'' := (1/4, 3/4)^d \subset\subset Q$: one could of course, if needed, obtain similar results in larger cubes, by changing/scaling the function φ, yet then the constants depend on φ and will degenerate to $+\infty$ as the inner cube tends to Q.

In this section, we need to assume $p > 1$.[2] We consider ω such that (4) and (5) hold,[3] and assume without loss of generality that $c = 0$. We define, as in the statement, $v = u\chi_{Q\setminus\omega}$, and we start as in Sect. 4.3, with the set ω_1^* defined as before, and recalling (10).

We let $w_1 = u\chi_{Q\setminus\omega_1^*}$ and since the set $\omega_1^* = (0,1) \times \Pi_1(\omega_1^*)$ are cylindrical, one may check that the distributional derivative $\partial_1 w_1$ is in fact $(\partial_1 u)\chi_{Q\setminus\omega_1^*} \in L^p(Q)$. In particular, $\partial_1(\varphi * w_1) = \varphi * \partial_1 w_1 = \varphi * ((\partial_1 u)\chi_{Q\setminus\omega_1^*})$ in $Q'' := (1/4, 3/4)^d$. We decompose, for a point in Q'':

$$\partial_1(\varphi * v) - \varphi * (\partial_1 u) = \partial_1\varphi * (v - w_1) + \varphi * (\partial_1 w_1 - \partial_1 u)$$

and we estimate both parts. We write

$$\|\varphi * (\partial_1 w_1 - \partial_1 u)\|_{L^p(Q'')} \le \|\varphi\|_p \int_{\omega_1^*} |\partial_1 u| dx \le C|\omega_1^*|^{1-\frac{1}{p}} \|\nabla u\|_p$$

(where now the constant C depends on d, p and $\|\varphi\|_p$). Similarly,

---

[2] The computations remain valid for $p = 1$ yet the statements deduced carry no relevant information.

[3] Actually, it could be the original set in Proposition 2, as well as any of the sets built later on in Proposition 3 or Proposition 4.

$$\|\partial_1\varphi * (v - w_1)\|_{L^p(Q'')} \le \|\partial_1\varphi\|_p \int_Q |u||\chi_{Q\setminus\omega} - \chi_{Q\setminus\omega_1^*}|dx$$
$$= \|\partial_1\varphi\|_p \int_Q |u|(\chi_{\omega_1^*\setminus\omega} + \chi_{\omega\setminus\omega_1^*})dx.$$

Then:

$$\int_Q |u|\chi_{\omega_1^*\setminus\omega}dx \le |\omega_1^*|^{1-\frac{1}{p}}\|u\|_{L^p(Q\setminus\omega)} \le C|\omega_1^*|^{1-\frac{1}{p}}\|\nabla u\|_p,$$

$$\int_Q |u|\chi_{\omega\setminus\omega_1^*} \le C|\omega|^{1-\frac{1}{p}}\|\nabla u\|_p,$$

the latter thanks to (10). We deduce:

$$\|\partial_1(\varphi * v) - \varphi * (\partial_1 u)\|_{L^p(Q'')}$$
$$\le C(|\omega_1^*|^{1-\frac{1}{p}} + |\omega|^{1-\frac{1}{p}})\|\nabla u\|_p \le C\mathcal{H}^{d-1}(K)^{1-\frac{1}{p}}\|\nabla u\|_p$$

for a constant C which now depends on d, p, $\|\varphi\|_p$ and $\|\nabla\varphi\|_p$. Upon appropriate translation and rescaling, we have shown.

Proposition 5 *In the setting of proposition 4, if $\varphi \in C_c^\infty(B_{1/4})$ is a smooth mollifier with $\varphi_r(x) = r^{-d}\varphi(x/r)$ and if we let $v = u\chi_{Q\setminus\omega} + c\chi_\omega$, then $\varphi_r * v$ obeys:*

$$\int_{Q''} |\nabla(\varphi_r * v) - \varphi_r * \nabla u|^p dx \le C\left(\frac{\mathcal{H}^{d-1}(K)}{r^{d-1}}\right)^{p-1} \int_Q |\nabla u|^p dx$$

where $Q'' = (-3r/4, 3r/4)^d$; the constant C depends only on p, d, φ.

Remark 1 (*Set ω from Proposition* 3) If we perform the previous construction starting from the set ω built in Sect. 4.2, then the exponent $p - 1$ above is to be replaced with $d(p-1)/(d-1)$.

Acknowledgements The author warmly thanks the scientific comittee of the ICIAM 2023 for having invited him. He also thanks his many co-authors and colleagues, and in particular Sergio Conti (Bonn), Gilles Francfort (Paris-Nord), Blaise Bourdin (McMaster University, Hamilton), Jean-François Babadjian (Paris-Saclay) and Vito Crismale (Roma) for their comments and many suggestions for improving previous versions of this note. Many thanks also to Blaise Bourdin for providing the simulations in Fig. 1. Thanks also to the anonymous referee for the many useful remarks and corrections.

References

Abuaisha, M.S., Eaton, D., Priest, J., Wong, R.: Hydraulic fracture simulation using the GEOS code. In: Microseismic Industry Consortium: Annual Research Report, vol. 5 (2015)

Almi, S., Davoli, E., Friedrich, M.: Non-interpenetration conditions in the passage from nonlinear to linearized Griffith fracture. J. Math. Pures Appl. **9**(175), 1–36 (2023)

Almi, S., Tasso, E.: A new proof of compactness in G(S)BD. Adv. Calc. Var. **16**(3), 637–650 (2023)

Ambrosio, L., Braides, A.: Energies in SBV and variational models in fracture mechanics. In: Homogenization and Applications to Material Sciences (Nice, 1995). GAKUTO International Series. Mathematical Sciences and Applications, vol. 9, pp. 1–22. Gakkōtosho, Tokyo (1995)

Ambrosio, L., Coscia, A., Maso, G.D.: Fine properties of functions with bounded deformation. Arch. Rat. Mech. Anal. **139**(3), 201–238 (1997a)

Ambrosio, L., Fusco, N., Pallara, D.: Functions of Bounded Variation and Free Discontinuity Problems. Oxford Mathematical Monographs. The Clarendon Press, Oxford University Press, New York (2000)

Ambrosio, L., Fusco, N., Pallara, D.: Partial regularity of free discontinuity sets. II. Ann. Scuola Norm. Sup. Pisa Cl. Sci. (4) **24**(1), 39–62 (1997b)

Ambrosio, L., Pallara, D.: Partial regularity of free discontinuity sets. I. Ann. Scuola Norm. Sup. Pisa Cl. Sci. (4) **24**(1), 1–38 (1997)

Ambrosio, L.: A compactness theorem for a new class of functions of bounded variation. Boll. Un. Mat. Ital. B (7) **3**(4), 857–881 (1989)

Amestoy, M., Leblond, J.-B.: Crack paths in plane situations. II. Detailed form of the expansion of the stress intensity factors. Int. J. Solids Struct. **29**(4), 465–501 (1992)

Amor, H., Marigo, J.-J., Maurini, C.: Regularized formulation of the variational brittle fracture with unilateral contact: numerical experiments. J. Mech. Phys. Solids **57**(8), 1209–1229 (2009)

Argatov, I.I., Nazarov, S.A.: Energy release in the kinking of a crack in a plane anisotropic body. Prikl. Mat. Mekh. **66**(3), 502–514 (2002)

Babadjian, J.-F., Chambolle, A., Lemenant, A.: Energy release rate for non-smooth cracks in planar elasticity. J. Éc. Polytech. Math. **2**, 117–152 (2015)

Babadjian, J.-F., Giacomini, A.: Existence of strong solutions for quasi-static evolution in brittle fracture. Ann. Sc. Norm. Super. Pisa Cl. Sci. (5) **13**(4), 925–974 (2014)

Babadjian, J.-F., Iurlano, F., Lemenant, A.: Partial regularity for the crack set minimizing the two-dimensional Griffith energy. J. Eur. Math. Soc. (JEMS) **24**(7), 2443–2492 (2022)

Bellettini, G., Coscia, A., Dal Maso, G.: Compactness and lower semicontinuity properties in $SBD(\Omega)$. Math. Z. **228**(2), 337–351 (1998)

Bonnet, A., David, G.: Cracktip is a global Mumford-Shah minimizer. Astérisque (274), vi+259 (2001)

Bonnet, A.: On the regularity of edges in image segmentation. Ann. Inst. H. Poincaré C Anal. Non Linéaire **13**(4), 485–528 (1996)

Bourdin, B., Francfort, G.A., Marigo, J.-J.: Numerical experiments in revisited brittle fracture. J. Mech. Phys. Solids **48**(4), 797–826 (2000)

Bourdin, B., Marigo, J.-J., Maurini, C., Sicsic, P.: Morphogenesis and propagation of complex cracks induced by thermal shocks. Phys. Rev. Lett. **112**, 014301 (2014)

Cagnetti, F., Chambolle, A., Scardia, L.: Korn and Poincaré-Korn inequalities for functions with a small jump set. Math. Ann. **383**(3–4), 1179–1216 (2022)

Cagnetti, F., Dal Maso, G., Scardia, L., Zeppieri, C.I.: Γ-convergence of free-discontinuity problems. Ann. Inst. H. Poincaré C Anal. Non Linéaire **36**(4), 1035–1079 (2019)

Chambolle, A., Conti, S., Francfort, G.: Korn-Poincaré inequalities for functions with a small jump set. Indiana Univ. Math. J. **65**(4), 1373–1399 (2016)

Chambolle, A., Conti, S., Iurlano, F.: Approximation of functions with small jump sets and existence of strong minimizers of Griffith's energy. J. Math. Pures Appl. **9**(128), 119–139 (2019)

Chambolle, A., Crismale, V.: A density result in $GSBD^p$ with applications to the approximation of brittle fracture energies. Arch. Rat. Mech. Anal. **232**(3), 1329–1378 (2019a)

Chambolle, A., Crismale, V.: A general compactness theorem in G(S)BD. Indiana Univ. Math. J. **74**(1), 233–249 (2025)

Chambolle, A., Crismale, V.: Compactness and lower semicontinuity in $GSBD$. J. Eur. Math. Soc. (JEMS) **23**(3), 701–719 (2021)

Chambolle, A., Crismale, V.: Equilibrium configurations for nonhomogeneous linearly elastic materials with surface discontinuities. Annali della Scuola Normale Superiore di Pisa, Classe di Scienze **24**(3), 1575–1610 (2023)

Chambolle, A., Crismale, V.: Existence of strong solutions to the Dirichlet problem for the Griffith energy. Calc. Var. Part. Differ. Equ. **58**(4), Paper No. 136, 27 (2019b)

Chambolle, A., Francfort, G.A., Marigo, J.-J.: Revisiting energy release rates in brittle fracture. J. Nonlinear Sci. **20**(4), 395–424 (2010)

Chambolle, A., Francfort, G.A., Marigo, J.-J.: When and how do cracks propagate? J. Mech. Phys. Solids **57**(9), 1614–1622 (2009)

Chambolle, A., Giacomini, A., Ponsiglione, M.: Crack initiation in brittle materials. Arch. Rat. Mech. Anal. **188**(2), 309–349 (2008)

Chambolle, A.: Addendum to: "An approximation result for special functions with bounded deformation". J. Math. Pures Appl. (9) **83**(7), 929–954 (2004); mr2074682; J. Math. Pures Appl. (9), **84**(1), 137–145 (2005)

Chambolle, A.: An approximation result for special functions with bounded deformation. J. Math. Pures Appl. (9) **83**(7), 929–954 (2004)

Chambolle, A.: A density result in two-dimensional linearized elasticity, and applications. Arch. Rat. Mech. Anal. **167**(3), 211–233 (2003)

Chambolle, A., Conti, S., Francfort, G.A.: Approximation of a brittle fracture energy with a constraint of non-interpenetration. Arch. Rat. Mech. Anal. **228**(3), 867–889 (2018)

Ciarlet, P.G.: Mathematical Elasticity Volume I. Three-Dimensional Elasticity. Studies in Mathematics and its Applications, vol. 20. North-Holland Publishing Co., Amsterdam (1988)

Conti, S., Focardi, M., Iurlano, F.: Existence of minimizers for the 2D stationary Griffith fracture model. C. R. Math. Acad. Sci. Paris **354**(11), 1055–1059 (2016)

Conti, S., Focardi, M., Iurlano, F.: Existence of strong minimizers for the Griffith static fracture model in dimension two. Ann. Inst. H. Poincaré C Anal. Non Linéaire **36**(2), 455–474 (2019)

Conti, S., Focardi, M., Iurlano, F.: Integral representation for functionals defined on SBD^p in dimension two. Arch. Rat. Mech. Anal. **223**(3), 1337–1374 (2017)

Costabel, M., Dauge, M.: Crack singularities for general elliptic systems. Math. Nachr. **235**, 29–49 (2002)

Crismale, V., Friedrich, M., Solombrino, F.: Integral representation for energies in linear elasticity with surface discontinuities. Adv. Calc. Var. **15**(4), 705–733 (2022)

Dal Maso, G., Morel, J.-M., Solimini, S.: A variational method in image segmentation: existence and approximation results. Acta Math. **168**(1–2), 89–151 (1992)

Dal Maso, G., Negri, M., Percivale, D.: Linearized elasticity as Γ-limit of finite elasticity. Set-Valued Anal. **10**(2–3), 165–183 (2002). Calculus of variations, nonsmooth analysis and related topics

Dal Maso, G., Toader, R.: A model for the quasi-static growth of brittle fractures based on local minimization. Math. Model. Methods Appl. Sci. **12**(12), 1773–1799 (2002a)

Dal Maso, G., Toader, R.: A model for the quasi-static growth of brittle fractures: existence and approximation results. Arch. Rat. Mech. Anal. **162**(2), 101–135 (2002b)

Dal Maso, G.: Generalised functions of bounded deformation. J. Eur. Math. Soc. (JEMS) **15**(5), 1943–1997 (2013)

Dal Maso, G., Zeppieri, C.I.: Homogenization of fiber reinforced brittle materials: the intermediate case. Adv. Calc. Var. **3**(4), 345–370 (2010)

De Giorgi, E., Carriero, M., Leaci, A.: Existence theorem for a minimum problem with free discontinuity set. Arch. Rat. Mech. Anal. **108**(3), 195–218 (1989)

De Giorgi, E., Ambrosio, L.: New functionals in the calculus of variations. Atti Accad. Naz. Lincei Rend. Cl. Sci. Fis. Mat. Nat. (8) **82**(2), 199–210 (1989)

Del Nin, G.: Rectifiability of the jump set of locally integrable functions. Ann. Sc. Norm. Super. Pisa Cl. Sci. (5) **22**(3), 1233–1240 (2021)

Destuynder, P., Djaoua, M.: Sur une interprétation mathématique de l'intégrale de Rice en théorie de la rupture fragile. Math. Methods Appl. Sci. **3**(1), 70–87 (1981)

Evans, L.C., Gariepy, R.F.: Measure Theory and Fine Properties of Functions. Studies in Advanced Mathematics. CRC Press, Boca Raton, FL (1992)

Federer, H.: Geometric Measure Theory. Die Grundlehren der mathematischen Wissenschaften, Band 153. Springer-Verlag New York Inc., New York (1969)

Francfort, G.A., Marigo, J.-J.: Cracks in fracture mechanics: a time indexed family of energy minimizers. In: Argoul, P., Frémond, M., Nguyen, Q.S. (eds.) IUTAM Symposium on Variations of Domain and Free-Boundary Problems in Solid Mechanics, Paris, April 1997, pp. 197–202. Springer Netherlands, Dordrecht (1999)

Francfort, G.A., Marigo, J.-J.: Revisiting brittle fracture as an energy minimization problem. J. Mech. Phys. Solids **46**(8), 1319–1342 (1998)

Francfort, G.A., Marigo, J.-J.: Stable damage evolution in a brittle continuous medium. Eur. J. Mech. Ser. A Solids **12**(2), 149–189 (1993)

Francfort, G.A., Larsen, C.J.: Existence and convergence for quasi-static evolution in brittle fracture. Commun. Pure Appl. Math. **56**(10), 1465–1500 (2003)

Freddi, F., Royer-Carfagni, G.: Regularized variational theories of fracture: a unified approach. J. Mech. Phys. Solids **58**(8), 1154–1174 (2010)

Friedrich, M., Labourie, C., Stinson, K.: An epsilon-regularity result for Griffith almost-minimizers in the plane (2023). arXiv:2310.07670

Friedrich, M., Perugini, M., Solombrino, F.: Γ-convergence for free-discontinuity problems in linear elasticity: homogenization and relaxation. Indiana Univ. Math. J. **72**, 1949–2023 (2023)

Friedrich, M., Solombrino, F.: Quasistatic crack growth in 2D-linearized elasticity. Ann. Inst. H. Poincaré C Anal. Non Linéaire **35**(1), 27–64 (2018)

Friedrich, M.: A compactness result in $GSBV^p$ and applications to Γ-convergence for free discontinuity problems. Calc. Var. Part. Differ. Equ. **58**(3), Paper No. 86, 31 (2019)

Friedrich, M.: A Korn-type inequality in SBD for functions with small jump sets. Math. Models Methods Appl. Sci. **27**(13), 2461–2484 (2017)

Friedrich, M.: A piecewise Korn inequality in SBD and applications to embedding and density results. SIAM J. Math. Anal. **50**(4), 3842–3918 (2018)

Friedrich, M.: Griffith energies as small strain limit of nonlinear models for nonsimple brittle materials. Math. Eng. **2**(1), 75–100 (2020)

Griffith, A.A.: The theory of rupture. In: Biereno, C.B., Burgers, J.M. (eds.) Proceedings of the 1st International Congress on Applied Mechanics Held in Delft, pp. 54–63 (1924)

Griffith, A.A.: VI. The phenomena of rupture and flow in solids. Phil. Trans. Roy. Soc. A **221**(586), 163–198 (1920)

Grisvard, P.: Elliptic Problems in Nonsmooth Domains. Monographs and Studies in Mathematics, vol. 24. Pitman (Advanced Publishing Program), Boston, MA (1985)

Hakim, V., Karma, A.: Laws of crack motion and phase-field models of fracture. J. Mech. Phys. Solids **57**(2), 342–368 (2009)

Iurlano, F.: A density result for $GSBD$ and its application to the approximation of brittle fracture energies. Calc. Var. Part. Differ. Equ. **51**(1–2), 315–342 (2014)

Karma, A., Kessler, D.A., Levine, H.: Phase-field model of mode III dynamic fracture. Phys. Rev. Lett. **87**, 045501 (2001)

Kristensen, P.K., Niordson, C.F., Martínez-Pañeda, E.: An assessment of phase field fracture: crack initiation and growth. Philos. Trans. R. Soc. A **379**(2203), 20210021 (2021)

Labourie, C., Lemenant, A.: Epsilon-regularity for griffith almost-minimizers in any dimension under a separating condition. Arch. Rat. Mech. Anal. **247**(6), 105 (2023)

Lancioni, G., Royer-Carfagni, G.: The variational approach to fracture mechanics. A practical application to the French *Panthéon* in Paris. J. Elast. **95**(1–2), 1–30 (2009)

Larsen, C.J.: Epsilon-stable quasi-static brittle fracture evolution. Commun. Pure Appl. Math. **63**(5), 630–654 (2010)

Larsen, C.J.: Local minimality and crack prediction in quasi-static Griffith fracture evolution. Discret. Contin. Dyn. Syst. Ser. S **6**(1), 121–129 (2013)

Lazarus, V., Leblond, J.-B.: Three-dimensional crack-face weight functions for the semi-infinite interface crack. II. Integrodifferential equations on the weight functions and resolution. J. Mech. Phys. Solids **46**(3), 513–536 (1998b)

Lazarus, V., Leblond, J.-B.: Three-dimensional crack-face weight functions for the semi-infinite interface crack. I. Variation of the stress intensity factors due to some small perturbation of the crack front. J. Mech. Phys. Solids **46**(3), 489–511 (1998a)

Leblond, J.-B., Lazarus, V., Mouchrif, S.: Crack paths in three-dimensional elastic solids. II. Three-term expansion of the stress intensity factors—applications and perspectives. Int. J. Solids Struct. **36**(1), 105–142 (1999)

Leblond, J.-B.: Crack paths in plane situation—I, General form of the expansion of the stress intensity factors. Int. J. Solids Struct. **25**, 1311–1325 (1989)

Leblond, J.-B.: Crack paths in three-dimensional elastic solids. I. Two-term expansion of the stress intensity factors—application to crack path stability in hydraulic fracturing. Int. J. Solids Struct. **36**(1), 79–103 (1999)

Loomis, L.H., Whitney, H.: An inequality related to the isoperimetric inequality. Bull. Am. Math. Soc. **55**, 961–962 (1949)

Luigi Ambrosio and Vincenzo Maria Tortorelli: Approximation of functionals depending on jumps by elliptic functionals via Γ-convergence. Commun. Pure Appl. Math. **43**(8), 999–1036 (1990)

Maddalena, F., Solimini, S.: Lower semicontinuity properties of functionals with free discontinuities. Arch. Rat. Mech. Anal. **159**(4), 273–294 (2001)

Miehe, C., Hofacker, M., Welschinger, F.: A phase field model for rate-independent crack propagation: robust algorithmic implementation based on operator splits. Comput. Methods Appl. Mech. Eng. **199**(45), 2765–2778 (2010)

Moës, N., Dolbow, J., Belytschko, T.: A finite element method for crack growth without remeshing. Int. J. Numer. Methods Eng. **46**(1), 131–150 (1999)

Mumford, D., Shah, J.: Optimal approximations by piecewise smooth functions and associated variational problems. Commun. Pure Appl. Math. **42**(5), 577–685 (1989)

Negri, M., Ortner, C.: Quasi-static crack propagation by Griffith's criterion. Math. Models Methods Appl. Sci. **18**(11), 1895–1925 (2008)

Podgórski, J., Gontarz, J.: Simulation of the Griffith's crack using own method of predicting the crack propagation. Adv. Sci. Technol. Res. J. **15**(4), 1–13 (2021)

Sack, R.A.: Extension of griffith's theory of rupture to three dimensions. Proc. Phys. Soc. **58**(6), 729 (1946)

Scardia, L.: Damage as the Γ-limit of microfractures in linearized elasticity under the non-interpenetration constraint. Adv. Calc. Var. **3**(4), 423–458 (2010)

Schreiber, C., Ettrich, T., Kuhn, C., Müller, R.: A phase field modeling approach of crack growth in materials with anisotropic fracture toughness. In: 2nd International Conference of the DFG International Research Training Group 2057–Physical Modeling for Virtual Manufacturing (iPMVM 2020). Schloss Dagstuhl-Leibniz-Zentrum für Informatik (2021)

Shao, Y., Xu, X., Meng, S., Bai, G., Jiang, C., Song, F.: Crack patterns in ceramic plates after quenching. J. Am. Ceram. Soc. **93**(10), 3006–3008 (2010)

Spatschek, R., Brener, E., Karma, A.: Phase field modeling of crack propagation. Philos. Mag. **91**(1), 75–95 (2011)

Suquet, P.-M.: Sur un nouveau cadre fonctionnel pour les équations de la plasticité. C. R. Acad. Sci. Paris Sér. A-B **286**(23), A1129–A1132 (1978)

Temam, R., Strang, G.: Functions of bounded deformation. Arch. Rat. Mech. Anal. **75**(1), 7–21 (1980/81)

Zeppieri, C.I.: Homogenisation of high-contrast brittle materials. Math. Eng. **2**(1), 174–202 (2020)

Ziemer, W.P.: Weakly Differentiable Functions: Sobolev Spaces and Functions of Bounded Variation. Graduate Texts in Mathematics, vol. 120. Springer-Verlag, New York (1989)

Open Access This chapter is licensed under the terms of the Creative Commons Attribution 4.0 International License (http://creativecommons.org/licenses/by/4.0/), which permits use, sharing, adaptation, distribution and reproduction in any medium or format, as long as you give appropriate credit to the original author(s) and the source, provide a link to the Creative Commons license and indicate if changes were made.

The images or other third party material in this chapter are included in the chapter's Creative Commons license, unless indicated otherwise in a credit line to the material. If material is not included in the chapter's Creative Commons license and your intended use is not permitted by statutory regulation or exceeds the permitted use, you will need to obtain permission directly from the copyright holder.

A Tutorial on the Dynamic Laplacian

Gary Froyland

Abstract Spectral techniques are popular and robust approaches to data analysis. A prominent example is the use of eigenvectors of a Laplacian, constructed from data affinities, to identify natural data groupings or clusters, or to produce a simplified representation of data lying on a manifold. This tutorial concerns the dynamic Laplacian, which is a natural generalisation of the Laplacian to handle data that has a time component and lies on a time-evolving manifold. In this dynamic setting, clusters correspond to long-lived "coherent" collections. We begin with a gentle recap of spectral geometry before describing the dynamic generalisations. We also discuss computational methods and the automatic separation of many distinct features through the SEBA algorithm. The purpose of this tutorial is to bring together many results from the dynamic Laplacian literature into a single short document, written in an accessible style.

Keywords Dynamic Laplacian · Dynamic cheeger constant · Finite-time coherent set · Inflated dynamic Laplacian

1 Introduction

Spectral geometry is concerned with relationships between the geometry of manifolds on the one hand, and spectral information from differential operators on manifolds on the other, most commonly the Laplace operator. While so-called direct problems seek spectral information from manifold geometry, inverse problems attempt to learn manifold geometry from the operator spectrum. Inverse problems are of great interest in practical areas such as spectral approaches to dimension reduction of high-dimensional manifolds and geometric clustering of data.

This brief tutorial focusses on the inverse problem, and its generalisation to answer questions concerning long-lived structurs in dynamical systems. In this dynamic set-

G. Froyland (✉)
School of Mathematics and Statistics, UNSW Sydney, Sydney, NSW 2052, Australia
e-mail: g.froyland@unsw.edu.au

© The Author(s) 2026

S. Oishi et al. (eds.), *Recent Developments in Industrial and Applied Mathematics*, ICIAM2023 Springer Series 1, https://doi.org/10.1007/978-981-95-1446-5_5

ting, the manifold is subjected to nonlinear transformations representing the evolution of the manifold over time. The inverse problem in this setting concerns learning the average geometry of the evolved manifold. Corresponding to this dynamic inverse problem is a natural dynamic generalisation of the Laplace operator to a *dynamic Laplace operator*. Using both the spectrum and eigenfunctions of the dynamic Laplacian we create a dynamic spectral geometry, and illustrate its use in identifying *finite-time coherent sets* in dynamical systems. These sets are the natural generalisations of Cheeger sets to the dynamic setting.

2 Recap on Properties of the Laplace Operator

We begin by recapping some basic spectral geometry for manifolds in the absence of dynamics; see for example Lablée (2015). Consider a full-dimensional manifold[1] $M \subset \mathbb{R}^d$. Expressed in local coordinates, the Laplace operator Δ acts on functions $f : M \to \mathbb{R}$ by $\Delta f = \sum_{i=1}^{d} \frac{\partial^2}{\partial x_i^2} f$. The eigenproblem is

$$\Delta f = \lambda f, \tag{1}$$

with $\langle \nabla f(x), n(x) \rangle = 0$ for all $x \in \partial M$, where $n(x)$ is a vector normal to the boundary of M at $x \in \partial M$, and $\langle \cdot, \cdot \rangle$ is the usual Euclidean inner product. With these Neumann boundary conditions, the eigenvalues consist of a countable sequence $0 = \lambda_1 \geq \lambda_2 \geq \cdots$, of negative numbers, diverging to $-\infty$. The leading eigenvalue $\lambda_1 = 0$ and the corresponding eigenfunction is constant on M. The eigenfunctions $f_1, f_2, \ldots$ form a complete orthonormal set for $L^2(M)$.

2.1 Optimality Properties of Laplace Eigenvalues and Eigenfunctions

There are variational expressions for the Laplace eigenvalues, for example, for each $n \geq 1$,

$$\sum_{k=1}^{n} \lambda_k = \min \left\{ \sum_{k=1}^{n} \int_M |\nabla f_k(x)|^2 \, dx : f_1, \ldots, f_n \text{ are orthonormal} \right\}. \tag{2}$$

Moreover, for each $n \geq 1$, the minimizing collection $\{f_1, \ldots, f_n\}$ is precisely the n leading Laplace eigenfunctions. Thus, denoting $L_n = \text{span}\{f_1, \ldots, f_n\}$, we see that

[1] All of the properties discussed here hold on general Riemannian manifolds, but for simplicity of presentation we will consider flat manifolds M endowed with the Euclidean metric.

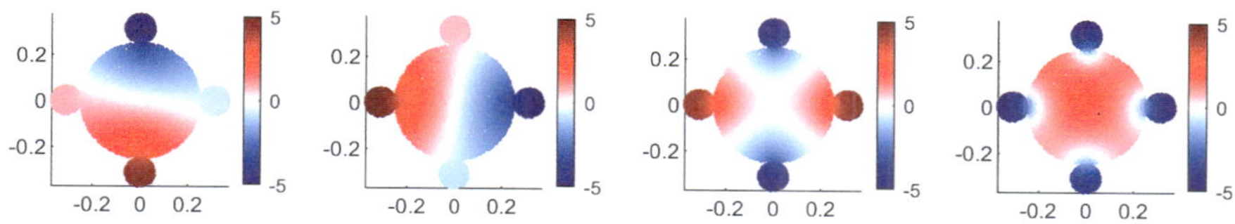

Fig. 1 Laplace eigenfunctions $f_2, \ldots, f_5$ of a two-dimensional manifold. The corresponding eigenvalues are $\lambda_2 \approx -33.42$, $\lambda_3 \approx -33.80$, $\lambda_4 \approx -49.99$, $\lambda_5 \approx -82.42$

the Laplace eigenfunctions are the most regular (with respect to the L^2 norms of gradients) orthonormal basis for L_n. Equivalently to (2) we have for each $n \geq 1$

$$\lambda_n = \min \left\{ \int_M |\nabla f_n(x)|^2 \, dx : \|f_n\|_2 = 1, \, f_n \perp L_{n-1} \right\}, \tag{3}$$

where L_0 is the trivial subspace. Figure 1 shows the first four nontrivial eigenfunctions of the Laplace operator on a two-dimensional manifold consisting of a large disk with four smaller disks attached near the boundary of the large disk. The decreasing regularity–left to right in Fig. 1–as indicated in (2) and (3) is apparent.

2.2 *The Heat Equation and Geometry*

The reader may also have noticed relationships between the eigenfunctions in Fig. 1 and the geometry of the manifold. Indeed there are well-known links between the Laplace eigendata of a manifold and the geometric shape of the manifold. Firstly, consider the heat equation on a manifold M, where the scalar value $u(t, x)$ denotes the temperature at position x and time t:

$$\frac{\partial u}{\partial t} = \Delta u. \tag{4}$$

We impose the same Neumann boundary conditions as above; this means that there is no heat lost at the boundary due to a zero derivative in the temperature field normal to the boundary. If we initialise a heat flow with $u(0, x) = f_k(x)$ then we obtain $u(t, x) = e^{\lambda_k t} f_k(x), k = 1, 2, \ldots$. The minimality properties (2) of the eigenvalues λ_k corresponds to a minimal decay of an orthonormal set of initial conditions $f_1, \ldots, f_n$ toward the heat equilibrium of $f_1 \equiv 1/\sqrt{\text{vol}(M)}$. In particular, in the two left panels of Fig. 1 we see that positive "heat" is concentrated to one side of the manifold and negative heat is concentrated on the other side, with the greatest concentration in the opposing small disks. Such an initialisation will lead to a lower heat relaxation toward equilibrium. Similarly, in the right two panels of Fig. 1 the heat is initialised predominantly in the small disks around the periphery of M. Because of the narrow necks joining the small disks to the main disk, the heat flow through these necks will contribute to slow relaxation to equilibrium.

2.3 Cheeger Problems and Laplace Eigenfunctions

Similar notions can be constructed in a purely geometric fashion through *Cheeger problems* Cheeger (1970) and Buser (1982). Let $\mathring{M}$ denote the interior of M and $|A|$ denote the full-dimensional Lebesgue volume of A. For a set $A \subset M$ with smooth boundary, we call

$$H(A) := \frac{|\partial A \cap \mathring{M}|}{|A|} \tag{5}$$

the *Neumann Cheeger ratio* of A, which quantifies the boundary size of A relative to its interior size. We call a collection $A_1, \ldots, A_n$ of n disjoint subsets of M an *n-packing* and define $H(\{A_1, \ldots, A_n\}) = \max_{1 \le k \le n} H(A_k)$. Cheeger (or isoperimetric) problems are concerned with decomposing manifolds into subsets with small boundaries relative to their interiors. To this end, we define the nth *Neumann Cheeger constant* of M as

$$h_n := \inf\{H(\{A_1, \ldots, A_n\}) : A_1, \ldots, A_n \subset M \text{ are disjoint}\}; \tag{6}$$

see Froyland and Rock (2025). It is not hard to show that h_n is increasing in n. Cheeger inequalities for n-packings of boundaryless manifolds have been studied in e.g. Miclo (2015), with analogous definitions.

In general, it is difficult to find the minimising collection $A_1, \ldots, A_n$, however the Laplace eigenfunctions can provide good candidate collections. A *nodal domain* of f_k is a maximal connected component of M where the sign of f_k is constant. We have the following result

Theorem 1.1 (Theorem 3.7 Froyland and Rock (2025)) *Consider an eigenpair* (λ, f).

1. *If f has at least n nodal domains then $h_n \le \sqrt{-2\lambda}$.*
2. *For each $k = 1, \ldots, n$, let $N_k \subset M$ denote the kth nodal domain and let $A_k(c_k) := \{x \in N_k : f(x)^2 \ge c_k\} \subset M$ denote the superlevel set of f^2 restricted to N_k. Then for a positive measure set of $(c_1, \ldots, c_n) \in \mathbb{R}^n$, one has*

$$H(A_1(c_1), \ldots, A_n(c_n)) \le \sqrt{-2\lambda}. \tag{7}$$

Part 1 says that the closer λ is to zero, the more easily M can be decomposed into n pieces, each with a small Cheeger ratio. This is consistent with our discussion of the heat equation because the closer λ is to zero, the more slowly the initial condition $u(0, x) = f(x)$ will relax to equilibrium. When f has n nodal domains, there are

n subsets of M containing obstacles that slow down this equilibriation. In Fig. 1, the left two panels show that f_2 and f_3 have two nodal domains (separated by the white strip), each roughly half of M, f_4 has four nodal domains, and f_5 has five nodal domains. Thus, by part 1 of Theorem 1.1, using the eigenvalues quoted in the caption of Fig. 1 we have $h_2 \leq 8.18$, $h_3 \leq h_4 \leq 10$, and $h_5 \leq 12.84$.

Part 2 of Theorem 1.1 provides a recipe for constructing n-packings of M with low Cheeger ratios, namely level sets of f^2 individually restricted to each nodal domain. In this regard, note that the nodal lines of f_2 and f_3 in Fig. 1 are slightly offset from the horizontal and vertical, respectively; this is not an accident. If the nodal lines were horizontal and vertical, they would split the small disks at the periphery of M, leading to larger Cheeger ratios for the corresponding blue and red nodal domains. Cheeger inequalities without nodal domain counts are available for boundaryless manifolds Miclo (2015), but they are much weaker.

2.4 *Sparse Eigenbasis Approximation (SEBA)*

Finding good candidate n-packings of a manifold with superlevel sets of individual nodal domains of f^2 can be practically challenging. Computing *individual* nodal domains can be nontrivial because of the connectivity requirement in the definition. Moreover, because the irregularity of the $\{f_k\}$ grows with k, we expect superior n-packings to be obtained from as few f_k as possible. The nth eigenfunction f_n contains *at most n* nodal domains (by the Courant Nodal Domain Theorem, Courant and Hilbert (1953)), and may contain fewer.

By taking linear combinations of $\{f_1, \ldots, f_n\}$ in combination with soft thresholding we can create a sparse collection of functions $\{s_1, \ldots, s_n\}$ that approximately span span$\{f_1, \ldots, f_n\}$. Each function s_k will support an individual element of an n packing; see Froyland and Rock (2025) for a discussion of SEBA in the manifold setting, including a direct analogue to Part 2 of Theorem 1.1 for sets $A_k(c_k) := \{x \in M : s_k(x) \geq c_k\}$. Implementational details, further discussion of constructing n-packings, and the short code may be found in Froyland et al. (2019) and at https://github.com/gfroyland/SEBA. Figure 2 illustrates this approach using eigenfunctions $f_1, \ldots, f_4$ (i.e. the constant function f_1 along with the left 3 panels of Fig. 1). One can be sure that SEBA has worked well when the sparse basis

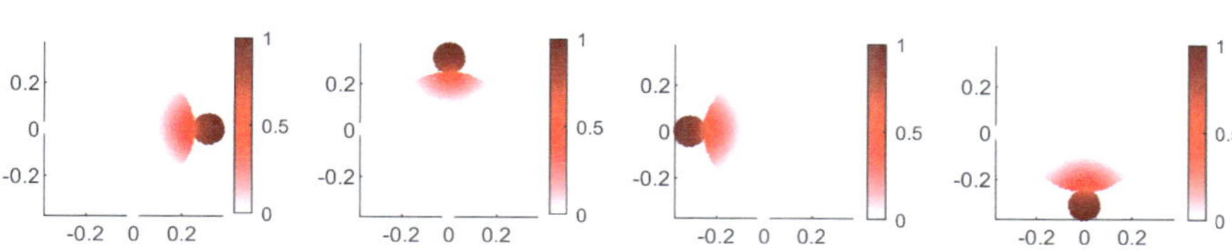

Fig. 2 Sparse functions $s_1, \ldots, s_4$ output by the SEBA algorithm applied to $f_1, \ldots, f_4$. One has span$\{s_1, \ldots, s_4\} \approx$ span$\{f_1, \ldots, f_4\}$ and may define A_k as a superlevel set of s_k, $k = 1, \ldots, 4$

functions $s_1, \ldots, s_n$ are all nonnegative. For example, applying SEBA to the collections of eigenfunctions $\{f_1, \ldots, f_n\}, 2 \le n \le 5$ in Fig. 1, we obtain nonnegative basis functions for $n = 2, 4, 5$, but not $n = 3$ because there is no natural encoding of three sets in the leading three eigenfunctions.

Finally, we remark that the k-means algorithm is often used to post-process spectral clustering output. Our Laplacian eigenanalysis is precisely spectral clustering on the manifold M and SEBA provides a more robust post-processing step for discrete data than k-means. This is because SEBA will zero out data points that do not naturally belong to a cluster and SEBA provides clear guidelines for rejecting spurious sparse vectors (s_k with large negative values should be rejected); see Froyland et al. (2019) for details.

3 Dynamical Systems and Finite-Time Coherent Sets

We now introduce dynamics on manifolds and explain how dynamic versions of the spectral geometric properties of the Laplacian discussed in the previous section can identify coherent sets in dynamical systems. We consider a domain[2] M and one-parameter family[3] of smooth transformations $\Phi^t : M \to \Phi^t(M)$ denoting the evolution of points in M for a duration $t \in [0, \tau]$. For example, Φ^t could be the flow generated by a time-dependent ODE $\dot{x} = F(t, x)$, i.e. $\frac{d}{ds}\Phi^s|_{s=t}(\cdot) = F(t, \cdot)$. For simplicity of presentation we will continue to assume that M is d-dimensional compact submanifold of $\mathbb{R}^d$ endowed with the Euclidean metric, and that Φ^t preserves Euclidean volume. All of the constructions below naturally generalise to evolving families of Riemannian manifolds endowed with metrics independent of the dynamics, and to non-Riemannian-volume-preserving dynamics Φ^t Froyland and Kwok (2020).

3.1 *Chaotic Advection, Mixing, and Coherence*

The above finite-time evolution is often highly heterogeneous in space, and an important characterisation of the dynamics is the decomposition of the phase space into coherent regions that are relatively non-dispersive and mixing regions that rapidly disperse. We will say that a subset A of M is *coherent over the interval* $[0, \tau]$ if the boundary of A remains small relative to its volume. In the presence of small diffusion–for example considering an SDE with drift F as above and small diffusion, $dX_t = F(t, X_t) + (\epsilon^2/2)dW_t$–the points in the interior of A that escape from A are those that are near to the boundary, and any escape is by diffusion alone; see Fig. 3. This approach is consistent with methods of quantification of advective mix-

[2] In dynamical systems, the domain is often called phase space or state space.

[3] All of the constructions developed here have obvious analogues if time is discrete.

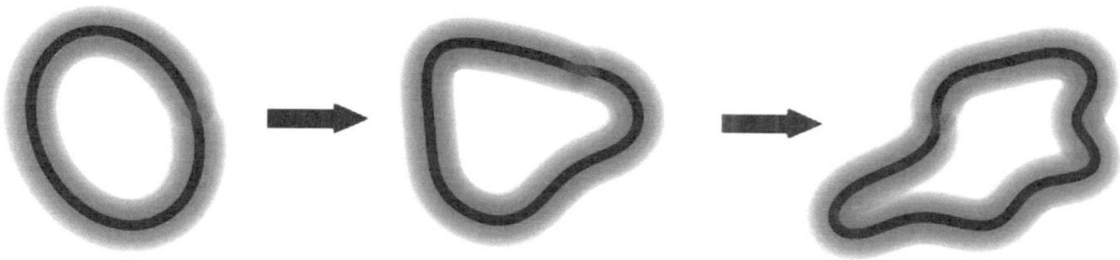

Fig. 3 Advective mixing in the presence of small diffusion. Left: a two-dimensional boundary shown as a black curve with ϵ-scale band of diffusion over the boundary. Centre and Right: under advective evolution with Φ^t the boundary evolves in a deterministic way, with the amount of mixing proportional to the size of the boundary

ing Ottino (1989), Pierrehumbert (1994), Mathew et al. (2005), Thiffeault (2012) and Haller and Beron-Vera (2013).

We briefly illustrate this idea in a simple example, based on the rotating double gyre Mosovsky and Meiss (2011). Let $M = [0, 1]^2$ and Φ^t, $t \in [0, 1]$ be generated by the time-dependent ODE given by Eq. (3.1) Mosovsky and Meiss (2011). Figure 4(left) shows two sets $A_1, A_2 \subset M$ with small Neumann Cheeger ratio (see (5)): $H(A_1) = H(A_2) = 1/(1/2) = 2$. In fact, these A_1, A_2 form an optimal 2-packing of M, with $h_2 = 2$. But when these sets are evolved under the dynamics Φ^t, their boundaries grow significantly: $H(\Phi^1(A_1)) = H(\Phi^1(A_2)) \approx 8.3057/(1/2) = 16.6114$. Thus, in a short period of time, exchange of fluid between the two shaded regions in Fig. 4 rapidly increases.

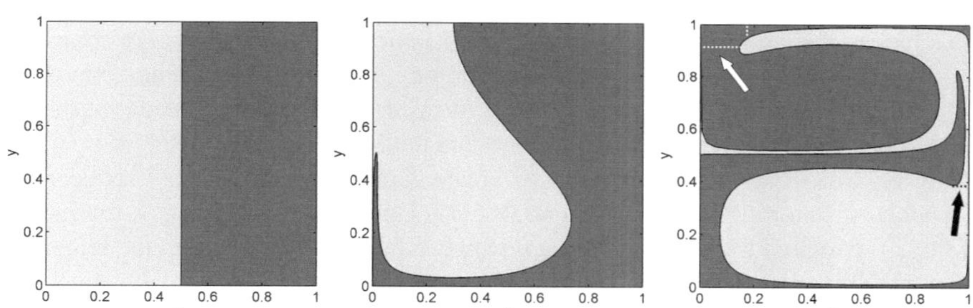

Fig. 4 Illustration of the dynamics of the rotating double gyre flow. Left: at time $t = 0$ we initialise with two sets A_1, A_2 given by the left and right halves of the domain. Centre: at time $t = 1/2$ the boundaries of A_1, A_2 begins to grow. Right: at the final evolution time $t = 1$ the boundaries have grown significantly, leading to advective mixing in the neighbourhoods of the boundaries. The arrows indicate asymmetries in the evolved sets. Image from Froyland and Padberg-Gehle (2014)

3.2 Dynamic Cheeger Ratios and Coherent Sets

In view of Fig. 3 and the above discussion on advective mixing, coherent subsets A of M are those for which the total exchange of trajectories from A to its complement, relative to the volume of A, is minimal. This total exchange is proportional to the average Cheeger ratio of $\Phi^t(A)$ over the time interval $[0, \tau]$. We therefore introduce a *dynamic Cheeger ratio* Froyland (2015) and Froyland and Rock (2025) in analogue to the Neumann Cheeger ratio (5):

$$H^D(A) := \frac{1}{\tau} \int_0^\tau \frac{|\partial(\Phi^t(A)) \cap \Phi^t(\mathring{M})|}{|\Phi^t(A)|} \, dt. \tag{8}$$

For $A = A_1$ in Fig. 4, the numerator of (8) is simply the average length of the common boundary of A_1 and A_2; in particular, we do not include the boundaries[4] of $\Phi^t(M)$. We therefore seek finite-time coherent sets as minimisers of $H^D(\cdot)$. Our approach will mirror our search for minimisers of $H(\cdot)$ in Sect. 2. Directly minimising the geometric expression (8) is difficult, so we instead develop a theory of *dynamic spectral geometry*, and find good candidates for minimisers with eigenfunctions of a *dynamic Laplace operator*.

3.3 Encoding Evolved Geometries with Pullback Metrics

If our dynamic Laplace operator is going to help us find finite-time coherent sets, it must incorporate the deformed geometry of M under the action of Φ^t. For example, in the right panel of Fig. 4 we computed the length of the evolved common boundary (let us denote this family of curves $\Phi^t(\Gamma)$)–with respect to the Euclidean metric–to be about 8.3057 at time $t = 1$. To encode this deformation using a single operator acting on functions on M, we need to create a geometry on M that reflects the Euclidean distances between points in $\Phi^t(M)$. Under such a geometry, the length of $\Phi^t(\Gamma)$–computed with the Euclidean metric e on $\Phi^t(M)$–would be the length of Γ on M. This geometry is described by the pullback metric $(\Phi^t)^*e =: g_t$. Because we have assumed that Φ^t preserves volume on M, the metric g_t has a particularly simple pointwise description in standard coordinates, namely it can be represented by the $d \times d$ positive-definite matrix $\mathbf{g}_t(x) := (D\Phi^t(x))^\top D\Phi^t(x)$ for $x \in M$, which is known as the Cauchy–Green tensor in fluid dynamics.

[4] In Sect. 3.6 we will discuss the Dirichlet dynamic Cheeger ratio and the circumstances in which to use each of the two types of dynamic Cheeger ratios.

Continuing with the example in Fig. 4, let $\gamma(s) = (1/2, s), 0 \le s \le 1$ be a parameterisation of Γ in the square domain M. The length of Γ according to the Euclidean metric on the square is

$$\int_0^1 \|\gamma'(s)\|_2 \, ds = \int_0^1 1 \, ds = 1. \tag{9}$$

We may similarly compute the length of $\Phi^t(\Gamma)$. A little notation before beginning: on the tangent space of M at $x \in M$, denoted $T_x M$, we define an inner product $\langle \cdot, \cdot \rangle_{g_t}$ by $\langle u, v \rangle_{g_t} = u^\top \mathbf{g}_t(x) v$ for $u, v \in T_x M$. We now may calculate the length of $\Phi^t(\Gamma)$ as:

$$\begin{aligned}
\int_0^1 \|D\Phi^t(\gamma(s)) \cdot \gamma'(s)\|_2 \, ds &= \int_0^1 \left(\gamma'(s)^\top (D\Phi^t(\gamma(s))^\top D\Phi^t(\gamma(s)))\gamma'(s)\right)^{1/2} ds \\
&= \int_0^1 \left(\gamma'(s)^\top \mathbf{g}_t(\gamma(s))\gamma'(s)\right)^{1/2} ds \\
&= \int_0^1 \langle \gamma'(s), \gamma'(s) \rangle_{g_t}^{1/2} ds \\
&= \int_0^1 \|\gamma'(s)\|_{g_t} \, ds.
\end{aligned} \tag{10}$$

Note that (10) is almost identical in form to the left-hand-side of (9); we have merely *changed the norm* in which we compute the length of the derivative vector γ'.

As discussed above, the metric g_t is the *pullback metric* we obtain by pulling back the Euclidean metric e with the diffeomorphism Φ^t. As Riemannian manifolds, Φ^t is an isometry between (M, g_t) and $(\Phi^t(M), e)$, because lengths computed in the tangent space $T_x M$ with g_t are identical to lengths computed in the tangent space $T_{\Phi^t x}\Phi^t(M)$ with the Eulidean metric. The concept of pullback metrics allows us to *transfer geometry on $\Phi^t(M)$ back onto M by applying a new metric on M.*

If M is d-dimensional, $d \ge 2$, we need to compute areas of $d-1$-dimensional hypersurfaces. We proceed in an analogous way to the two-dimensional case above, again using the same pullback metric g_t. A bounded $d-1$-dimensional hypersurface S can be parameterised by a function $\gamma : [0,1]^{d-1} \to \mathbb{R}^d$. To compute the $d-1$-dimensional area of the hypersurface $\Phi^t(S)$ one integrates

$$\int_{[0,1]^{d-1}} \left(\det(\mathrm{Gram}_{g_t}(\mathbf{s}))\right)^{1/2} d\mathbf{s},$$

where $d\mathbf{s} = ds_1\, ds_2 \cdots ds_{d-1}$ and $\mathrm{Gram}_{g_t}(\mathbf{s})_{ij} = \langle \partial\gamma(\mathbf{s})/\partial s_i, \partial\gamma(\mathbf{s})/\partial s_j \rangle_{g_t}$ is the $(d-1) \times (d-1)$ Gram matrix formed by partial derivatives of the parameterisation γ with respect to the $d-1$ parameters. In the case $d = 2$, the Gram matrix is scalar-valued and we obtain (10).

3.4 The Dynamic Laplacian

Consider a Riemannian manifold (M, g) with volume form ω_g. The divergence of a vector field X on M is the unique scalar-valued function $\mathrm{div}(X) : M \to \mathbb{R}$ satisfying $\mathrm{div}(X)\omega_g = X(\omega_g)$, where $X(\omega_g)$ is the Lie derivative of ω_g with respect to X. The gradient $\nabla_g f$ is the unique vector field on M satisfying $\langle \nabla_g f, X\rangle_g = X(f)$ for all vector fields X on M, where $X(f)$ is the Lie derivative of f with respect to X. The Laplace–Beltrami operator with respect to a metric g is defined as $\Delta_g f = \mathrm{div}(\nabla_g f)$. The dynamic Laplacian Froyland (2015) for Φ^t over the interval $[0, \tau]$ combines Laplace–Beltrami operators for the metrics g_t, $0 \le t \le \tau$: Let

$$\Delta^D f := \frac{1}{\tau}\int_0^\tau \Delta_{g_t} f \, dt, \tag{11}$$

where Δ_{g_t} is the Laplace–Beltrami operator on M with respect to the metric g_t. Recall that each of the pullback metrics is a metric on M and so (11) is an average of Laplace–Beltrami operators, each of which is defined on M.

We now consider the (Neumann) eigenproblem for the dynamic Laplacian

$$\Delta^D f = \lambda f, \tag{12}$$

with $\int_0^\tau \langle \nabla_{g_t} f(x), n(x)\rangle_e \, dt = 0$ for all $x \in \partial M$. Because Δ^D is a linear combination of Laplace–Beltrami operators, it inherits the symmetry and ellipticity properties of these operators and therefore also many of their spectral properties. The eigenspectrum of Δ^D is a countable sequence $0 = \lambda_1^D \ge \lambda_2^D \ge \cdots$ of negative numbers diverging to $-\infty$, and the leading eigenfunction is constant. The eigenfunctions $f_1^D, f_2^D, \ldots$ form a complete orthonormal set for $L^2(M)$ Froyland (2015).

3.5 Dynamic Cheeger Problems and Dynamic Laplace Eigenfunctions

We now discuss the links between dynamic Cheeger problems and eigenfunctions of the dynamic Laplacian, in analogy to Sect. 2.3. Recall the dynamic Cheeger ratio $H^D(A)$ for a set $A \subset M$ in Eq. (8). When $H^D(A)$ is low, this means that the relative evolved boundary size of $\Phi^t(A)$ is low on average over the time interval $[0, \tau]$. Given a dynamical evolution of a manifold M under Φ^t, $0 \le t \le \tau$, to find n finite-time coherent sets we search for dynamic n-packings $\{A_1, \ldots, A_n\}$ of M with low $H^D(\{A_1, \ldots, A_n\}) := \max_{1\le k\le n} H^D(A_k)$. To quantify how well a choice can be

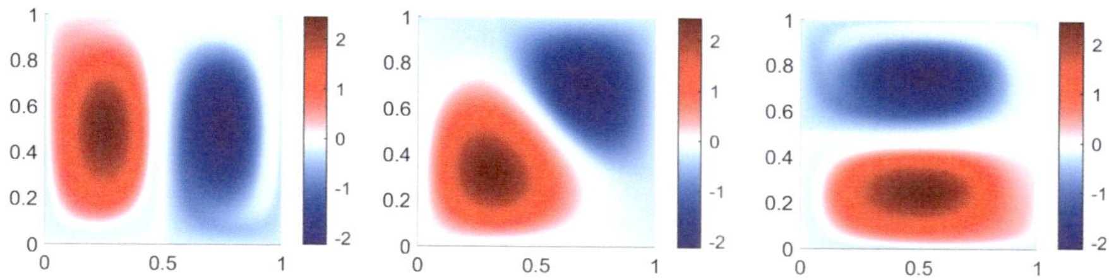

Fig. 5 Left: Leading nontrivial eigenfunction f_2^D for the rotating double gyre shown in Fig. 4. Centre: Pushforward of f_2^D by 1/2 a time unit to $t = 1/2$, namely $f_2^D \circ (\Phi^{1/2})^{-1}$. Right: Pushforward of f_2^D by 1 time unit to $t = 1$, namely $f_2^D \circ (\Phi^1)^{-1}$

made for a given finite-time dynamics we define the *dynamic Cheeger constant* of (Φ^t, M) for $t \in [0, \tau]$ as[5]

$$h_n^D := \inf\{H^D(\{A_1, \ldots, A_n\}) : A_1, \ldots, A_n \subset M \text{ are disjoint}\}. \tag{13}$$

We will use eigenfunctions of the dynamic Laplace operator to find good candidate dynamic n-packings.

Theorem 1.2 (Theorem 3.19 Froyland and Rock (2025)) *Consider an eigenpair* (λ^D, f^D) *for the dynamic Laplacian* Δ^D.

1. *If* f^D *has at least n nodal domains then* $h_n^D \le \sqrt{-2\lambda^D}$.
2. *For each* $k = 1, \ldots, n$, *let* $N_k \subset M$ *denote the* k^{th} *nodal domain and let* $A_k(c_k) := \{x \in N_k : f^D(x)^2 \ge c_k\} \subset M$ *denote the superlevel set of* $(f^D)^2$ *restricted to* N_k. *Then for a positive measure set of* $(c_1, \ldots, c_n) \in \mathbb{R}^n$, *one has*

$$H^D(A_1(c_1), \ldots, A_n(c_n)) \le \sqrt{-2\lambda^D}. \tag{14}$$

The interpretation of this theorem follows the discussion of Theorem 1.1. By (8), part 1 says that the closer the eigenvalue λ^D is to zero, the more easily we can decompose M into n disjoint sets $A_1, \ldots, A_n$ such that the average boundary size of each evolved family $\{\Phi^t(A_k)\}_{0\le t\le \tau}$ is small. Part 2 provides an ansatz for identifying such families using the dynamic Laplacian eigenfunctions.

Figure 5(left) shows the leading nontrivial eigenfunction f_2^D of the dynamic Laplacian for the rotating double gyre dynamics discussed in Sect. 3.1. If we use superlevel sets of $(f_2^D)^2$ as per Part 2 of Theorem 1.2 we will obtain two sets A_1, A_2, which we can visualise as the deeper red and blue areas in Fig. 5(left). We may push these red and blue regions forward under the dynamics Φ^t in the usual way, to obtain a function $f_2^D \circ (\Phi^t)^{-1}$ on $\Phi^t(M)$; this is shown at $t = 1/2$ and $t = 1$ in Fig. 5(centre & right). These are snapshots of the evolving coherent families of sets $\Phi^t(A_1)$, $\Phi^t(A_2)$ for $t \in [0, \tau]$. The red and blue structures that we see follow the

[5] See Froyland (2015) for $n = 2$, and Froyland and Rock (2025) for general n.

dynamics visible in Fig. 4, but the evolved boundaries of our sets A_1, A_2 are much smaller than the gray regions in Fig. 4, and therefore much more coherent.

3.6 Dirichlet Boundary Conditions

We have discussed the eigenproblem for the dynamic Laplace operator with Neumann boundary conditions. These boundary conditions are particularly appropriate when the domains $\Phi^t(M)$ have a physically meaningful boundary and the finite-time coherent sets one seeks may touch the boundary. In situations where one explicitly wishes to exclude coherent sets that touch the boundary, or where the boundary of $\Phi^t(M)$ may be arbitrary because the manifolds $\Phi^t(M), 0 \leq t \leq \tau$ live inside a larger ambient space, then Dirichlet boundary conditions–i.e. the eigenfunctions are zero on ∂M–may be applied. The main changes are as follows Froyland and Junge (2018) and Froyland and Rock (2025):

1. λ_1^D is now strictly negative and f_1^D may be chosen to be strictly positive, except on ∂M where it vanishes.
2. Equation (8) becomes

$$h^D(A) := \frac{1}{\tau} \int_0^\tau \frac{|\partial(\Phi^t(A))|}{|\Phi^t(A)|}\, dt, \tag{15}$$

   so that the entire boundaries of $\Phi^t(A)$ contribute, even if part of these boundaries coincide with the boundaries of $\Phi^t(M)$. With this adjustment Theorem 1.2 holds for eigenpairs (λ^D, f^D) satisfying Dirichlet boundary conditions.
3. Dirichlet boundary conditions force the superlevel sets $A_k(s_k)$ in Theorem 1.2 to not touch the boundary of M.

4 Computation

Our starting point will be several trajectories of a time-dependent dynamical system $x_{i,t_l}, 0 \leq i \leq N, 1 \leq l \leq T$, where $x_{i,t_l} \in \Phi^{t_l}(M)$ is the location of the ith trajectory at time t_l. These trajectories may be numerically generated from a known dynamical system Φ^t or they may be state observations made from some unknown dynamics. The basic building block we need to approximate numerically is the Laplace–Beltrami operator for the pullback metric g_t. Estimating $g_t = (\Phi^t)^* e$ numerically requires high-resolution data in order to accurately estimate the matrix $(D\Phi^t)^\top D\Phi^t$ (see Sect. 3.3). To avoid this high computational cost, we instead use the important relationship

$$\Delta_{g_t} f = \Delta_e(f \circ \Phi^{-t}) \circ \Phi^t = (\Phi^t)^*(\Delta_e(\Phi^t_* f)), \tag{16}$$

where $(\Phi^t)_* f = f \circ \Phi^{-t}$ is the pushforward of f under Φ^t, and $(\Phi^t)^* f = f \circ \Phi^t$ is the pullback of f under Φ^t.

Because we wish to estimate Laplace–Beltrami operators, a natural established numerical framework is the finite-element method. This has been adapted to the dynamic Laplacian in Froyland and Junge (2018). We wish to solve the eigenproblem $\Delta^D f = \lambda f$ with the dynamic Neumann boundary condition. We begin by integrating both sides of this equality against a test function $\tilde{f}$ using the volume measure V_e for the metric e, i.e. Lebesgue measure. This yields the weak form of the dynamic Laplacian eigenproblem, namely

$$\int_M (\Delta^D f)\tilde{f}\, dV_e = \lambda \int_M f\tilde{f}\, dV_e \text{ for all test functions } \tilde{f}. \tag{17}$$

The weak form has the immediate advantage of eliminating the Neumann[6] boundary condition Froyland and Junge (2018). By (11) we see that the LHS of (17) is a time integral of $\int_M (\Delta_{g_t} f)\tilde{f}\, dV_e$ over the time interval $[0, \tau]$. Taking one of these terms and applying (16) we obtain

$$\int_M (\Phi^t)^*(\Delta_e(\Phi^t_* f))\tilde{f}\, dV_e.$$

Changing variables with $\Phi^t(M)$ and performing integration by parts yields

$$-\int_{\Phi^t(M)} \nabla(\Phi^t_* f) \cdot \nabla(\Phi^t_* \tilde{f})\, dV_e. \tag{18}$$

We now introduce the FEM discretisation, which is a basis of nodal hat functions $\{\varphi_i^{t_l}\}_{j=1}^N$ defined on a mesh of $\Phi^t(M)$ with N nodes $\{\Phi^{t_l} x_i\}_{i=1}^N$, namely the trajectories we were initially supplied with. As a real-world example, Fig. 6 shows a mesh of around 3000 deep-ocean ARGO floats at December 2013. Each $\varphi_i^{t_l}$ is locally supported on mesh elements containing the node $\Phi^{t_l} x_i$ as a vertex, and $\varphi_i^{t_l}(\Phi^{t_l} x_i) = 1$, $i = 1, \ldots, N$. We form a (sparse) matrix $D_{ij}^{t_l}$ by evaluating (18) with $\Phi^t_* f = \varphi_i^{t_l}$ and $\Phi^t_* \tilde{f} = \varphi_j^{t_l}$, $1 \le i, j, \le N$:

$$D_{ij}^{t_l} := -\int_{\Phi^{t_l}(M)} \nabla\varphi_i^{t_l} \cdot \nabla\varphi_j^{t_l}\, dV_e, \tag{19}$$

which is a standard FEM integral. We have thus handled the left hand side of (17), obtaining the matrices $D_{ij}^{t_l}$ for t in a discrete set of points $\{t_1, \ldots, t_T\}$ sampled from the time interval $[0, \tau]$.

---

[6] Dirichlet boundary conditions are also eliminated by simply restricting the functions f to those that vanish at the boundary of M.

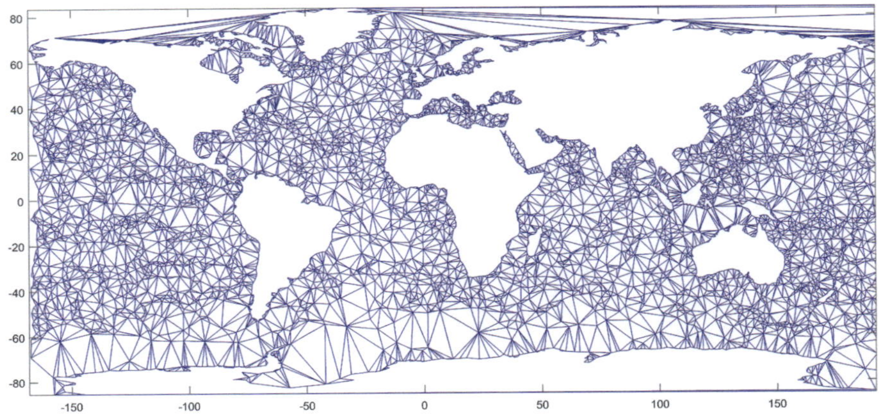

Fig. 6 Delaunay mesh of approximately 3000 Argo float positions (nodes) at December 2013. Abernathey et al. (2022) identified the major long-term coherent sets at 1000 m depth by meshing Argo float locations $x_i^{t_l}$ at each month t_l over six years

For the right hand side of (17) we substitute $f = \varphi_i^0$, $\tilde{f} = \varphi_j^0$ to obtain the $N \times N$ sparse matrix $M_{ij} := \int_M \nabla\varphi_i^0 \cdot \nabla\varphi_j^0 \, dV_e$. One now solves the discretised problem

$$\left(\frac{1}{T}\sum_{l=1}^{T} D^{t_l}\right) v = \lambda M v \tag{20}$$

to find eigenvalues λ and corresponding basis coefficient vectors v, and reconstructs eigenfunctions on M as $f = \sum_{j=1}^{N} v_j \varphi_j^0$.

This FEM approach is very robust and accurate, and is numerically efficient for dimensions 2 and 3; it has been used for all eigenfunction images in this chapter. It has been tested in many situations on sparse trajectory data Froyland and Junge (2018), Froyland et al. (2019) and Abernathey et al. (2022) and is insensitive to highly varying densities of trajectory data sampling (see e.g. Fig. 10 Froyland and Junge (2018)). The method also copes very well with situations where observed trajectories are incomplete and many observations are missing. Figure 12 Froyland and Junge (2018) illustrates this for the rotating double gyre eigenfunction in Fig. 5.

The identification of coherent regions in the global ocean at 1000 m depth from Argo drifter observations is an example of a real-world computation requiring a method than can handle sparsity, varying data density, and missing data. Figure 6 illustrates the sparsity and the varying data density. Figure 7 displays eigenfunctions of the dynamic Laplacian built from around 8000 individual deep-sea Argo float trajectories, where only 3000 floats are providing location signals in any given month, and only 10% of floats continue to transmit over the six years from 2011–2017. In Fig. 7 the dynamic Laplacian was estimated in a single computation constructed from the many partial and non-overlapping-in-time trajectories of Argo floats.

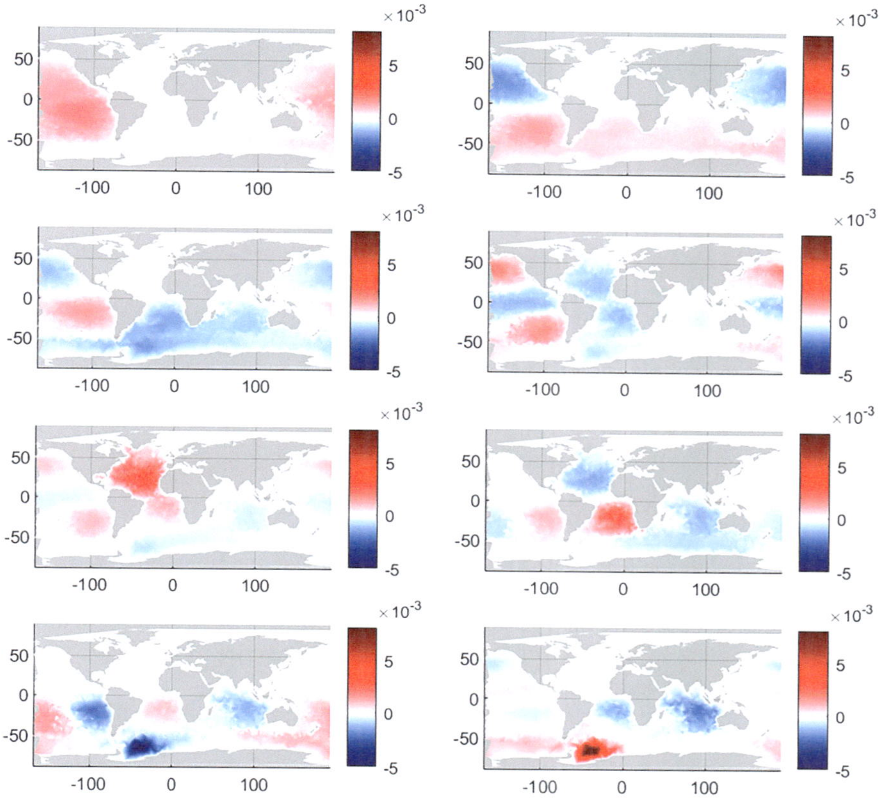

Fig. 7 Leading eight eigenfunctions $f_1^D, \ldots, f_8^D$ of the dynamic Laplacian computed from Argo trajectory data over the six years 2011–2017. The eigenfunctions are represented as scalar fields at January 2011. Image from Abernathey et al. (2022)

Finally, we note that there are no free parameters to choose and the method preserves the symmetric structure of the original eigenproblem. MATLAB code and demos are available at https://github.com/gaioguy/FEMDL. Alternate FEM-based strategies and generalisations to weighted manifolds are discussed in detail in Froyland and Junge (2018).

4.1 Alternate Approaches to Estimate Δ^D and Related Objects

Ulam and finite-difference

In Froyland (2015) the pushforwards and pullbacks Φ^t_* and $(\Phi^t)^*$ in (16) were estimated with Ulam-discretisations and the Laplacian Δ_e was discretised by finite-difference. This approach was generalised to the measure-preserving setting on weighted manifolds in Froyland and Kwok (2020), and we refer the reader to Froyland (2015) and Froyland and Kwok (2020) for a full explanation of this approach. A large number of points were used in Froyland (2015) and Froyland and Kwok (2020) to estimate the Ulam matrices representing Φ^t_* and $(\Phi^t)^*$, but similar results can be obtained with many fewer points. Because this approach is grid-based, a reasonably uniform sampling of initial conditions is needed.

Radial basis function collocation

The paper Froyland and Junge (2015) replaces the indicator function bases of the Ulam approach with smooth radial basis functions and proceeds to approximate Φ^t_*, $(\Phi^t)^*$ and Δ^D by collocation. The shape parameter–which determines the width of the basis functions–needs to be selected and the method requires a suitable positioning of basis function centres.

Related objects

We noted earlier that the dynamic Laplacian is a type of spectral clustering on manifolds. A lower-order approach, compared to FEM, is to construct a graph Laplacian based on initial conditions of trajectories Hadjighasem et al. (2016), where edges are weighted according to a dynamic distance introduced in Froyland and Padberg-Gehle (2015). In a similar direction, Padberg-Gehle and Schneide (2017) apply constant edge weights that record if one trajectory comes within a threshold of another. In these approaches, cutoffs for weights or distances need to be selected. Varying point density can influence the results, whereas this is automatically adjusted for in the FEM approach, which can itself be interpreted as a graph Laplacian, see the discussion in Sect. 3.4 Froyland and Junge (2018).

5 Finding Multiple Coherent Sets: SEBA

In Sect. 2.4 we discussed how the SEBA algorithm could create sparse linear combinations of Laplace eigenfunctions to automatically separate individual nodal domains and create additional candidate sets for n-packings. The same methodology can be applied to the eigenfunctions $f_1^D, \ldots, f_n^D$ of the dynamic Laplace operator to cre-

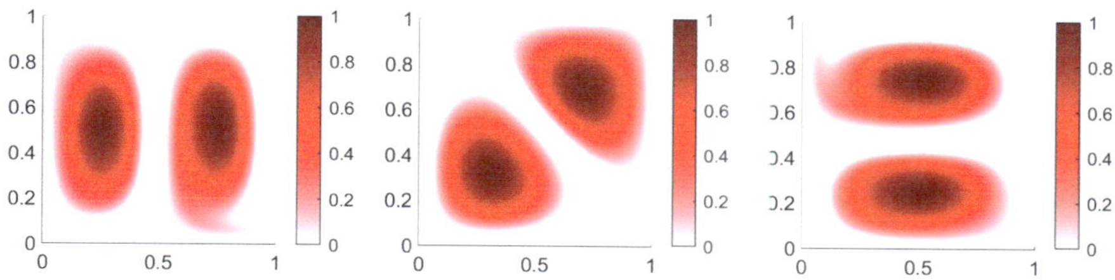

Fig. 8 Left: Sum of two SEBA vectors arising from SEBA applied to the dynamic Laplacian eigenfunctions $f_1^D \equiv \mathbf{1}$ and f_2^D from Fig. 5(left) for the rotating double gyre. Centre: Pushforward of the left panel by 1/2 a time unit with $\Phi_*^{1/2}$ to $t = 1/2$. Right: Pushforward of the left panel by 1 time unit with Φ_*^1 to $t = 1$

ate a sparse basis $\{s_1^D, \ldots, s_n^D\}$ whose span approximates $\mathrm{span}\{f_1^D, \ldots, f_n^D\}$. Each sparse function s_k^D supports a single coherent set and in this way we disentangle individual coherent sets from multiple eigenfunctions. Performance guarantees for candidate coherent sets $A_k(c_k) := \{x \in M : s_k^D(x) \geq c_k\}$ are available in direct analogue to Part 2 of Theorem 1.2; see Froyland and Rock (2025). By applying SEBA to the leading eigenfunctions $f_1^D \equiv \mathbf{1}$ and f_2^D from Fig. 5(left), we obtain two (sparse) SEBA vectors s_1^D, s_2^D taking values in $[0, 1]$. The sum of these two vectors is shown in red in Fig. 8(left) and their pushforwards under Φ^t shown in the centre and right panels.

Similarly, we can apply SEBA to the eight eigenfunctions shown in Fig. 7 to separate the eight dominant coherent regions in the global ocean at 1000 m depth; see Fig. 9. With the strongest coherent regions clearly separated, it has been noted in Abernathey et al. (2022) that they are largely aligned with the major ocean boundaries, but with a bias toward the east of the basin due to the stronger-mixing currents on their western boundaries. The Weddell gyre off the Antarctic coast is highlighted. The North and South Atlantic is separated by the equatorial countercurrent. The subpolar and subtropical gyres in the North Pacific are separated and the South Pacific Gyre splits into eastern and western components. The reader is referred to Abernathey et al. (2022) for more detailed analysis. These results are broadly consistent with coarser results Froyland and Padberg-Gehle (2015) and Banisch and Koltai (2017) using floating drifters on the ocean surface.

Sometimes the trajectory data is such that it does not support exactly n coherent sets in the leading n eigenfunctions. We have already seen this issue in Sect. 2.4 where the eigenfunctions in Fig. 1 supported decompositions into 2, 4, and 5 regions, but not 3 regions. The SEBA vectors themselves contain a simple "reliability" test, namely the magnitude of the most negative value; see Sects. 4.1 and 4.2 in Froyland et al. (2019)). A discussion of assessing spectral gaps is in Sect. 4.2 Froyland et al. (2019).

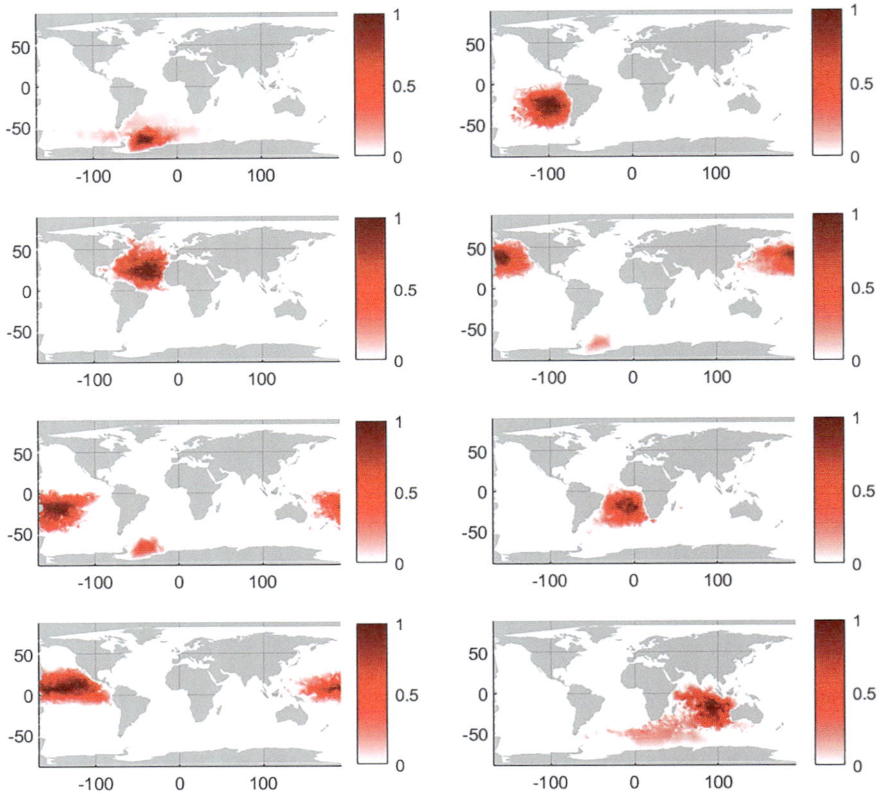

Fig. 9 Result of SEBA applied to the eight eigenfunctions shown in Fig. 7 to construct eight SEBA vectors $s_1^D, \ldots, s_8^D$. Each SEBA vector is supported on an individual coherent region of the ocean at January 2011, each of which remains coherent under ocean currents at 1000 m depth over a six-year duration from 2011–2017. Image from Abernathey et al. (2022)

6 Weights, Dynamic Metrics, and Zero-Noise Limits

Measure-preserving dynamics on weighted Riemannian manifolds

We have mainly discussed the special situation where we have a flat manifold M embedded in Euclidean space and Φ^t preserves Euclidean volume. We may also consider a dynamic Laplacian on a sequence of weighted Riemannian manifolds (M_t, h_t, m_t), where $M_t = \Phi^t(M)$, m_t is a Riemannian metric on M_t and $h_t : M_t \to \mathbb{R}^+$ is a positive weight function. The measure of $A \subset M$ is $\int_A h_0 \, dV_{m_0}$, where V_{m_0} is the volume measure induced by the metric m_0. The dynamics Φ_t can be made *measure preserving* by defining positive weight functions h_t so that the measure of a set $A \subset M$ equals the measure of $\Phi^t(A)$ on M_t: $\int_{\Phi^t A} h_t \, dV_{m_t} = \int_A h_0 \, dV_{m_0}$, for $0 < t \leq \tau$. One now defines

$$\Delta^D f := \frac{1}{\tau} \int_0^\tau \Delta_{h_t, g_t} f \, dt, \tag{21}$$

where $\Delta_{h,g} f = \frac{1}{h} \text{div}(h \nabla_g f)$ is the h-weighted Laplace–Beltrami operator for g.

In the context of fluid flow, this weighted setting is useful for finding finite-time coherent sets in compressible fluid flows, where Euclidean volume is not preserved, but fluid mass is conserved. The weight functions h_t track the evolution of an initial fluid density h_0 (with respect to V_{m_0}) under Φ^t. This weighted setup also covers the situation where one is interested in coherence of a passive tracer carried by the flow and this passive tracer is initialised with a spatial concentration different to the fluid density; e.g. in the volume-preserving case, the fluid density would be uniform. One sets h_0 to be the initial density of the passive tracer, computed with respect to V_{m_0}, and future passive tracer densities are tracked by the weight functions h_t. We refer the reader to Froyland and Kwok (2020) for the relevant constructions and Froyland and Junge (2018) for numerical computations.

Is the dynamic Laplacian a Laplace–Beltrami operator?

One may ask whether there is a dynamic Riemannian metric g^D so that $\Delta^D = \Delta_{g^D}$. Let us return to the simplified setting of a flat manifold $M \subset \mathbb{R}^d$ endowed with the Euclidean metric e and Φ^t Lebesgue volume preserving; i.e. $V_{g_t} = V_e$ for all $t \in [0, \tau]$. The appropriate construction for g^D is to set $g^D = \left(\frac{1}{\tau} \int_0^\tau g_t^{-1} \, dt\right)^{-1}$, where g^{-1} represents the dual metric, which is a symmetric, positive-definite, bilinear form acting on pairs of covectors in cotangent space Karrasch and Keller (2020) and Karrasch and Schilling (2021). One then obtains that Δ^D is the weighted Laplace–Beltrami operator Δ_{θ, g^D}, where θ is the Radon-Nikodym derivative dV_e/dV_{g^D}. This operator acts on the weighted manifold (M, θ, g^D), where the role of the weight θ is to transform the weighted volume of a set $A \subset M$, namely $\int_A \theta \, dV_{g^D}$, into the Lebesgue volume of A, $\int_A \mathbf{1} \, dV_e$. Similar results hold for the measure-preserving case in the previous paragraph. See Karrasch and Keller (2020), Karrasch and Schilling (2021) and Froyland and Lee (2025) for details.

The dynamic Laplacian as a zero-noise limit

In Sect. 3.1 we discussed the connection between dynamic isoperimetric problems and finite-time coherent sets by considering advective dynamics with small-noise diffusive dynamics. In fact, there is a formal connection in the following sense. Finite-time coherent sets were initially introduced and motivated by minimally mixing behaviour of subsets in the presence of small noise Froyland et al. (2010), and Froyland (2013). The classical transfer operator approach Froyland (2013) uses leading singular vectors of a diffusive transfer operator $\mathcal{L}_\epsilon = \mathcal{D}'_\epsilon \circ \Phi^\tau_* \circ \mathcal{D}_\epsilon$, where $\mathcal{D}'_\epsilon$ and $\mathcal{D}_\epsilon$ are isotropic diffusion (of scale ϵ) operators on M and $\Phi^\tau(M)$, respectively. One may ask what happens in the $\epsilon \to 0$ limit. The naive approach of considering leading eigenvectors of $\lim_{\epsilon \to 0} \mathcal{L}^*_\epsilon \mathcal{L}_\epsilon$ is uninformative because this limit is the identity operator. However, a finer analysis Froyland (2015) shows that the dynamic Laplacian emerges in an appropriate limit. For $f \in C^3(M)$

$$\lim_{\epsilon \to 0} \frac{(\mathcal{L}_\epsilon^* \mathcal{L}_\epsilon - I) f}{\epsilon^2} = C \cdot (\Delta_e + (\Phi^\tau)^* (\Delta_e (\Phi_*^t f)) = C \cdot \left(\Delta_{g_0} + \Delta_{g_\tau}\right)/2,$$

for a constant C depending on the common ϵ-scaled kernel of the diffusion operators $\mathcal{D}_\epsilon$. The RHS above is a dynamic Laplace operator where the average of Laplace–Beltrami operators is taken at the end points of the time interval $[0, \tau]$. This result was proved for general weighted Riemannian manifolds Froyland and Kwok (2020) and the limit shown to be uniform in the C^0-topology for f ranging over $C^3(M)$. Related calculations are in Banisch and Koltai (2017) and Karrasch and Schilling (2021).

7 When Finite-Time Coherent Sets Come and Go: the Inflated Dynamic Laplacian

The dynamic Laplacian tracks dynamics over a time interval $[0, \tau]$ and is an average of Laplace–Beltrami operators Δ_{g_t} with respect to metrics that are pullbacks under the dynamics. Because of this averaging property, the dynamic Laplacian succeeds in constructing dynamic isoperimetric properties of sets, and consequently finite-time coherent sets for the interval $[0, \tau]$. In several real-world situations, one may have complex dynamics in which many finite-time coherent sets emerge and disappear, each with their own distinct lifetimes. The dynamic Laplacian may be deployed on distinct time intervals to either separately find these sets and/or stitch distinct computations together across time Schneide et al. (2022). A recent alternative is to create an *inflated dynamic Laplace operator* on the spacetime manifold $[0, \tau] \times M$ for a perhaps larger τ than many of the lifetimes of the coherent sets. On each time fibre $\{t\} \times M$, one applies Δ_{g_t} (acting in space), while the distinct time fibres are tied together with a one-dimensional Laplace operator on $[0, \tau]$. The eigenfunctions of this inflated dynamic Laplacian encode the appearance and disappearance of multiple finite-time coherent sets, whose individual lifetimes can be extracted by applying the SEBA algorithm in spacetime. Much of the theory of the dynamic Laplacian can be naturally modified for the inflated dynamic Laplacian; see Froyland and Koltai (2023) and Atnip et al. (2024) for details.

7.1 Clusters in Time-Varying Networks

The dynamic Laplacian and the inflated dynamic Laplacian have counterparts on time-varying networks. For example, one may consider permutation dynamics on the nodes of the network and develop a corresponding dynamic spectral theory for dynamic graph cuts Froyland and Kwok (2015). The concept of a supra-Laplacian Gómez et al. (2013) and Sole-Ribalta et al. (2013) is related to a spacetime network version of the inflated dynamic Laplacian and recent work Froyland et al. (2024) has shown how spacetime spectral clustering in time-varying networks can automatically detect the appearance and disappearance of clusters.

Acknowledgements The author thanks Yat-Long Lee for a careful reading of the manuscript. GF's research is partially supported by an Australian Research Council Discovery Project.

References

Abernathey, R., Bladwell, C., Froyland, G., Sakellariou, K.: Deep Lagrangian connectivity in the global ocean inferred from Argo floats. J. Phys. Oceanogr. **52**(5), 951–963 (2022)

Atnip, J., Froyland, G., Koltai, P.: An inflated dynamic Laplacian to track the emergence and disappearance of semi-material coherent sets (2024). arXiv:2403.10360

Banisch, R., Koltai, P.: Understanding the geometry of transport: Diffusion maps for Lagrangian trajectory data unravel coherent sets. Chaos Interdiscip. J. Nonlinear Sci. **27**(3), (2017)

Buser, P.: A note on the isoperimetric constant. Annales scientifiques de l'École Normale Supérieure **Ser. 4, 15**(2), 213–230 (1982)

Cheeger, J.: A lower bound for the smallest eigenvalue of the Laplacian. In: Problems in Analysis: a Symposium in Honor of Salomon Bochner, pp. 195–199. Princeton University Press (1970)

Courant, R., Hilbert, D.: Methods of Mathematical Physics, vol. 1. Interscience Publishers, New York (1953)

Froyland, G., Junge, O.: On fast computation of finite-time coherent sets using radial basis functions. Chaos **25**(8), 087409 (2015)

Froyland, G., Kalia, M., Koltai, P.: Multi-set spectral clustering of time-evolving networks using the supra-Laplacian (2024). arXiv:2409.11984

Froyland, G., Lee, Y.L.: In preparation

Froyland, G., Padberg-Gehle, K.: A rough-and-ready cluster-based approach for extracting finite-time coherent sets from sparse and incomplete trajectory data. Chaos Interdiscip. J. Nonlinear Sci. **25**(8), 087406 (2015)

Froyland, G., Padberg-Gehle, K.: Almost-invariant and finite-time coherent sets: directionality, duration, and diffusion. In: Ergodic Theory, Open Dynamics, and Coherent Structures, pp. 171–216 (2014)

Froyland, G., Rock, C.P.: Higher Cheeger ratios of features in Laplace-Beltrami eigenfunctions. Appl. Comput. Harmon. Anal. **74**, 101710 (2025)

Froyland, G., Santitissadeekorn, N., Monahan, A.: Transport in time-dependent dynamical systems: Finite-time coherent sets. Chaos **20**, 043116 (2010)

Froyland, G.: An analytic framework for identifying finite-time coherent sets in time-dependent dynamical systems. Phys. D **250**, 1–19 (2013)

Froyland, G.: Dynamic isoperimetry and the geometry of Lagrangian coherent structures. Nonlinearity **28**, 3587–3622 (2015)

Froyland, G., Junge, O.: Robust FEM-based extraction of finite-time coherent sets using scattered, sparse, and incomplete trajectories. SIAM J. Appl. Dyn. Syst. **17**(2), 1891–1924 (2018)

Froyland, G., Koltai, P.: Detecting the birth and death of finite-time coherent sets. Commun. Pure Appl. Math. **76**(12), 3642–3684 (2023)

Froyland, G., Kwok, E.: Partitions of networks that are robust to vertex permutation dynamics. Spec. Matrices **3**, 22–42 (2015)

Froyland, G., Kwok, E.: A dynamic Laplacian for identifying Lagrangian coherent structures on weighted Riemannian manifolds. J. Nonlinear Sci. **30**, 1889–1971 (2020)
Froyland, G., Rock, C.P., Sakellariou, K.: Sparse eigenbasis approximation: multiple feature extraction across spatiotemporal scales with application to coherent set identification. Commun. Nonlinear Sci. Numer. Simul. **77**, 81–107 (2019)
Gómez, S., Díaz-Guilera, A., Gómez-Gardeñes, J., Pérez-Vicente, C.J., Moreno, Y., Arenas, A.: Diffusion dynamics on multiplex networks. Phys. Rev. Lett. **110**(2), (2013)
Hadjighasem, A., Karrasch, D., Teramoto, H., Haller, G.: Spectral-clustering approach to Lagrangian vortex detection. Phys. Rev. E **93**(6), 063107 (2016)
Haller, G., Beron-Vera, F.J.: Coherent Lagrangian vortices: the black holes of turbulence. J. Fluid Mech. **731**, R4 (2013)
Karrasch, D., Schilling, N.: A Lagrangian perspective on nonautonomous advection-diffusion processes in the low-diffusivity limit (2021). arXiv:2102.04777
Karrasch, D., Keller, J.: A geometric heat-flow theory of Lagrangian coherent structures. J. Nonlinear Sci. **30**(4), 1849–1888 (2020)
Lablée, O.: Spectral Theory in Riemannian Geometry. EMS Textbooks in Mathematics. European Mathematical Society (2015)
Mathew, G., Mezić, I., Petzold, L.: A multiscale measure for mixing. Phys. D **211**(1–2), 23–46 (2005)
Miclo, L.: On hyperboundedness and spectrum of Markov operators. Invent. Math. **200**(1), 311–343 (2015)
Mosovsky, B.A., Meiss, J.D.: Transport in transitory dynamical systems. SIAM J. Appl. Dyn. Syst. **10**(1), 35–65 (2011)
Ottino, J.M.: The Kinematics of Mixing: Stretching, Chaos, and Transport, vol. 3. Cambridge University Press (1989)
Padberg-Gehle, K., Schneide, C.: Network-based study of Lagrangian transport and mixing. Nonlinear Process. Geophys. **24**(4), 661–671 (2017)
Pierrehumbert, R.T.: Tracer microstructure in the large-eddy dominated regime. Chaos Solitons Fractals **4**(6), 1091–1110 (1994)
Schneide, C., Vieweg, P.P., Schumacher, J., Padberg-Gehle, K.: Evolutionary clustering of Lagrangian trajectories in turbulent Rayleigh–Bénard convection flows. Chaos **32**(1), (2022)
Sole-Ribalta, A., De Domenico, M., Kouvaris, N.E., Diaz-Guilera, A., Gomez, S., Arenas, A.: Spectral properties of the Laplacian of multiplex networks. Phys. Rev. E **88**(3), 032807 (2013)
Thiffeault, J.L.: Using multiscale norms to quantify mixing and transport. Nonlinearity **25**(2), R1 (2012)

Open Access This chapter is licensed under the terms of the Creative Commons Attribution 4.0 International License (http://creativecommons.org/licenses/by/4.0/), which permits use, sharing, adaptation, distribution and reproduction in any medium or format, as long as you give appropriate credit to the original author(s) and the source, provide a link to the Creative Commons license and indicate if changes were made.

The images or other third party material in this chapter are included in the chapter's Creative Commons license, unless indicated otherwise in a credit line to the material. If material is not included in the chapter's Creative Commons license and your intended use is not permitted by statutory regulation or exceeds the permitted use, you will need to obtain permission directly from the copyright holder.

A Consideration of Scientific-Technical Aspects and Artistic Aspect of Origami Engineering-Aiming to Create New Big Industries and a New Fan Culture

Ichiro Hagiwara

Abstract Washi (Japanese paper) is virtually the world's first foldable paper. It is inevitable that origami is born and nurtured in Japan. Origami consists of 4 types such as courtesy origami dedicated to god, play origami such as cranes and turtles, modern origami such as floor gate fold and biomimetics origami for origami engineering. All of these four types of origami have been born from Japan. Thanks to Japan's isolation during the Edo period (1603–1868A.D.) and a peaceful and prosperous era, art and culture such as haiku (poetry), kabuki (theater), noh (dance-drama) and ogi (folding fans) peculiar to Japan were nurtured. Origami has long been a precious art and was not considered to be a target for money in Japan. But during World War II, British engineers developed a mass production method for honeycombs, inspired by Japanese Tanabata decorations which developed into a trillion yen industry. This fact inspired us to start origami engineering in 2002 of which possibility will be discussed here.

1 From Origami to Origami Engineering

1.1 Birth of Paper and History of Japanese Origami

Papyrus, the origin of the word paper, began to be used for paper in Egypt around 2500 BC. Since Egyptian and Western writing instruments were designed to scratch, the papyrus-based paper of the time had to be thicker and could not be folded. In Japan and China, on the other hand, brushes were used as writing tools, so the thinner the paper, the better. Such paper for brushes was first invented in China in 105AD, and these paper- and ink-making methods were introduced to Japan in 610 AD by a Korean monk. Japanese don't just accept it as is but they have the characteristic of not being satisfied unless they try making everything themselves. At last, other

I. Hagiwara (✉)
Institute for Advanced Study of Mathematical Sciences (MIMS) & Institute of Autonomous Driving and ELSI (MIAD), Meiji University, Tokyo, Japan
e-mail: ihagi@meiji.ac.jp

© The Author(s) 2026

S. Oishi et al. (eds.), *Recent Developments in Industrial and Applied Mathematics*, ICIAM2023 Springer Series 1, https://doi.org/10.1007/978-981-95-1446-5_6

paper-making techniques developed in Japan around 800 A.D., and the Japanese created the world's first foldable paper Washi using tenacious elm. Subsequently, it was discovered that Washi became more rigid and produced a beautiful lustre when folded. Thus was the birth of etiquette (courtesy) origami dedicated to god.

During the Edo period, when Japan adopted a policy of national seclusion, thanks to a peaceful and prosperous era, art and culture such as haiku (poetry), kabuki (theater), noh (dance-drama) and ogi (folding fans) peculiar to Japan were nurtured. Origami is one of them. With the popularization of paper in the Edo period, a more playful form of origami, so called play origami to create animals such as cranes and turtles was developed. It was also during this period that the world's oldest book on origami was produced. Later, immediately after World War II, Kamon-ori (floor gate fold) was founded by Mitsuhiro Uchiyama who gave birth to modern origami where the author's name is attached to origami works. In such way, origami has long been a precious art and was not considered to be a target for money in Japan. While doing so, during World War II, British engineers developed a method to mass-produce honeycomb, inspired by the Japanese Tanabata festival decorations. Honeycomb, which has the highest bending stiffness per weight, is now used in rockets and bullet trains and has become a multi-trillion-yen industry. This realization changed the Japanese mindset to origami such that we are also permitted to earn money with origami.

1.2 Birth of Origami Engineering Based on Biomimetics Origami, the Fourth Origami by Nojima

This invention of mass production for honeycomb core mentioned above inspired us. Nojima proposed the concept of origami engineering (Nojima 2002) to further apply its advantages to industrial applications in 2002. In response to this, Hagiwara set up study groups on origami engineering in both JSME (The Japan Society of Mechanical Engineers) and JSIAM (The Japan Society for Industrial and Applied Mathematics). Nojima founded biomimetics origami immediately as follows.

The origin of biomimetic origami began with the study of plants: The arrangement of leaves and flowers is called the phyllotaxis or inflorescence. These leaves grow one by one in a spiral pattern so that each leaf receives an equal amount of light from the sun. As shown in Fig. 1c, after three rotations, the eighth leaf will be directly above the first leaf which is called 3/8 phyllotaxis. It is described the 1/3, 2/5 and 3/8 phyllotaxis schematically in Fig. 1a–c respectively. These numbers follow the Fibonacci sequence (where every number is the sum of the previous two numbers, i.e. 1, 1, 2, 3, 5, 8, etc.). These leaves are evenly spaced on this spiral. This interval is expressed as an angle and is called the opening degree φ'. For an example, in the case of 3/8 phyllotaxis, $\varphi' = 3 \times 360°/8 = 135°$ so that φ' will be very close to golden angle $\varphi_0(137.5077°)$. Assuming the stems in Fig. 1a–c as cylinders, by cutting at the busbars, the developed views of Fig. 1a–c are obtained. The group of straight lines in the figure that gradually rise to the right is a single spiral called the generative spiral that wraps around the cylinder in Fig. 1. Based on these development views, Nojima designed foldable cylindrical models, by furthermore, considering two conditions. One is the folding condition at the node where the folding lines meet such that sum of discrete angles made by folding lines at a node is angle 180°. The other is that the condition for closing in the circumferential direction after folding holds true. Furthermore, Nojima considered a case in which primordia such as seeds are arranged spirally in a circular area like sun flower in Fig. 2a. At first, the development diagram of the phyllotaxis model in Fig. 2 changed to a conical shape is shown in Fig. 2b. Here vertical angle of developed view is 60°. The opening degree in Fig. 2 is golden angle φ. And so the opening degrees of this development view in Fig. 2b is (137.5/360)/6 ≅ 22.9166°. A conical origami model made from this development diagram is shown in Fig. 2c. If you fold this, it will look like Fig. 2d. Based on the above, a generalization of the circular membrane folding model has been obtained. For details, please refer to pages 68–76 of the book (Nojima and Hagiwara 2012).

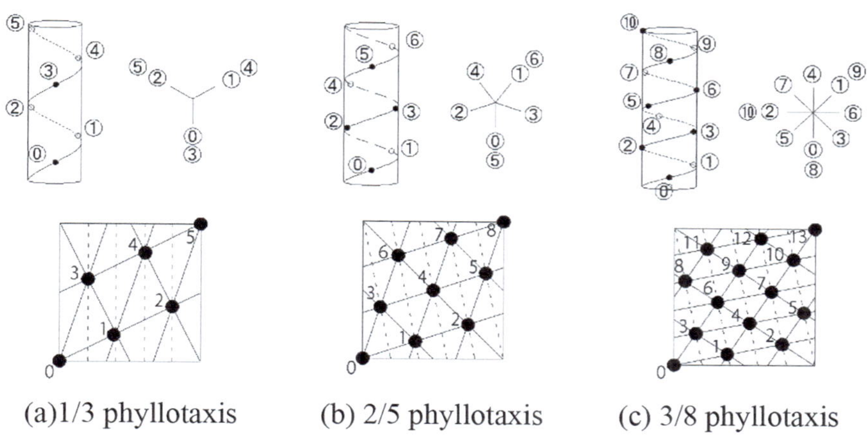

Fig. 1 How leaves form in a spiral according to fibonacci sequence and each development view (Nojima and Hagiwara 2012)

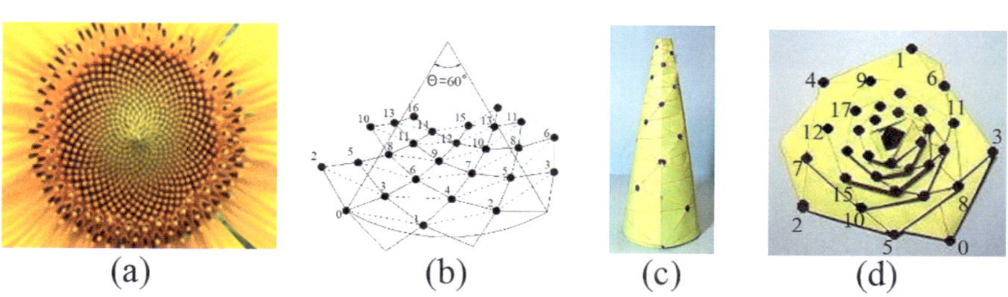

Fig. 2 The spiral pattern of sunflower seeds (**a**), developed view of a cone (**b**), origami model of the cone (**c**) and its folded state (**d**) (Nojima and Hagiwara 2012)

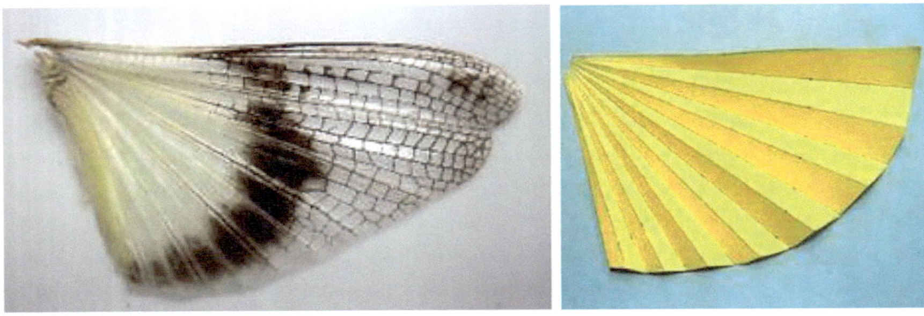

Fig. 3 Origami reproduction of an insect wing (Hagiwara and Kondoh 2022)

Nojima also succeeded in making insects out of origami, thus creating a new field known as biomimetics origami. As indicated, biomimetics origami is inspired by the natural world. For instance, various insects are able to fold their wings very compactly, and these complex folding patterns can be drawn using elementary geometric rules and then reproduced using origami as shown in Fig. 3.

As above, there are 4 types of origami such as etiquette (courtesy) origami, play origami which is sometimes called tradition origami because its creator is unknown, modern origami (floor gate origami) and biomimetics origami. And also as shown in Fig. 4, origami engineering has industrial aspects, artistic aspects, and game (geometric) aspects because of course it comes from origami (Fig. 5).

Here, we describe the current status and issues of the industrial aspect, where there are 3 types of directions such as origami design, design manufacturing for light but rigid and deployable and fordable origami structures. Below, we explain these three approaches with examples.

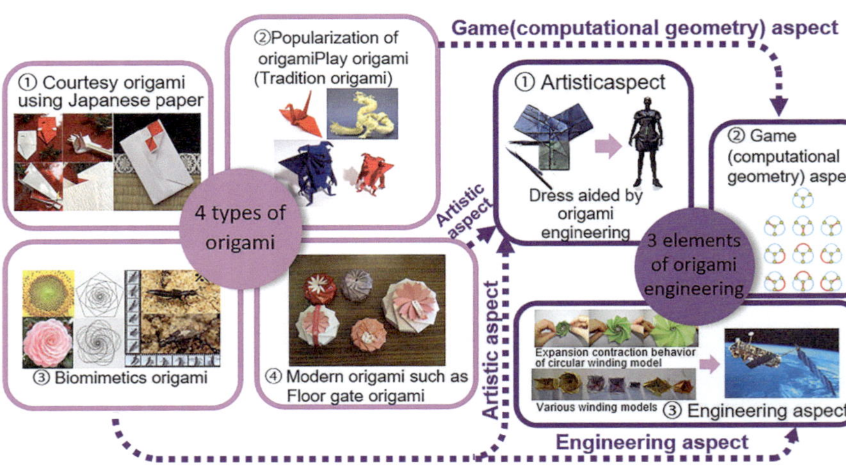

Fig. 4 4 types of origami and 3 elements of origami engineering (Hagiwara 2024)

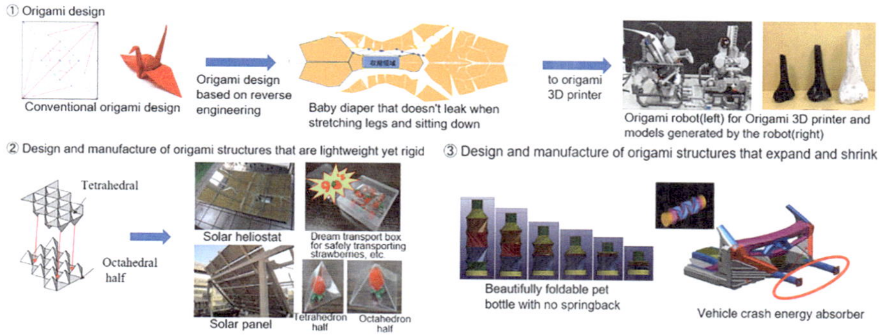

Fig. 5 Three pillars of industrial engineering aspect of origami engineering design manufacturing for light but rigid and for deployable and foldable origami structures

2 Creating an Origami Design for Anything

2.1 Proposed Origami Design System

Let's look at origami design, which attempts to make anything out of origami. It is a particular challenge to fold a complex polyhedral shape from a single sheet of paper without any gaps. To begin with we used reverse engineering (Hagiwara and Shinoda 2018), mainly used in manufacturing industry, to build a system that uses a photo of the product to create a development view with a glue allowance, and origami mountain and valley lines. Reverse engineering means working backwards in the design and manufacturing process. In practice, for instance, a point cloud is generated from the data of an actual vehicle or clay model; STL (Stereolithography) data is generated for conversion to CAD; and then a CAD or finite element method (FEM) model is generated simplified one (402 points) and size of each STL element, (c) Generation of development view, (d) Prototype by for simulation to verify the part. Based on this reverse engineering, we have developed origami design system Ori-3D (Hagiwara et al. 2017; Yu et al. 2016) as shown in Fig. 6. As far as origami design system, there are some splendid systems such as Tree Maker (Lang 2015), Origamizer (Demaine et al. 2017), ORI-REVO (Mitani 2008). But Ori3D is the most practical so that it has the potential comparable to 3D additive printer. Let's at this situation.

Because the degree of adhesion between diaper and skin is different when the legs are stretched and when the legs are seated, the problem of leaking "Loose poo" in the low-moon period has been solved based on photo image of baby doll and generating its development view. The baby diapers thus obtained are being sold with good reviews (http://www.unicharm.co.jp/company/news/2019/1211066_13296.html). Here, it is shown in Fig. 7 the procedure to make diapers for the elderly, similar to baby diapers.

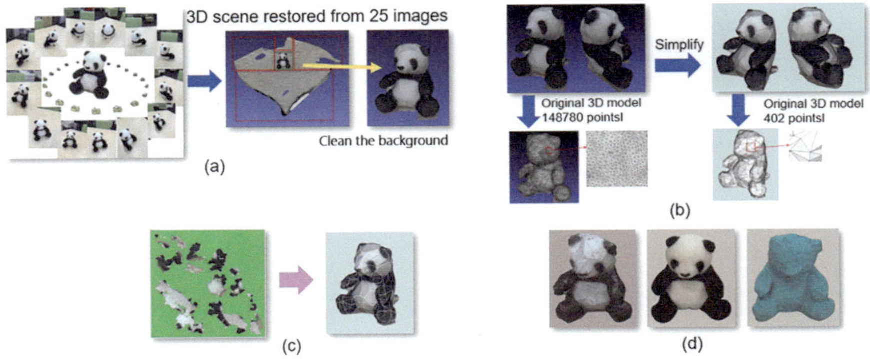

Fig. 6 Reconstruction based on development view with glue-margins from scattered points by photographic image along reverse-engineering routine: **a** 3D scene restored from 25 images without the background, **b** Simplification from image (148,780 points) to Ori3D (left), Real object (center), Prototype by Additive 3D (right)

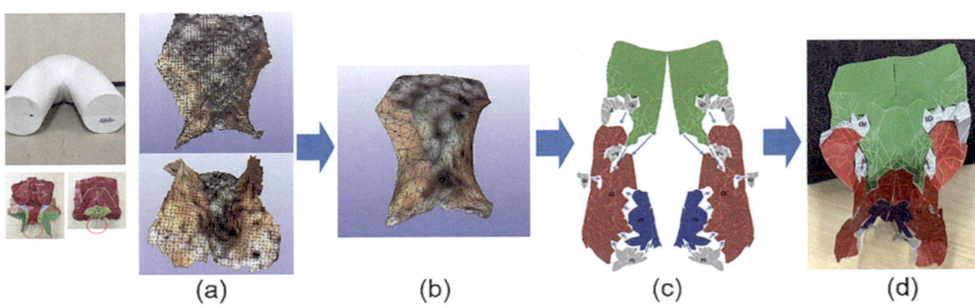

Fig. 7 Flow chart for diaper by Ori3D: **a** 3D surface mesh model generated from photographic image, **b** simplified model, **c** development view by Ori3D, **d** reconstructed model

2.2 *Required Items for Achieving Origami3D Printer*

Although the STL data is also the input data for additive 3D printer (abbreviated as Add-3D) currently used generally, itself, it is difficult to generate such diaper with Add-3D. From this, it can be considered that Ori-3D has the potential to complement Add-3d if Ori-3D is automatically operated with origami robot. The research of origami robot seems to have reached a dead end where its sensors and actuators are too sophisticated because origami robot was required to be able to imitate human folding behavior. There are things to note here that the tree structure of the development view of interested origami is almost close. Here, the tree structure is generated by randomly selecting one point in each triangle with valley and mountain lines on the development view and connecting them in order. Such tree structure gives humans moderate difficulty and a sense of accomplishment but this is considered to be a

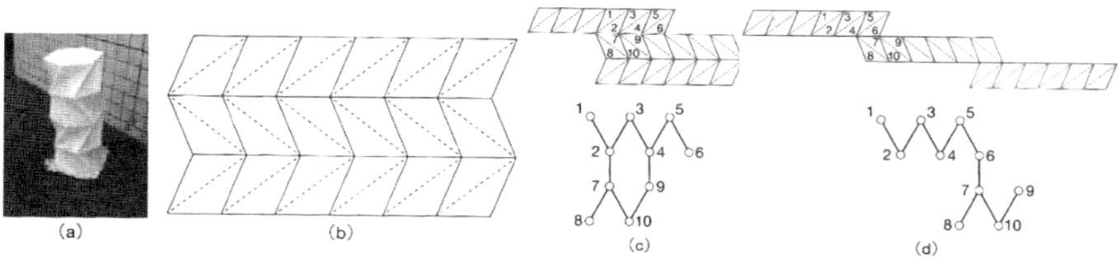

Fig. 8 Many patterns can be generated by Ori3D and some of them (if open tree structure) can be done by a robot that can only bend and glue (Hagiwara 2018)

difficulty beyond the capabilities of robots. For example, the firstly presented development view (Nojima 2000) of RSO (Reversed Spiral Origami Structure) shown in Fig. 8a is shown in Fig. 8b. We can also get Fig. 8c, d with Ori-3D. The tree structures of Fig. 8b, c are close but that of Fig. 7d is open. If you bend triangle 2 in Fig. 8c, triangles 1, 3 and 7 around triangle 2 are deformed. And when triangle 3 is deformed, triangle 4 is deformed. A routine like this transforms triangle 3 into triangles 4, 9, 10, 7. On the other hand, when you bend firstly only triangle 2 in Fig. 8c, triangle 7 is received transformations from two routines 2 → 7 and 2 → 3 → 4 → 9 → 10 → 7. This correspondence is difficult for robot. On the other hand, in the case of Fig. 8d, if you deform triangle 7, triangles 6, 8, 10 etc. receive deformation from only one routine so that it is easy for robot with only mountain-folding and valley-folding functions. And so, if the development view with close tree structure can be automatically transformed to the one with open tree structure (Hagiwara et al. 2019), origami robot (Romero et al. 2016) can be realized and the above mentioned diaper is obtained by only one system, also called Origami 3D printer which is being achieved by us (Hagiwara 2016, 2018).

3 Design and Manufacture of Origami Structures that Are Lightweight Yet Rigid

3.1 Kirigami Honeycomb

These days, Kirigami honeycomb is also in the spotlight as a treasure trove of metamaterial creation. As above, stimulated by the invention of honeycomb mass production method, Nojima advocated origami engineering and he developed firstly curved honeycomb with Japanese tradition Kirigami (Nojima 2002) against the top and bottom surfaces of the honeycomb core are flat. After that, Saito mainly extended and also had a patent (Saito 2017). Honeycomb obtains improvement in bending stiffness and acquisition of shear stiffness only after its sandwich structure with joining core material and top and bottom plates. And so Saito and subsequent authors start with the formula for top and bottom surfaces. It can be said that Harvard/MIT and

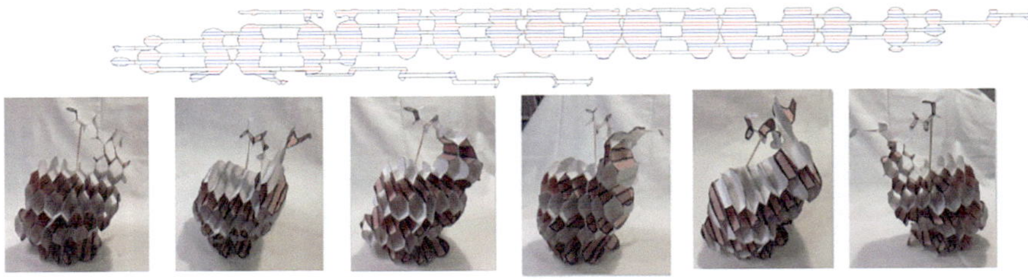

Fig. 9 Stanford Bunny Kirigami-honeycomb example (upper: the 2D pattern, lower: 6 directions photographs of the real model generated from 2D pattern) (Diago et al. 2023)

Fig. 10 Bicycle helmet made from kirigami honeycomb

University of Bristol are the two major bases of meta material creation based on Kirigami honeycomb and Kirigami also becomes an international language. On the other hand, Diago et al. has achieved completely 3D Kirigami-honeycomb by directly using CAD and STL themselves and by covering them with honeycombs and search for intersections between honeycombs and CAD or STL data (Diago et al. 2023). An example is shown in Fig. 9 which solves the drawbacks of not being able to reproduce complex three-dimensional shapes with Kirigami honeycomb. Here too, the development of origami robot is indispensable for its practical application. In Japan, bicycle users are also becoming obligated to wear helmets. We can say our recent Kirigami honeycomb helmet shown in Fig. 10 is the most safety because each column direction of all the cells is perpendicular to the surface.

3.2 Qubic Core

Another influential thing is the qubic core panel (QP) (Hagiwara and Zhao 2022). It is compared with honeycomb core panel (HP) in Fig. 11 where HP has the line contact and QP has the surface contact between the core material and the upper/lower panels. For this reason, HP cannot be used for the structures which are subjected to long-term vibration or shear. In the case of HP, walls and walls are glued and also are glued between core material and the upper/lower panels so that HP becomes powerless in fire. QP eliminates these drawbacks of HP by surface contact and forming without glue. As shown in Fig. 11, by the red dotted line, the central widths of both equivalent cross sections can be approximated to be the same. The face plate thickness of the equivalent cross section of the cubic core can be approximated to 1.5 times the original plate thickness. And so QP is also superior to HP in terms of bending stiffness per weight as shown in Fig. 12 (Tian et al. 2023). The solid red line indicates the rate of increase of the area moment of inertia of the cubic core panel by the honeycomb panel, and the black dotted line indicates the increase rate of the thin plate volume of the cubic core panel by the honeycomb panel. The horizontal axis indicates the ratio between the side length and the thickness of the panel.

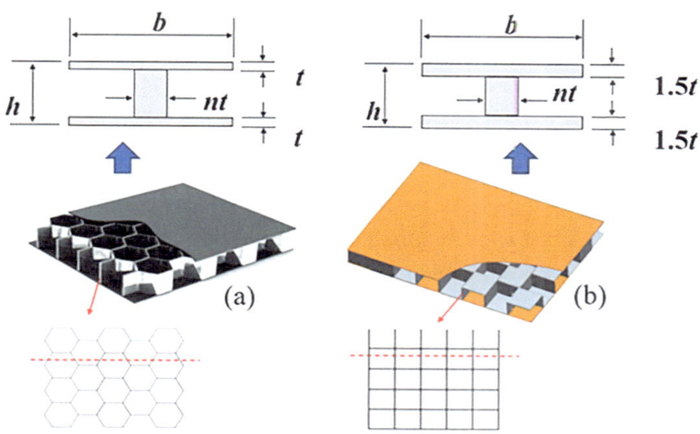

Fig. 11 Equivalent cross section of honeycomb core panel and cubic core panel

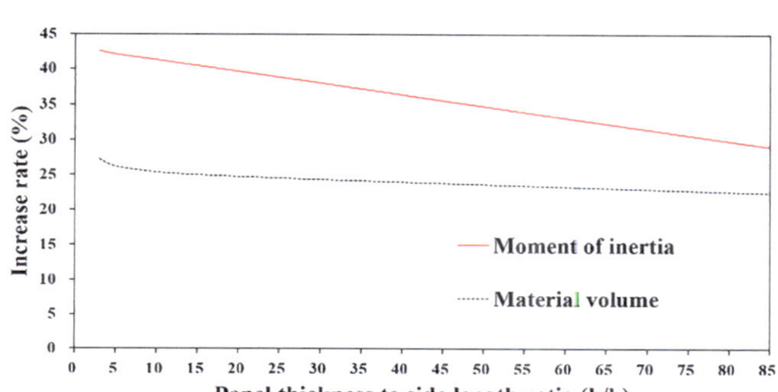

Fig. 12 Comparison of bending properties between honeycomb and cubic core panels (Tian et al. 2023)

3.3 Truss Core

Now let's move on to truss core (Hagiwara 2021) which is generated by space filling of tetrahedral and octahedral half as shown in Fig. 13. Bending stiffness per weight of truss core is about the same as honeycomb panel and 6–7 times greater than flat plate. So, when truss cores are applied to solar cell panels, it does not need to reinforce eaves because it does not need to reinforce truss cores to maintain a wide flat surface. Note here, when the cores are arranged uniformly as shown in Fig. 13, if it is made of resin, it can be molded without any problems but if it is of metal, it is not easy. Therefore, we are solving this problem by developing multi-step forming method (Tokura and Hagiwara 2010) which first creates hemispherical cores and then creates pyramidal cores in the next step as shown in Fig. 14. As this result, well-shaped truss cores are formed and they are already applied to solar cell panels and solar heliostats. However, as the height of the cores is limited when constructing multiple cores from a single panel by multi-step press forming method, the concept of the Assembly Core Truss Panel (ACTP) (Hagiwara and Terada 2022), in which the cores are built and assembled one by one, was created. For example, a space-filled shipping box with 24 hollow octahedral halves and 25 hollow tetrahedral can support a person weighing 60 kg with no cores pain as shown in Fig. 15. Even if some cores are damaged, you only need to replace the damaged cores. If the core is made of 0.1 mm steel plate, it can withstand 6 tons. The flange length that does not require gluing is calculated here. And this is one of the appeal points of the patent (Hagiwara and Terada 2022). The properties of these ACTPs can be applied to pentagonal and cubic octahedral semi-shapes, for example to create effective packaging materials for transporting foods such as strawberries and eggs, which are prone to breakage during transport, as well as for pluripotent stem cells (ips) or blood, which degrade easily during transport. If the length of one side of an equilateral triangle of the core is 100 mm, an egg can be stored in the core. As shown in Fug. 16, the egg is held still in the core with a 0.5 mm pp (Polypropylene) sheet ①, the egg is put into the octahedral half core ②, It is prepared 5 pieces of this specification ③, the prototype with pentagon-shaped outer box ③ is 137 g. Assuming a fall accident during transportation, an experiment was conducted ④ in which the object was dropped from a height of 1 m. As the result, All five eggs had no damage such as cracks. From the above, it can be seen that the pp pentagon can also be used as an excellent cushioning material and shipping box (Fig. 16).

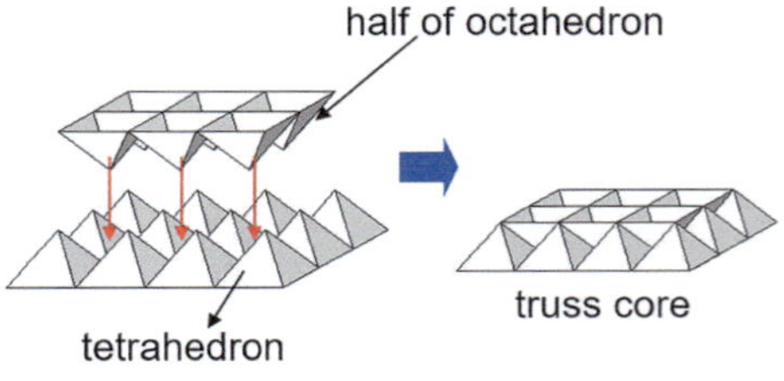

Fig. 13 Structure with octahedral halves and tetrahedra that interlock

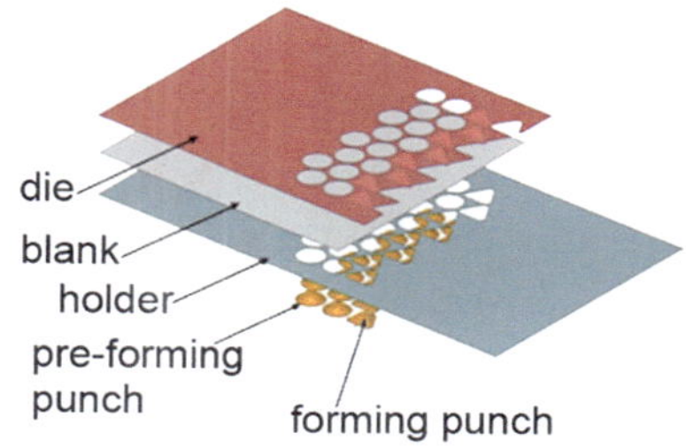

Fig. 14 Multi-step press forming process (Tokura and Hagiwara 2010)

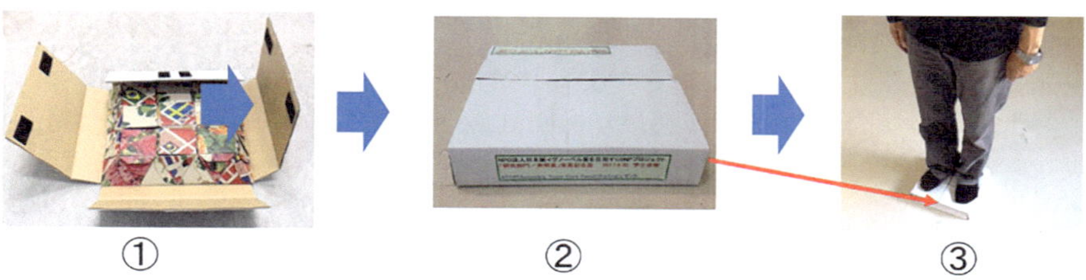

Fig. 15 Assembly Truss Core Panel (ATCP): ① Store 49 cores (25 tetrahedrons and 24 octahedrons) in cardboard box, ② Close cardboard box and stop with Velcro, ③ cardboard box has not damage for 60 kg iron box has not damage for 6 ton!

Fig. 16 5 eggs do not break even if they fall from 1 m above: ① Spread the PP sheet to stand the egg, ② Store the eggs in the octahedron half-core, ③ Store in ATCP pentagon type consisting of regular octahedron 5 half cores, ④ Drop from 1 m above

4 The Design and Manufacture of Origami Structures that Expand and Shrink

4.1 Crash Energy-Absorbing Characteristics of Origami Structure

Because the energy absorbing material is a column structure, the most important thing to ensure the safety of vehicle occupants during a collision is that the column structures are designed to collapse in the axial direction as long as possible without bending on the way. In order for the column structure to obtain the above ideal deformation mode, there are beads installed in the appropriate position of the energy absorber of automobile manufacturers around the world, this effectiveness was firstly demonstrated by us (Hagiwara et al. 1991; Kitagawa et al. 1991). Bead is a general term for notches and protrusions. However, even though the ideal deformation mode is realized, always the initial peak load is too high. I judged that the only way to solve this problem is to use origami structures. This was also the reason why the author promoted origami engineering.

The graph in Fig. 17c compares the load of the absorber. When the conventional rectangular type is used, the initial peak load is high, as shown by the red line, and this can sometimes harm the occupants. With this type of absorber, the amount of deformation is only about 70% of its length. On the other hand, when the reverted spiral origami structure (RSO) is used for a new energy-absorbing material (Hagiwara and Nadayoshi 2003; Hagiwara et al. 2004; Wu et al. 2007; Zhao et al. 2011), the load value remains almost constant without changing with the deformation, as shown by the black line in Fig. 17c, while the amount of deformation reaches 90% of its length due to the folding characteristic. Here, α, β in development view of RSO are decided by the conditions that the left and right end lines must match as shown in Eq. (1) and they must not come apart during expansion and contraction after the two lines are connected as shown in Eq. (2) (Nojima 2000). Consequently, research is now being conducted on the use of the new origami structure for absorbing materials that are able to absorb high impact energy, reduce the impact load as gently as a cushion. For this reason, we developed a hydro-forming simulation (Kong et al. 2017) for equipment and made a prototype of it. But due to its high manufacturing cost, RSO is not yet commercialized.

$$\alpha = \pi/n \tag{1}$$

$$\pi/4 - \pi/2n < \beta < \pi/2 - \pi/n \tag{2}$$

10 years later, we have developed the reversed torsional origami structure (RTO) by the partial heating torsion forming system (Hagiwara and Zhao 2021) shown in Fig. 18 cheaply. In Fig. 18, IH means induced heat. As shown in Fig. 19, there is good

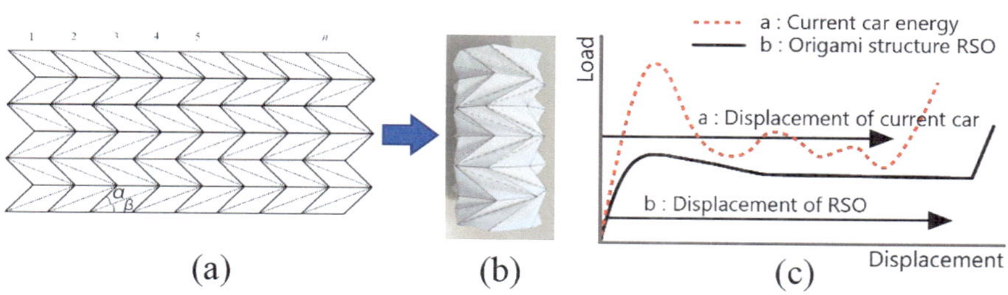

Fig. 17 **a** Development viewof RSO, **b** RSO, **c** comparison of load–displacement curves of RSO and present vehicle crash absorber

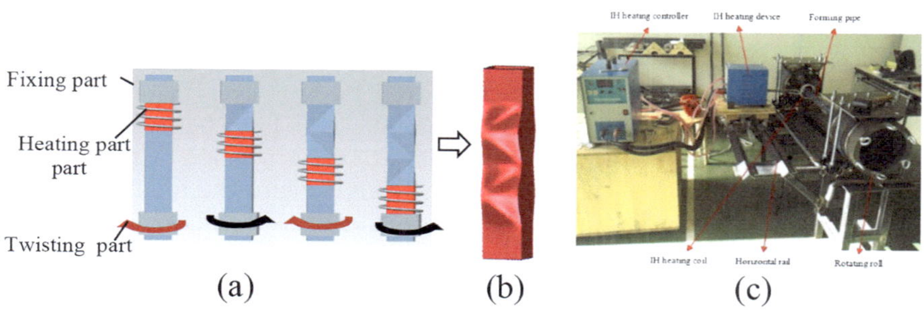

Fig. 18 **a** RTO (right end) obtained by partially heated rotational Forming method, **b** Forming system for RTO (Zhao et al. 2022)

Fig. 19 Comparison between simulation and experiment for RTO (Zhao et al. 2022)

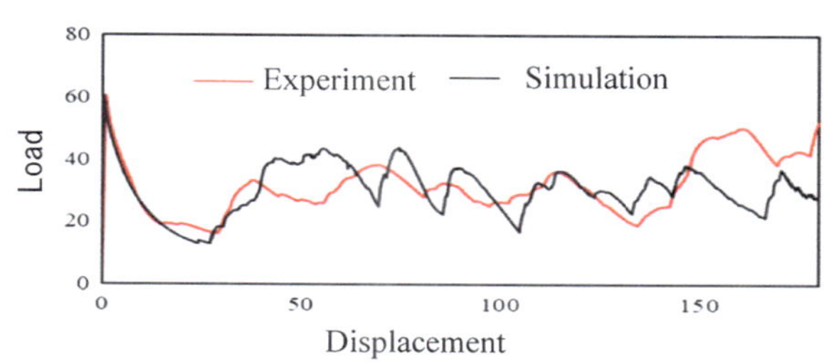

correspondence between the prototype crushing experiment and the simulation. The performance of RTO is superior to the one of RSO (Zhao et al. 2022) and so that there is a good possibility that it will be applied to the actual structure.

4.2 *Foldable Pet Bottle*

Another application is a polyethylene terephthalate (PET) bottle. PET bottle is expected that it can be easily made into small chunks by human hands after drinking.

What is more desired is that it can be folded up to the height of the drink in the direction of its pillar. As for the circumferential compacting, it has been obtained but it is not easy to fold in the direction of its pillar and has finally succeeded recently.

We developed a PET bottle that easily folds axially without bending. For the object to easily fold, we used an FSO (Forward Spiral Origami Structure) or RSO structure for the foldable parts as shown in Fig. 20. With these FSO or RSO structures, the maximum height of each layer is decided according to the perimeter. The overall structure has plural layers. We began with seven layers from an aesthetic perspective as shown in Fig. 20. In the case of adopting spiral structures except for the lid and bottom, the bottle can fold rather easily but it has remarkable spring back. We found that to suppress spring back, a folded layer can be inserted into the non-foldable layer and the groove between the folded and non-foldable layers deforms in synchronization with the deformation of the folded layer. The feasibility of models with two spiral layers based on this idea was investigated; that is, the number of layers was reduced from seven to six while two of the four layers that were not the lid or bottom were set as cylindrical shells. The model in Fig. 20 is easily foldable but it readily returns to its original height through the generation of spring back. Figure 21 shows a reversed spiral model for which the third and fourth layers are spiral so as to be foldable. This model has six rather than seven layers. Points 1–6 in Fig. 21a shift to the same points in Fig. 21b after folding. It is thus seen that the folded third layer is inserted into the not folded conical shell second layer and the folded fourth layer is inserted into the not folded cylindrical shell fifth layer after slight spring back. These results are achieved in two important ways. First, as shown in Fig. 22, truss-shaped lines are applied to the second layer for strength as to be inserted by the folded third layer. Second, grooves are set between layers. Figure 23 compares a deformation mode with (a) and without (b) grooves. Figure 23(I) shows that the spiral third layer inserts into the conical shell second layer. With or without grooves, the spiral layer inserts into the conical shell layer in Fig. 23a and b-(I). With grooves, if the grooves deform in conjunction with the spiral layer, once the folded part inserts into the not folded layer, it does not get out as shown in Fig. 23a-(II). Meanwhile, without grooves, the folded part is prone to get out from the not folded part because the spring back is large in Fig. 23b-(II). In such a way, the grooves suppress the spring back by deforming in conjunction with the spiral structure. Here, though this is due to the software system used, the first layer is yellow, the second layer is brown, and the third layer is red. The spiral layer with the groove is therefore red in Fig. 23a. Meanwhile, the spiral layer is brown in Fig. 23b. The groove has a dent on the inside so that the crushed groove suppresses the spring back by entwining with the groove and the folded spiral that is inserted. This explains how the grooves suppress the spring back. Figure 24 shows a prototype with two spiral layers and its folded status in the axial direction without bending (Hagiwara et al. 2022).

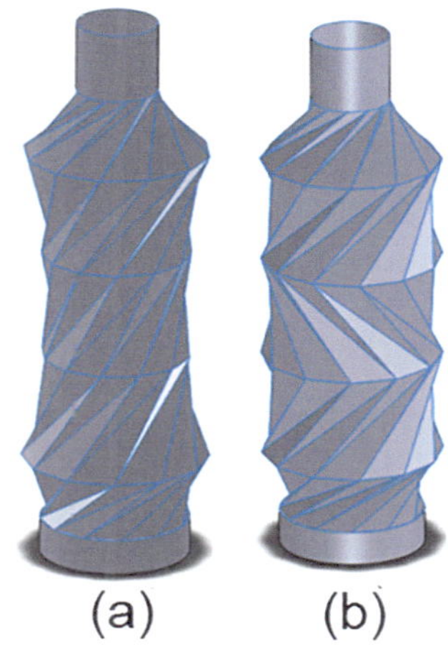

Fig. 20 FSO model (**a**) and RSO model (**b**) (Hagiwara et al. 2022)

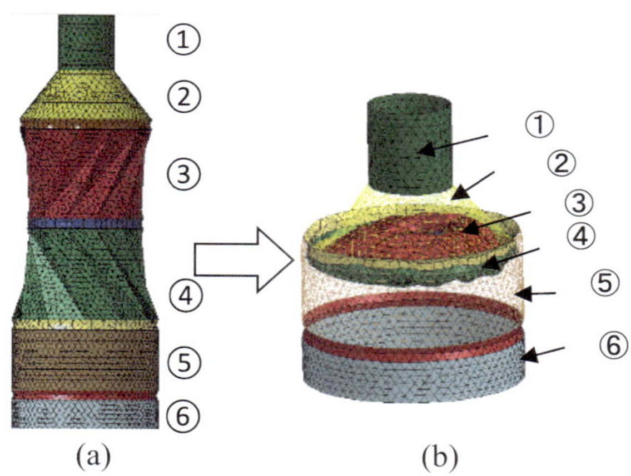

Fig. 21 2-layer model without spring back (Hagiwara et al. 2022)

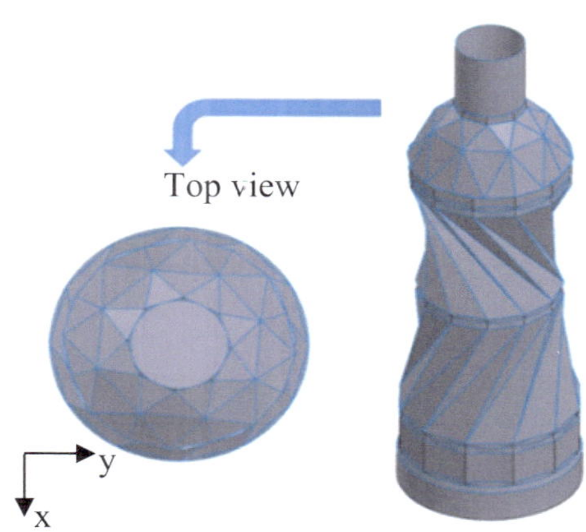

Fig. 22 2nd layer strengthened by truss shape (Hagiwara et al. 2022)

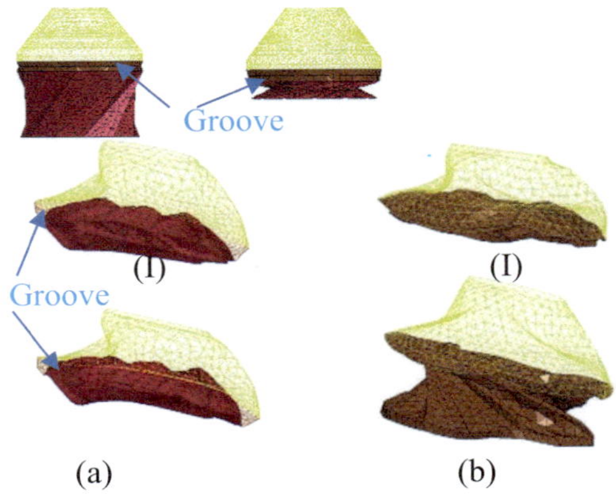

Fig. 23 Comparison of deformation mode between with (**a**) and-without (**b**) grooves (Hagiwara et al. 2022)

Fig. 24 Prototype of 2 layers model and its beautifully folded one without spring back (Hagiwara et al. 2022)

4.3 Origami Hydraulic Damper

With the shift to EV cars, the internal combustion engine is replaced by an electric motor. This results that vehicle interior noise becomes lower so that demands for the sound quality of speakers are increasing. On the other hand, as the result of further pursuit of vehicle body weight reduction, the door outer panel of vehicle body is reduced to 0.8 mm. So a problem arises in which the sound through a speaker and the vibration of the door outer panel affect each other. Based on this, we applied for a patent of an origami-type hydraulic damper made of polypropylene in order to be able to use it not only for the conventional heavy and long size hydraulic cylinder of iron but also for light, thin, short and small sizes (Hagiwara et al. 2023). The proposed basic concept is shown in Fig. 25b. The conventional hydraulic cylinder in

Fig. 25a does not deform even half of its length and so it's not suitable for frivolous, short and small. On the other hand, RSO damper of pp deforms as much as 90% of its own length. Structure overview of RSO origami hydraulic damper is shown in Fig. 26a and its mathematical model is in Fig. 26b. Based on Fig. 26, the kinetic equation was derived as in Eq. (3) (Gao et al. 2020).

$$M\ddot{x} + c\dot{x} + C_{\text{damper}}\dot{x}^2 + Kx = -m\ddot{x}_e \tag{3}$$

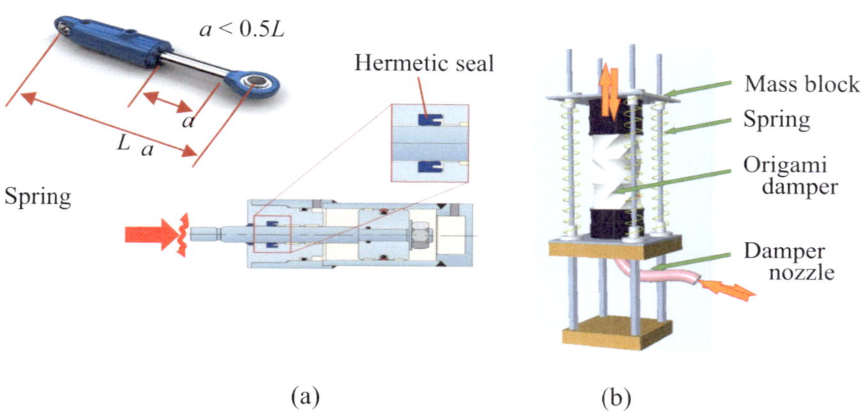

Fig. 25 Proposed inverted spiral origami hydraulic damper (Gao et al. 2020)

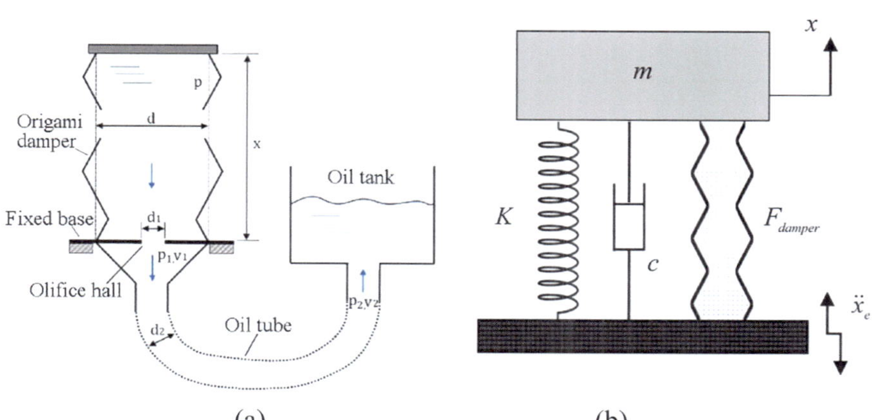

Fig. 26 Structure overview of RSO origami hydraulic damper (**a**) and its mathematical model (**b**) (Gao et al. 2020)

$$C_{\text{damper}} = \frac{\pi \rho d^6}{8}\left(\frac{1}{c^2 {d_1}^4} + \frac{\lambda L}{{d_2}^5} + \frac{K_c}{{d_2}^4}\right)$$

Here, $\ddot{x}_e$ is external force acceleration, λ is the friction coefficient of the oil pipe, L is the length of the oil tube and other parameters are in Fig. 26a. In such way, we found out for the first time that the origami damper is related to $\dot{x}^2$. We solved the equation and confirmed the effect in Figs. 27 and 28. In other word, the effect of the difference in orifice diameter is greater than the difference between water and oil. If it is made of iron, it will rust due to water, but if it is made of pp material, it will not rust due to water. And so what is noteworthy is that it is convenient that water can be used with pp material. In order to cope with cases where the direction of the load is arbitrary, we made prototype for an arc shape stroke origami type hydraulic damper as shown in Fig. 29. The kinetic equation was derived as in Eq. (4) (Guan et al. 2023).

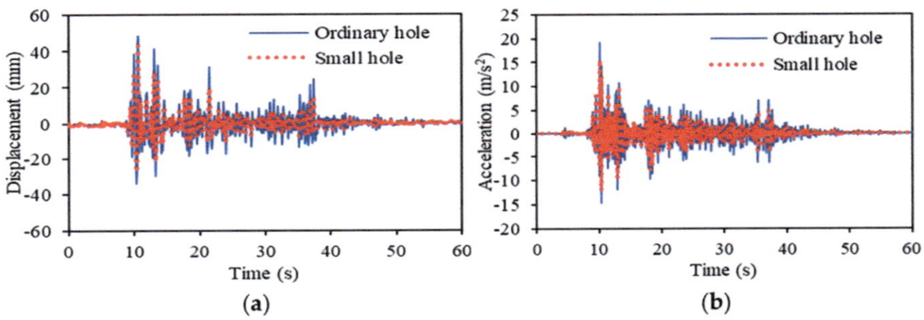

Fig. 27 Effect of attenuation hole diameter when excited by EI Centro NS seismic wave. **a** Response displacement. **b** Response acceleration (Gao et al. 2020)

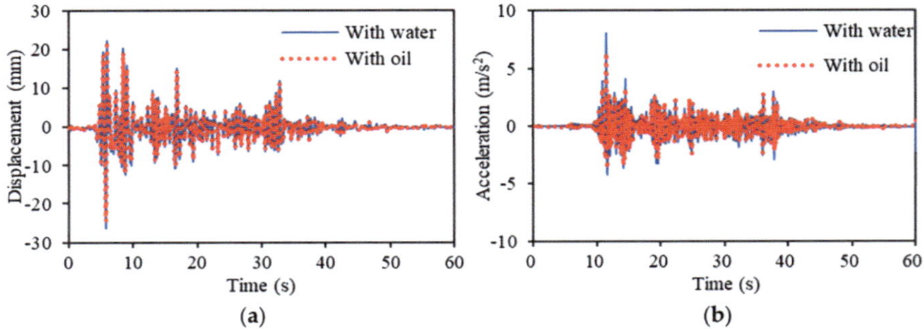

Fig. 28 Effect of type of liquid when excited by EI Centro NS seismic wave. **a** Response displacement. **b** Response acceleration

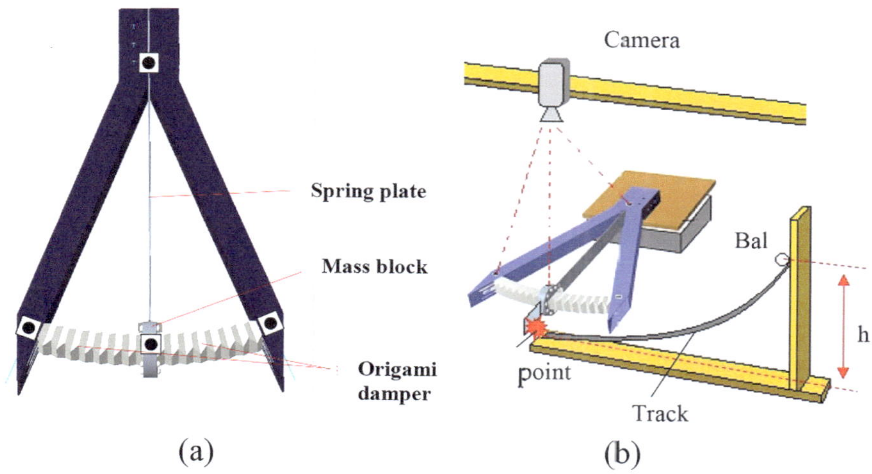

Fig. 29 Arc shape origami type hydraulic damper prototype (**a**), its experiment situation (**b**) (Guan et al. 2023)

$$MR\ddot{\theta} + cR\dot{\theta} + F_{\text{damper}} + KR\theta = -mR\ddot{\theta}_e \tag{4}$$

Here, θ is angular displacement of mass block in Fig. 29 and $\ddot{\theta}_e$ is angular acceleration of external force. This effect has also been confirmed (Guan et al. 2023). But I omit it here.

5 Japanese Folding Fan as Bellows Folding

At last, Japanese fan which combines functionality and beauty that are characteristic of Japanese manufacturing. This bellows folded fan with bamboo bones behaves differently than single sheet bellows folding.

(1) Fan used to repel insects and start fires was born all over the world at about the same time. However, folding fan was only born in Japan around 800 AD.
(2) The famous painters such as Hokusai etc. of the Edo period tried to create stories on fan by taking advantage of the fact that characters and backgrounds look different with change of perspective. But it was difficult to create a composition because each point on fan (3D) is distorted such as a perfect circle on fan painting (2D) becomes an ellipse on fan because the center of the arc of the fan painting is different from the pivot of the arc of the fan.
(3) Since strain mathematics on fan has become clear (Hagiwara et al. 2020; Yamazaki et al. 2021), we succeeded in bringing out characteristics of fan by taking advantage of the fact that characters and backgrounds look different with change of perspective as shown in Fig. 30.

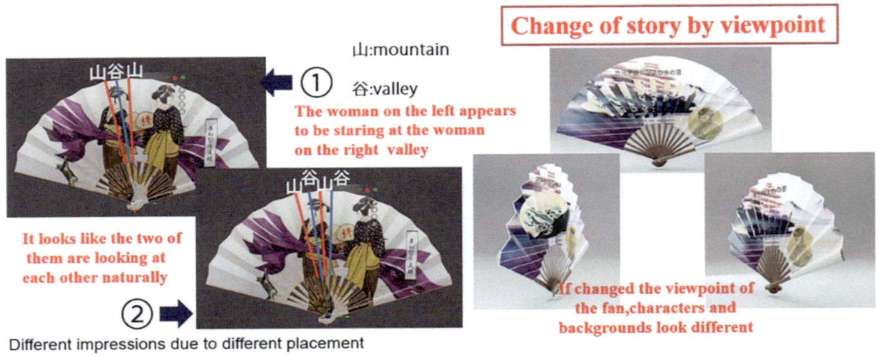

Fig. 30 Examples of actively utilizing Fan feature (Hagiwara 2024)

Now, the most outstanding feature of Japanese culture is beautiful simplification such as ① Simplification of characters: Simplification of rough-looking Chinese kanji to beautiful Japanese hiragana, ② Simplification of story using letters: 31 character tanka and 17 character haiku, and ③ Simplification of shape: origami.

Fan was originally intended to simplify the story using picture. Because we have developed mathematics (Yamazaki et al. 2021) that can make this possible, we will contribute to this realization as part of the activities of the art side of origami engineering.

6 Conclusion

It has been 20 years since we started origami engineering with the aim of industrializing origami structures beyond honeycomb cores. First, we started by building biomimetics origami anew.

And as mentioned above, a wide range of results have been obtained from dress to large scale structures.

Origami is customarily made from paper; when using cardboard, plastic, or metal, however, one has to confront many issues, such as a lack of stability after folding and unfolding, difficulty in reproducing complex folds, and the challenge of manufacturing inexpensive objects while maintaining their original functions.

For these reasons, we still haven't achieved the objective of obtaining further industrialization than honeycomb core. However, industries such as qubic core, RSO, RTO, and fully three dimensional Kirigami honeycomb are becoming much more than honeycomb aided by modelling and numerical simulation techniques. It is also thought that origami 3D printer also has the potential to achieve this objective.

This origami engineering will be more and more important because interests in science and technology span both micro and macro extremes with remarkable progress in measurement technology where there is no manufacturing equipment and the self-folding function of the origami structure is only relied on.

We would like to continue contributing to the SDGs with origami structures.

Acknowledgements The results introduced here were achieved through joint research by many collaborators. The author would like to take this opportunity to express my gratitude to Dr. Taketoshi Nojima, Dr. Zhao Xile and Dr. Diago Luis in particular.

References

Demaine, E., Tachi, T.: Origamizer: a practical algorithm for folding any polyhedron. In: Proceedings of the 33rd International Symposium on Computational Geometry (SoCG 2017), Brisbane, Australia, July 4–7, 2017

Diago, L., Shinoda, J., Yamazaki, K., Hagiwara, I.: Study on Kirigami honeycomb generation for complicated structures. Trans. Jpn. Soc. Mech. Eng. **89**(917), 22–00273 (2023). https://doi.org/10.1299/transjsme.22-00273.(InJapanese)

Gao, H., Chi, M., Dai, L., Yang, J., Zhou, X.: Mathematical modelling and computational simulation of the hydraulic damper during the orifice-working stage for railway vehicles. Math. Probl. Eng. **2020**, 1830150 (2020)

Guan, J., Yao, Y., Zhao, W., Hagiwara, I., Xilu Zhao, X.: Development of an impact energy absorption structure by an arc shape stroke origami type hydraulic damper. Shock Vib. Article ID 4578613, 11 pages (2023). https://doi.org/10.1155/2023/4578613

Hagiwara, I.: Cheaper than honeycomb: welding two embossed panel steel plates. Autom. Technol. 96–101 (2011)

Hagiwara, I.: Aiming for a 3D origami printer that surpasses 3D additive printer. Bull. Jpn. Soc. Ind. Appl. Math. Invited Paper **26**(1), 22–28 (2016)

Hagiwara, I.: Birth of an "origami robot" with a new idea-https-any three-dimensional object can be reproduced with origami, Asahi newspaper WEBRONZA. https://webronza.asahi.com/science/articles/2018100900012.html (2018). (In Japanese)

Hagiwara, I.: A consideration of scientific-technical aspects and artistic aspect of origami engineering-aiming to create a new big industry. SIAM Newsl. Res. I (2024). https://sinews.siam.org/Details-Page/reflecting-on-the-science-technology-and-artistry-of-origami-engineering

Hagiwara, I., Kondoh, A.: Origami engineering: inspired by Japanese folding culture Kirigami and fan folds represent new opportunities. In: Futurities, The Simulation Based Engineering & Sciences Magazine, pp. 7–12. Springer (2022), First issue

Hagiwara, I., Nadayoshi, S.: Consideration on crash characteristics of cylindrical structures using origami engineering. Trans. Soc. Autom. Eng. Jpn. **34**(4), 145–149 (2003). (In Japanese)

Hagiwara, I., Shinoda, J.: Reverse engineering based on computer graphics technique. In: The 28th Volume of Gakuto International Series, Mathematical Sciences and Applications, pp. 53–62 (2008)

Hagiwara, I., Terada, K.: Core panel and cushioning material. Patent No. 7161180-2022 (2022). (In Japanese)

Hagiwara, I., Zhao, X.: Crash absorber manufacturing equipment, crash absorber, manufacturing method of crash absorber. Patent application 2020-215790 (2021)

Hagiwara, I., Zhao, X.: Core panel, manufacturing method and core manufacturing equipment. Patent application 2022-177019 (2022). (In Japanese)

Hagiwara, I., Zhao, X.: Shock absorber, Patent Application Number 2023-037761 (2023). (In Japanese)

Hagiwara, I., Chie Nara, C., Yang, Y.: Development of new foldable polyethylene terephthalate bottles. J. Adv. Simul. Sci. Eng. (JASSE) **9**(2), 247–262 (2022)

Hagiwara, I., Nguyen, H.T.T., Yu, B., Thai, P.T.: Developed diagram creation device. Developed diagram creation method and Program, Patent No. 6529025 (2019). (In Japanese)

Hagiwara, I., Savchenko, M., Yu, B., Shinoda, J.: Method, equipment and program for manufacturing three-dimensional structures Patent No. 6198107 (2017). (In Japanese)

Hagiwara, I., Tsuda, M., Kitagawa, I.: Determination method for beads placement. Patent No. 2727680 (1991). (In Japanese)

Hagiwara, I., Yamamoto, C., Tao, X., Nojima, T.: Optimization for crash characteristics of cylindrical origami structure using reversed spiral model. JSME Ser. A **70**(689), 36–42 (2004). (in Japanese)

Hagiwara, I., Yamazaki, K., Abe, F.: Fan drawings conversion device, conversion method for fan drawings and Fan picture parts. Patent Application Number 2020-149060 (2020). (In Japanese)

http://www.unicharm.co.jp/company/news/2019/1211066_13296.html

Kitagawa, Y., Hagiwara, I., Tsuda, M.: Development of a collapse mode control method for side members in vehicle collisions. SAE910809 1991 Transaction Section 6 (1992–4 Yuè), pp. 1101–1107

Kong, C., Zhao, X., Hagiwara, I.: Hydroforming process of manufacturing for reverse spiral origami structure. Int. J. Veh. Perform. **3**(4), 347–364 (2017)

Lang, R.J.: TREEMAKER. http://www.langorigami.com/science/computational/treemaker (2015)

Mitani, J.: ORI-REVO. http://mitani.cs.tsukuba.ac.jp/ori_revo/ (2008)

Nojima, T.: Modeling of folding patterns in flat membranes and cylinders by using origami. JSME Int. J. Ser. C **66**(543), 1050–1056 (2000). (In Japanese)

Nojima, T.: Structure model based on mathematical origami-Proposition of origami engineering. Kyoto university IIC fair, The Kyoto Shimbun, November (2002). (In Japanese)

Nojima, T., Hagiwara, I.: Mathematics for Origami, Kyoritsu Shuppan (2012) (In Japanese)

Romero, J., Diago, L., Nara, C., Shinoda, J., Hagiwara, I.: Norigami folding machines for complex 3D shapes. In: Proceeding of the ASME 2016 IDETC/CIE 2016

Saito, K.: Manufacturing device for honeycomb core and manufacturing method. Patent number JP6075006B2 (2017). (In Japanese)

Tian, Z., Guan, J., Wei Zhao, W., Fukuchi, A.B., Zhao, X., Hagiwara, I.: Development of a cubic core lightweight panel using Origami-Kirigami engineering. Int. J. Mech. Eng. Appl. **11**(2), 38–48 (2023). http://www.sciencepublishinggroup.com/j/ijmea https://doi.org/10.11648/j.ijmea.20231102.11. ISSN: 2330-023X (Print); ISSN: 2330-0248 (Online)

Tokura, S., Hagiwara, I.: Forming process simulation of truss core panel. J. Comput. Sci. Technol. **4**(1), 25–35 (2010) (Release Date: March 30, 2010)

Wu, Z., Hagiwara, I., Tao, X.: Optimization of crash characteristics of the cylindrical origami structure. Int. J. Veh. Design **43**(1–4), 66–81 (2007)

Yamazaki, K., Abe, F., Hagiwara, I.: Mathematical elucidation of the traditional Japanese fan focusing on its structure. Trans. JSME **87**(898) (2021). (In Japanese). https://doi.org/10.1299/transjsme.21-00042

Yu, B., Savchenko, M., Shinoda, J., Diago, L., Hagiwara, I., Savchenko, V.: Producing physical copies of the digital models via generating 2D patterns for "Origami 3D Printer" system. J. Adv. Simulat. Sci. Eng. **3**(1), 58–77 (2016)

Zhao, X., Hu, Y., Hagiwara, I.: Shape optimization to improve energy absorption ability of cylindrical thin-walled origami structure. J. Comput. Sci. Technol. 148–162, Release Date: November 30, 2011

Zhao, X., Kong, C., Yang, Y., Hagiwara, I.: Reversed-torsion-type crush energy absorption structure and its inexpensive partial-heating torsion manufacturing method based on origami engineering. J. Manuf. Sci. Eng. (2022). https://doi.org/10.1115/1.4052438(2021-9)

Open Access This chapter is licensed under the terms of the Creative Commons Attribution 4.0 International License (http://creativecommons.org/licenses/by/4.0/), which permits use, sharing, adaptation, distribution and reproduction in any medium or format, as long as you give appropriate credit to the original author(s) and the source, provide a link to the Creative Commons license and indicate if changes were made.

The images or other third party material in this chapter are included in the chapter's Creative Commons license, unless indicated otherwise in a credit line to the material. If material is not included in the chapter's Creative Commons license and your intended use is not permitted by statutory regulation or exceeds the permitted use, you will need to obtain permission directly from the copyright holder.

Bayesian Experimental Design for Symbolic Discovery

Kenneth L. Clarkson, Cristina Cornelio, Sanjeeb Dash, Joao Goncalves, Lior Horesh, and Nimrod Megiddo

Abstract This study concerns the formulation and application of Bayesian optimal experimental design to symbolic discovery, which is the inference from observational data of predictive models taking general functional forms. We apply constrained first-order methods to optimize an appropriate selection criterion, using Hamiltonian Monte Carlo to sample from the prior. A step for computing the predictive distribution, involving convolution, is computed via either numerical integration, or via fast transform methods.

1 Motivation

This study concerns work done on Bayesian Optimal Experimental Design (OED) in the context of symbolic discovery (Cornelio et al. 2021). The latter, also called symbolic regression, seeks to explain observational data using a model chosen from

K. L. Clarkson (✉) · N. Megiddo
IBM Research, San Jose, CA, USA
e-mail: klclarks@us.ibm.com

N. Megiddo
e-mail: megiddo@us.ibm.com

C. Cornelio · S. Dash · J. Goncalves · L. Horesh
IBM Research, Yorktown Heights, NY, USA
e-mail: corneliocristina@gmail.com

S. Dash
e-mail: sanjeebd@us.ibm.com

J. Goncalves
e-mail: jpgoncal@us.ibm.com

L. Horesh
e-mail: lhoresh@us.ibm.com

C. Cornelio
Samsung AI Research, Cambridge, UK

© The Author(s) 2026

S. Oishi et al. (eds.), *Recent Developments in Industrial and Applied Mathematics*,
ICIAM2023 Springer Series 1, https://doi.org/10.1007/978-981-95-1446-5_7

a range of functional forms, that may include general algebraic expressions, trigonometric functions, and so on. The model is typically chosen to balance criteria involving fidelity to the data, the simplicity of the model by some measure, and prior knowledge.

The goal of optimal experimental design (OED) is to find the optimal design of a data acquisition system, so that the uncertainty in the inferred parameters, or some predicted quantity derived from them, is minimized with respect to a statistical criterion. In the context of model discovery, a large body of work addresses the question of experimental design for pre-determined functional forms, and another body of research addresses the selection of a model (functional form) out of a set of candidates. In the context of symbolic regression, our aim is to devise an experimental setup that attends to both the functional form and the continuous set of parameters that defines the model behavior. In realistic settings, experimental data may be restricted or costly, providing limited support for any given hypothesis as to the underlying functional form. Here we formulate a Bayesian framework for experimental design where joint model selection and parameter estimation are pursued. Within that formulation, we show relationships, sometimes equivalence, of a few known selection criteria. Following that we perform a preliminary validation study. A key computational challenge will be computation of derivatives, for which we have used symbolic differentiation. We also aim to explore both Markov Chain Monte Carlo (MCMC) as well more efficient sampling strategies as Hamiltonian Monte Carlo (HMC).

2 Background Art

The topic of experimental design, and in particular design in a Bayesian framework, is rich and well-studied; an overview is given by Ryan et al. (2016). We will develop the framework appropriate to our setting and implementation, and return in Sect. 6 to alternative choices and further work.

3 Formulation

Models, priors, and updates Denote the design (input) space as $\mathcal{X}$. We have a set of models $\mathcal{M}$, where each model $m \in \mathcal{M}$ has parameters $\theta_m \in \mathbb{R}^n$ for some n. We assume that there is some ground-truth model m_{true} with associated parameters θ_{true}, and the data we receive is

$$y(x) \triangleq m_{\text{true}}(x, \theta_{\text{true}}) + \epsilon, \tag{1}$$

where $\epsilon \sim \mathcal{N}(0, \sigma^2)$, for known variance σ^2. Letting $\phi(z; \mu, \sigma^2)$ denote the density function of a normal distribution $\mathcal{N}(\mu, \sigma^2)$, and writing $y(x)$ as y where this is clear, we can also write this as $p(y|m, \theta_m, x) = \phi(y; m(x, \theta_m), \sigma^2)$.

We consider $m \in \mathcal{M}$ to be random, with prior $p(m)$, and θ_m to be random, with prior $p(\theta_m)$. These distributions express our belief that a given m is m_{true} (in functional form), and our beliefs about the locations of the associated parameters.

An experiment here is a choice of x; given that choice, we learn $y(x)$ as in (1) above. For that given x and y, we can update $p(m)$ and $p(\theta_m)$ using Bayes rule, with posterior probability

$$
\begin{aligned}
\tilde{p}(m|x, y) &= p(m)\frac{p(y|m, x)}{p(y|x)}, \text{ where} \\
p(y|m, x) &= \mathbf{E}_{\theta_m} p(y|m, \theta_m, x) = \mathbf{E}_{\theta_m}\phi(y; m(x, \theta_m), \sigma^2), \text{ and} \\
p(y|x) &= \mathbf{E}_m p(y|m, x) = \sum_{m\in\mathcal{M}} p(m)p(y|m. x).
\end{aligned} \tag{2}
$$

Similarly,

$$
\tilde{p}(\theta_m) = p(\theta_m)\frac{p(y|m, \theta_m, x)}{p(y|x)} = p(\theta_m)\frac{\phi(y; m(x, \theta_m), \sigma^2)}{p(y|x)} \tag{3}
$$

We will use Monte Carlo methods to sample from $p(\theta_m)$, for each m, obtaining $S_m \subset \mathbb{R}^n$ for each $m \in \mathcal{M}$, such that (ideally) each member of S_m has distribution $p(\theta_m)$. For functions $f(\theta)$, we then estimate

$$
\mathbf{E}_{\theta_m} f(\theta_m) \approx \frac{1}{|S_m|}\sum_{\theta_m\in S_m} f(\theta). \tag{4}
$$

Selection criteria. We are looking for a design point $x \in \mathcal{X}$ that is "most informative," in some sense, about the model. Picking a point x^*, we then update the posterior distributions as in (2) and (3), and repeat. That is, we make the resulting posterior into the prior for the next choice of design point, so the *current* $p(m)$ and $p(\theta_m)$ are based on a sequence of choices $x^1, x^2, \ldots$, and corresponding $y^1, y^2, \ldots$, each y^t from (1) and using the corresponding x^t.

Maximizing mutual information. A natural version of this general goal is to find x that maximizes the mutual information $I(y; m|x)$ between the response y and the model m, conditioned on x. This approach, to find

$$
x^*_{\text{MI}} \triangleq \arg\max_x I(y; m|x),
$$

was proposed in this form in, for example, Drovandi et al. (2014).

Minimizing model entropy. We could also consider selecting a point x that minimizes the entropy $H(m|y, x)$; however, since $I(y; m|x) = H(m|x) - H(m|y, x) = H(m) - H(m|y, x)$, we have

$$x^*_{\mathrm{ME}} \triangleq \arg\min_x H(m|y,x) = x^*_{\mathrm{MI}}.$$

Maximizing Jensen-Shannon divergence. A criterion based on the multi-way Jensen-Shannon divergence was proposed by Vanlier et al. (2014). The form used was

$$D_{\mathrm{JS}}(x) \triangleq \mathbf{E}_m[D_{\mathrm{KL}}((p(y|m,x)\;\|p(y|x)\;)], \tag{5}$$

where $D_{\mathrm{KL}}(\;\|\;)$ is the Kullback-Liebler divergence. That is, we seek to find the design point of

$$x^*_{\mathrm{JS}} \triangleq \arg\max_x D_{\mathrm{JS}}(x), \tag{6}$$

that maximizes the expected (w.r.t. m) divergence of $p(y|m,x)$ from the expectations (w.r.t. m) of those divergences.

However, as is well-known (Wikipedia 2020; Shulkind et al. 2018), the mutual information between random variables W and Z satisfies $I(W;Z) = \mathbf{E}_W[D_{\mathrm{KL}}(p(Z|W)\;\|p(Z)\;)]$, so that taking $W = m$ and $Z = y$, and conditioning everywhere on x (and noting that m is independent of fixed x), the conditional mutual information $I(y;m|x) = \mathbf{E}_m[D_{\mathrm{KL}}((p(y|m,x)\;\|p(y|x)\;)]$, so we have also $x^*_{\mathrm{JS}} = x^*_{\mathrm{MI}}$.

Maximizing response entropy. Suppose the uncertainty of observation y at a given point x under model m is independent of m and x, that is, for any given m, x, m', x', $H(y|m,x) = H(y|m',x')$. This holds under our assumption of i.i.d. measurement noise, since for all m, x, we have $H(y|m,x) = H(\epsilon)$, where ϵ is the error as in (1) above. Then, since also $I(y;m|x) = H(y|x) - H(y|m,x)$, and $H(y|m,x)$ is fixed with respect to x, we have

$$\begin{aligned} x^*_{\mathrm{RE}} \triangleq \arg\max_x H(y|x) &= \arg\max_x H(y|x) - H(y|m,x) \\ &= \arg\max_x I(y;m|x) = x^*_{\mathrm{MI}}, \end{aligned} \tag{7}$$

a formulation apparently going back to Borth (1975). That is, the best x makes the variation in the response distribution $p(y|x)$ as large as possible.

Maximzing log det. So far, all selection criteria yield the same optimum $x^*_{\mathrm{MI}} = x^*_{ME} = x^*_{\mathrm{JS}} = x^*_{RE}$; the simplest of these to compute seems to be x^*_{RE}, from (7).

A different idea that we have also explored is again based on maximizing the dispersion of the response. Here we (conceptually) build a matrix $D(x)$ whose rows and columns are indexed by the set $\{(m,\theta) \mid m \in \mathcal{M}, \theta \in S_m\}$, recalling that the S_m are samples of $p(\theta_m)$, as used in (4). That is, this matrix $D(x)$ has $\sum_{m\in\mathcal{M}} |S_m|$ rows and columns. The entry of $D(x)$ for (m,θ) and (m',θ') is, using standard facts about the KL-divergence,

$$D_{\mathrm{KL}}(p(y|m,\theta,x)\;\|p(y|m',\theta',x)\;)$$

$$= D_{\mathrm{KL}}(\mathcal{N}(m(x,\theta),\sigma^2)\,\|\mathcal{N}(m'(x,\theta'),\sigma^2)\,)$$
$$= \frac{\left(m(x;\theta),\sigma^2) - m'(x;\theta'),\sigma^2)\right)^2}{2\sigma^2},$$

and the optimum is

$$x^*_{\mathrm{LD}} \triangleq \arg\min_x - \log\det D(x). \tag{8}$$

4 Implementation Considerations

We implemented algorithms for finding x^*_{JS}, which is the optimum for the design selection criterion based on Jensen-Shannon (6); the equivalent one x^*_{RE} from (7) based on response entropy; the "logdet" optimum x^*_{LD} from (8); and a variant of the logdet criterion. Below we will focus on the response entropy, which seems the most promising.

Estimating $p(y|x)$. For computing x^*_{RE}, we need an estimate of

$$H(y|x) = -\int_y p(y|x)\log p(y|x) = -\mathbf{E}_{y|x}[\log p(y|x)], \tag{9}$$

using the convention (from the limit) that $0\log 0 = 0$. Our estimate of $p(y|x)$ is, following (2) and (4),

$$\begin{aligned} p(y|x) &= \mathbf{E}_m\, p(y|m,x) = \mathbf{E}_m \mathbf{E}_{\theta_m} \phi(y; m(x,\theta_m),\sigma^2) \\ &\approx \frac{1}{|\mathcal{M}|}\sum_{m\in\mathcal{M}} p(m)\frac{1}{|S_m|}\sum_{\theta_m\in S_m}\phi(y; m(x,\theta_m),\sigma^2). \end{aligned} \tag{10}$$

We implemented two ways to estimate $H(y|x)$: either numerical integration of (9) using (10), or via convolution, noting that

$$p(y|m,x) \approx \frac{1}{|S_m|}\sum_{\theta_m\in S_m}\phi(y; m(x,\theta_m),\sigma^2) = \frac{1}{|S_m|}\mathbf{1}_{S_m} * \phi(z; 0,\sigma^2),$$

where "$*$" is the convolution operator, and $\mathbf{1}_{S_m}$ is the function having $\mathbf{1}_{S_m}(z) = 1$ when $z = m(x,\theta)$ for each $\theta \in S_m$, and zero otherwise. By representing $\mathbf{1}_{S_m}$ and the Gaussian mask $\phi(z; 0,\sigma^2)$ on a fine one-dimensional grid, and using fast convolution, we can obtain an estimate of $p(y|m,x)$ more quickly than via numerical integration. We can do better for accuracy than rounding $m(x,\theta)$ to the nearest grid value by distributing the weight for each such value across multiple grid points. This still allows fast convolution, but has the effect of interpolation. In our implementation we distribute weight to the grid values such that we have cubic interpolation of the Gaussian mask. The weights, convolution, and mask are shown in Fig. 1.

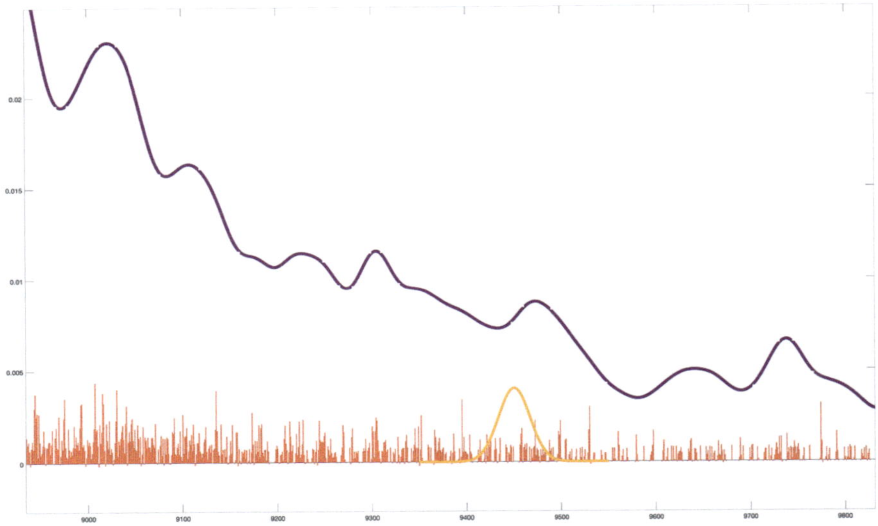

Fig. 1 Computing $p(y|m, x)$ via convolution

Such accuracy might not seem necessary, but we use `SQP` within `Matlab` to optimize our estimate $\hat{H}(y|x)$ of $H(y|x)$ with respect to x, subject to box constraints on x. This works best if $\hat{H}(y|x)$ is a smooth function of x, and if a relatively high-accuracy estimate of its gradient $\nabla_x \hat{H}(y|x)$ is provided. Some discussion of the gradient computation is given below.

Obtaining S_m. To obtain a sample S_m of $\theta \sim p(\theta_m)$, we use Hamiltonian Monte Carlo (HMC), as provided by `Matlab`. We change the sample only when $p(\theta_m)$ changes. HMC is relatively fast, but requires $p(\theta)$ to have unbounded support, and requires both $\log p(\theta)$ and $\nabla_\theta \log p(\theta)$ to be provided. Some discussion of the gradient computation is given below.

The requirement of unbounded support implies that the easiest approach is to allow e.g. parameters to be negative even when we know that θ_{true} has non-negative coordinates. This causes problems when some parameter is an exponent of an input coordinate close to zero: the response become unnaturally large. We ameliorate this by including a term in the prior for θ that makes it unlikely that $m(x_0, \theta)$ is extremely large, for some given input point x_0. More generally, we could use changes of variables, or rejection methods, to enforce constraints on the parameters and still use HMC.

Obtaining gradients. For HMC, we need $\nabla_{\theta_m} \log p(\theta_m)$, and for `SQP`, it is helpful to have $\nabla_x H(y|x)$. (That is, the gradients of our approximations to these functions.) The former is straightforward to derive and compute, and the latter is computed via either numerical integration or fast convolution, just as $H(y|x)$ is. Via the chain rule, for $\nabla_{\theta_m} \log p(\theta_m)$ we need $\nabla_\theta m(x, \theta)$, and for $\nabla_x H(y|x)$, we need $\nabla_x m(x, \theta)$. In our implementations, we describe the functional forms as symbolic expressions, and use `Matlab`'s symbolic toolbox to obtain functions for computing the models and their gradients. This is purely as a convenience, as we could write these functions manually, but this approach allows some scalability in implementation, and reduces errors.

5 Numerical Study

We tested and refined our implementation on a challenging-enough small example, equation (I.24.6) from Feynman's lecture notes, which is

$$E = cm^{e_1}(\omega^{e_2} + \omega_0^{e_3})z^{e_4}, \tag{11}$$

where $c = 1/4$, $e_1 = 1$ and $e_2 = e_3 = e_4 = 2$. This model has four inputs $x \triangleq (m, \omega, \omega_0, z)$ and five parameters $\theta \triangleq (c, e_1, e_2, e_3, e_4)$. We use three candidate models, the first of which has the same functional form as the ground-truth model m_{true} of (11). The other two models are

$$E = cm^{e_1}\omega^{e_2}\omega_0^{e_3}z^{e_4}, \tag{12}$$

$$E = cm^{e_1}(\omega^{e_2} + z^{e_4})\omega_0^{e_3}. \tag{13}$$

We encode the initial values of the parameters of each model, say θ_m for m, in the prior distribution $p(\theta_m) \sim \mathcal{N}(\mu_m, \Sigma_m^2)$, where vectors μ_m have coordinates of magnitude at most 2, and the matrices $\Sigma_m = I$. We generate 4,000 samples via HMC. The results of two computational experiments are shown in Fig. 2. Here the correct model has probability one after twelve trials, for small noise, and does not quite get to one, for large noise.

We also tracked the increased knowledge of the parameters, by way of the mean variance of each (the trace of the covariance, divided by the number of parameters, as shown in Fig. 3.

6 Other Choices, Further Work

We need to experiment our new formulation with more settings, and larger problems, although our current runtimes on modest workstations (laptops) are not unwieldy. There are a number of proposed techniques to accelerate Bayesian OED; some of them seem promising for our setting.

- Importance sampling of log posteriors (a.k.a. particle methods). As discussed by Drovandi et al. (2014) for example, these methods reduce the number of times Monte Carlo sampling methods are needed, by weighting existing samples to

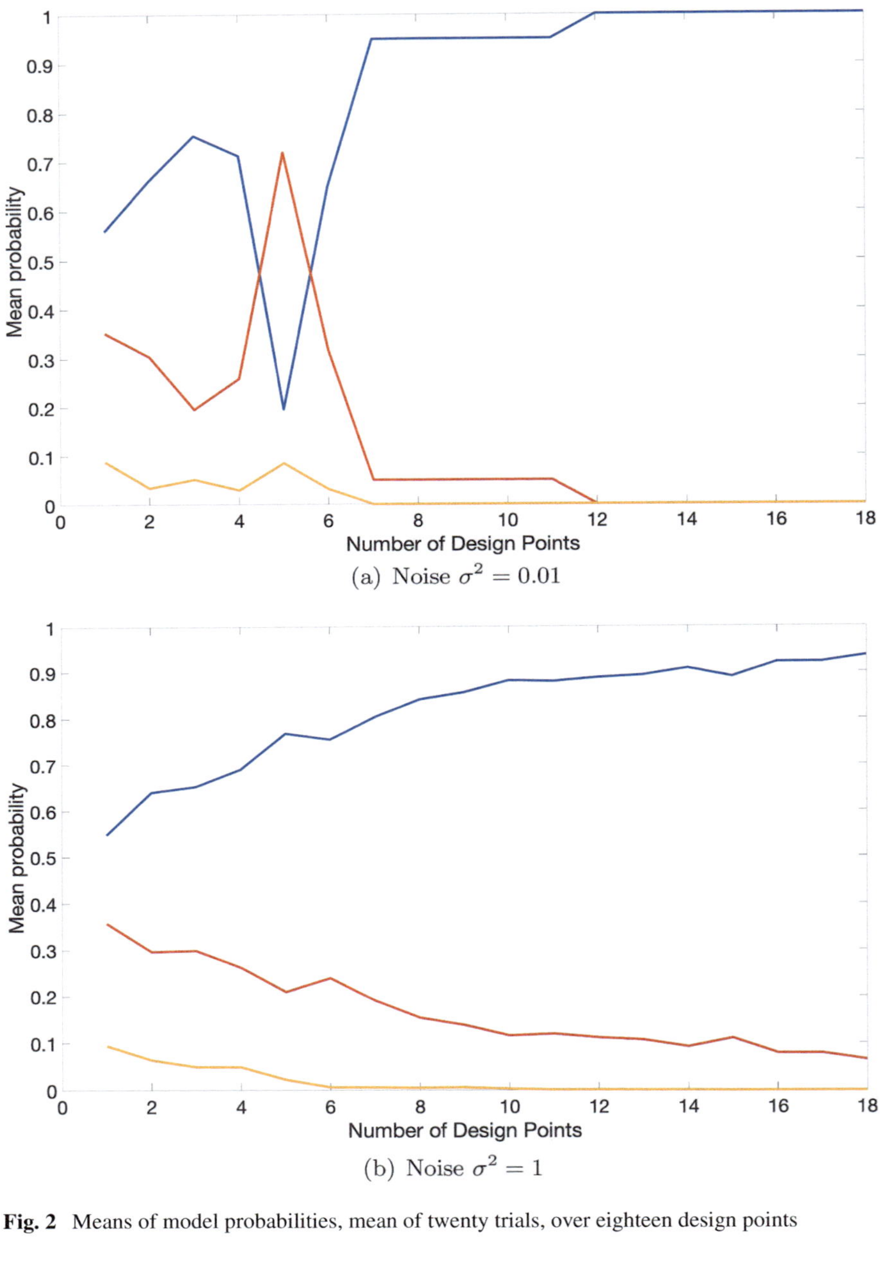

(a) Noise $\sigma^2 = 0.01$

(b) Noise $\sigma^2 = 1$

Fig. 2 Means of model probabilities, mean of twenty trials, over eighteen design points

reflect the current posterior distribution. So far HMC seems fast enough, but we need to scale up more to see if it becomes a bottleneck.

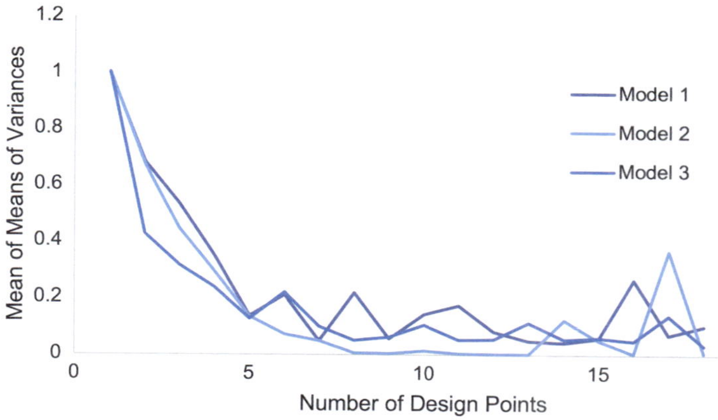

(a) Noise $\sigma^2 = 0.01$

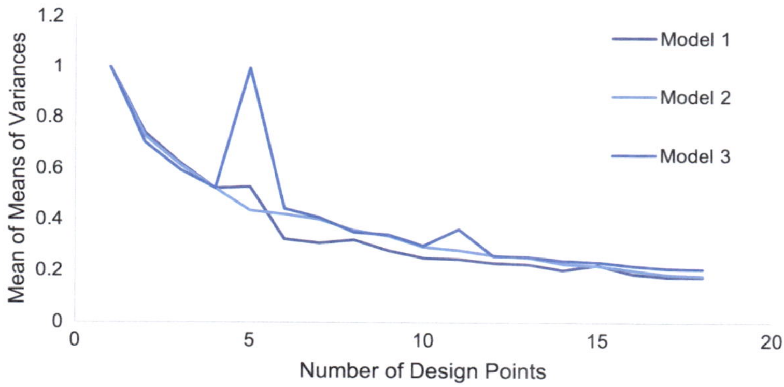

(b) Noise $\sigma^2 = 1$

Fig. 3 Means of per-model parameter variances, mean of twenty trials, over eighteen design points

- Laplace approximations to the posterior. These use quadratic approximations to the posterior log likelihood, in the neighborhood of its maximum. It seems unlikely that such an approximation would work well in our case, but this is to be determined.
- Approximate Bayesian methods (Haber et al. 2009; Ryan et al. 2016). These are appropriate in settings where the model evaluation itself is expensive; this is not the case here.
- Formulations where the design point itself is a random variable, and optimum point is the output of an MCMC process (Cavagnaro et al. 2010). We may explore this.

Patent Application. Aspects of this work are covered by U.S. Patent Application P201809327 "Experimental Design For Symbolic Model Discovery," filed April 2020.

Acknowledgements This work was supported in part by the Defense Advanced Research Projects Agency (DARPA) (PA-18-02-02). The U.S. Government is authorized to reproduce and distribute reprints for Governmental purposes notwithstanding any copyright notation thereon.

References

Borth, D.M.: A total entropy criterion for the dual problem of model discrimination and parameter estimation. J. R. Stat. Soc. Ser. B (Methodological) **37**(1), 77–87 (1975)

Cavagnaro, D.R., Myung, J.I., Pitt, M.A., Kujala, J.V.: Adaptive design optimization: a mutual information-based approach to model discrimination in cognitive science. Neural Comput. **22**(4), 887–905 (2010). PMID: 20028226

Cornelio, C., Dash, S., Austel, V., Josephson, T., Goncalves, J., Clarkson, K., Megiddo, N., El Khadir, B., Horesh, L.: AI Descartes: combining data and theory for derivable scientific discovery. arXiv:2109.01634 (2021)

Drovandi, C.C., McGree, J.M., Pettitt, A.N.: A sequential Monte Carlo algorithm to incorporate model uncertainty in Bayesian sequential design. J. Comput. Graph. Stat. **23**(1), 3–24 (2014)

Haber, E., Horesh, L., Tenorio, L.: Numerical methods for the design of large-scale nonlinear discrete ill-posed inverse problems. Inverse Prob. **26**(2), 025002 (2009)

Ryan, E.G., Drovandi, C.C., McGree, J.M., Pettitt, A.N.: A review of modern computational algorithms for Bayesian optimal design. Int. Stat. Rev. **84**(1), 128–154 (2016)

Shulkind, G., Horesh, L., Avron, H.: Experimental design for nonparametric correction of misspecified dynamical models. SIAM/ASA J. Uncertain. Quantif. **6**(2), 880–906 (2018)

Vanlier, J., Tiemann, C.A., Hilbers, P.A.J., van Riel, N.A.W.: Optimal experiment design for model selection in biochemical networks. BMC Syst. Biol. **8**(1), 20 (2014)

Wikipedia: Mutual information—Wikipedia, the free encyclopedia (2020). Accessed 27 March 2020

Open Access This chapter is licensed under the terms of the Creative Commons Attribution 4.0 International License (http://creativecommons.org/licenses/by/4.0/), which permits use, sharing, adaptation, distribution and reproduction in any medium or format, as long as you give appropriate credit to the original author(s) and the source, provide a link to the Creative Commons license and indicate if changes were made.

The images or other third party material in this chapter are included in the chapter's Creative Commons license, unless indicated otherwise in a credit line to the material. If material is not included in the chapter's Creative Commons license and your intended use is not permitted by statutory regulation or exceeds the permitted use, you will need to obtain permission directly from the copyright holder.

An Overview of Convergence Rates for Sum of Squares Hierarchies in Polynomial Optimization

Monique Laurent and Lucas Slot

Abstract In this chapter we consider polynomial optimization problems, asking to minimize a polynomial function over a compact semialgebraic set, defined by polynomial inequalities. This models a great variety of (in general, nonlinear nonconvex) optimization problems. Various hierarchies of (lower and upper) bounds have been introduced, having the remarkable property that they converge asymptotically to the global minimum. These bounds exploit algebraic representations of positive polynomials in terms of sums of squares and can be computed using semidefinite optimization. Our focus in this chapter lies in the performance analysis of these hierarchies of bounds, namely, in how far the bounds are from the global minimum as the degrees of the sums of squares they involve tend to infinity. We present the main state-of-the-art results and offer a gentle introductory overview over the various techniques that have been recently developed to establish them, stemming from the theory of orthogonal polynomials, approximation theory, Fourier analysis, and more.

1 Introduction

This paper offers a gentle introduction and overview over the *design and performance analysis of approximation hierarchies for polynomial optimization*. In this section we will first introduce polynomial optimization, and its use for modeling hard optimization problems arising within a broad range of fields and application domains. After that we will explain how to design hierarchies of approximations using sums of squares of polynomials as a tractable surrogate for polynomial nonnegativity, and how sums of squares of polynomials can be modeled using semidefinite optimization.

M. Laurent (✉)
CWI Amsterdam, Amsterdam, XG, The Netherlands
e-mail: M.Laurent@cwi.nl

Tilburg University, Tilburg, AB, The Netherlands

L. Slot
ETH, Zürich, Switzerland
e-mail: lucas.slot@inf.eth.ch

© The Author(s) 2026

S. Oishi et al. (eds.), *Recent Developments in Industrial and Applied Mathematics*, ICIAM2023 Springer Series 1, https://doi.org/10.1007/978-981-95-1446-5_8

1.1 Polynomial Optimization

Throughout we use the following notation. We let $\mathbb{R}[\mathbf{x}] = \mathbb{R}[x_1, \ldots, x_n]$ denote the ring of n-variate real polynomials in the variables $\mathbf{x} = (x_1, \ldots, x_n)$. For a multi-index $\alpha \in \mathbb{N}^n$, we let $x^\alpha = x_1^{\alpha_1} \cdots x_n^{\alpha_n}$ denote the associated monomial, whose degree is $|\alpha| := \|\alpha\|_1 = \alpha_1 + \ldots + \alpha_n$. We let $\mathbb{N}^n_d$ denote the set of sequences $\alpha \in \mathbb{N}^n$ with $|\alpha| \leq d$. Any polynomial $f \in \mathbb{R}[\mathbf{x}]$ can be expressed in the monomial basis as $f = \sum_\alpha f_\alpha x^\alpha$ for some coefficients $f_\alpha \in \mathbb{R}$. Here, the sum is finite and the largest value of $|\alpha|$ for which $f_\alpha \neq 0$ is the *degree* of f, denoted $\deg(f)$. For an integer $d \geq 0$, we let $\mathbb{R}[\mathbf{x}]_d$ denote the set of polynomials with degree at most d. It is a vector space, with $[x]_d = \{x^\alpha : \alpha \in \mathbb{N}^n_d\}$ as its standard monomial basis. As we will see later, it is sometimes convenient to use other polynomial bases, that are orthogonal with respect to some inner product induced by a selected measure on the space $\mathbb{R}^n$.

The general setting of polynomial optimization is as follows. We are given polynomials $f, g_1, \ldots, g_m \in \mathbb{R}[\mathbf{x}]$ and the goal is to find the minimum value that f takes on the feasible region defined by the polynomials g_j. In other words, the goal is to solve the following optimization problem.

The polynomial optimization problem

$$\begin{aligned} f_{\min} &= \min\{f(\mathbf{x}) : g_1(\mathbf{x}) \geq 0, \ldots, g_m(\mathbf{x}) \geq 0\} \\ &= \min\{f(\mathbf{x}) : \mathbf{x} \in \mathbf{X}\}, \end{aligned} \tag{1}$$

after setting

$$\mathbf{X} = \{\mathbf{x} \in \mathbb{R}^n : g_1(\mathbf{x}) \geq 0, \ldots, g_m(\mathbf{x}) \geq 0\}. \tag{2}$$

We also sometimes use the notation $f_{\min,\mathbf{X}}$ for the minimum value of (1) and the notation $\mathrm{Se}(\mathbf{g})$ for the set in (2).

Such a set $\mathbf{X}$ is known as a basic closed semialgebraic set. Typical instances include the unit sphere $\mathbb{S}^{n-1}$, the standard simplex Δ^n, the unit ball B^n, the box $[-1, 1]^n$ or $[0, 1]^n$, and the binary cube $\{\pm 1\}^n$ or $\{0, 1\}^n$, where

$$\mathbb{S}^{n-1} = \{\mathbf{x} \in \mathbb{R}^n : \|\mathbf{x}\| = 1\}, \; B^n = \{\mathbf{x} \in \mathbb{R}^n : \|\mathbf{x}\| \leq 1\},$$

$$\Delta^n = \Big\{\mathbf{x} \in \mathbb{R}^n : x_i \geq 0 \; (i \in [n]), \sum_{i=1}^n x_i = 1\Big\},$$

and $\|\mathbf{x}\|^2 = \sum_{i=1}^n x_i^2$ denotes the squared Euclidean norm. Throughout we assume $\mathbf{X}$ is compact. So, problem (1) always has a global minimizer.

The polynomial optimization problem (1) contains linear programming as a (very) special case, when all polynomials $f, g_1, \ldots, g_m$ are linear (i.e., of degree at most 1). In this case, efficient optimization algorithms exist (see, e.g., Roos et al. (2005)). But, it also models a much broader range of problems, which are generally nonlinear and nonconvex. This includes well-known NP-hard problems, already when restricting to seemingly simple feasible regions **X** in (2), such as the simplex, the box, or the sphere as we illustrate on some examples in the next section.

Polynomial optimization has received growing research interest in the past decades, when it was realized that algebraic and geometric properties of polynomials could be exploited to design dedicated methods, able to capture the *global minimum*, in contrast to general nonlinear optimization methods where one can often only gain information about local minima. In a nutshell, this research direction builds on combining real algebraic geometry results (about sums of squares of polynomials) and functional analytical results (about moments of measures) with semidefinite optimization. It roots in foundational works, in particular, by Shor (1987), Nesterov (2000), Lasserre (2001) and Parrilo (2000, 2003). The field has substantially grown[1] and has a broad literature. We mention some books and overviews that can serve as introduction to the topic and give further references to many applications and additional aspects that are not mentioned in the present paper; in particular, by Lasserre (2009), Lasserre (2015), Laurent (2009), Blekherman et al. (2012), Henrion et al. (2020), Magron and Wang (2023) and Nie (2023).

The present paper will focus on the performance analysis of various hierarchies of bounds that have been introduced for the polynomial optimization problem (1), based on using tailored sums of squares representations for positive polynomials. We will recall the definition of these hierarchies of (upper and lower) bounds and discuss the main state-of-the-art results that have been shown in recent years about their quantitative convergence properties (these are not covered in the literature mentioned above). We focus on offering a gentle overview of the main techniques that are needed to prove these quantitative results.

1.2 Examples and Applications

We begin with mentioning a few instances of the polynomial optimization problem (1) that capture well-known hard combinatorial optimization problems.

Consider a graph $G = (V, E)$, where $V = [n] = \{1, \dots, n\}$ is the set of vertices and the pairs in $E \subseteq V \times V$ correspond to the edges of G. A set $I \subseteq V$ is said to be independent (or stable) if it contains no edge and a fundamental combinatorial problem is determining the largest cardinality of an independent set, denoted $\alpha(G)$, a well-known NP-hard problem (see Garey and Johnson (1979)). Interestingly, this problem admits several equivalent reformulations as instances of polynomial optimization over the boolean cube, the box, the simplex, the sphere, respectively:

[1] This is witnessed, e.g., by the fact that it has received its own Mathematics Subject Classification number 90C23.

$$\alpha(G) = \max\left\{\sum_{i\in V} x_i : x_i x_j = 0 \text{ for } \{i,j\}\in E, x_i^2 = x_i \text{ for } i\in V\right\},$$

$$\alpha(G) = \max\left\{\sum_{i\in V} x_i - \sum_{\{i,j\}\in E} x_i x_j : x\in[0,1]^n\right\},$$

$$\frac{1}{\alpha(G)} = \min\left\{x^{\mathsf T}(I + A_G)x : x\in\mathbb{R}^n_+, \sum_{i\in V} x_i = 1\right\},$$

$$\frac{1}{\alpha(G)} = \min\left\{(x^{\circ 2})^{\mathsf T}(I + A_G)x^{\circ 2} : x\in\mathbb{R}^n, \sum_{i\in V} x_i^2 = 1\right\},$$

where we set $x^{\circ 2} = (x_1^2, \ldots, x_n^2)^{\mathsf T}$ and A_G is the adjacency matrix of graph G. An additional reformulation is in terms of linear optimization over copositive matrices:

$$\alpha(G) = \min\{\lambda : \lambda(I + A_G) - J \text{ is copositive}\},$$

where J denotes the all-ones matrix. We refer to de Klerk and Pasechnik (2002) for details about the above formulations.

Another fundamental combinatorial problem is the maximum cut problem, asking for the largest cardinality of a cut in a graph $G = (V, E)$, denoted as $\mathrm{mc}(G)$, that can be formulated as any of the following polynomial optimization problems:

$$\mathrm{mc}(G) = \max\left\{\sum_{\{i,j\}\in E} \frac{1}{2}(1 - x_i x_j) : x\in\{\pm 1\}^n\right\},$$

$$\mathrm{mc}(G) = \max\left\{\frac{1}{4}x^{\mathsf T} L_G x : x\in[-1,1]^n\right\},$$

where $L_G = D - A_G$ is the Laplacian matrix, with D the diagonal matrix having the degrees of the vertices as diagonal entries.

A fundamental question in analysis that can be formulated in terms of polynomial optimization is testing whether an n-variate polynomial f is convex. Indeed, f is convex if and only if its Hessian matrix $H_f(\mathbf{x}) = \left(\frac{\partial^2 f}{\partial x_i \partial x_j}(\mathbf{x})\right)_{i,j=1}^n$ is positive semidefinite at any $\mathbf{x}\in\mathbb{R}^n$ or, equivalently, if the $2n$-variate polynomial $F(\mathbf{x},\mathbf{y}) = \mathbf{y}^{\mathsf T} H_f(\mathbf{x})\mathbf{y}$ is nonnegative over $\mathbb{R}^n\times\mathbb{R}^n$ (i.e., its global minimum is 0). It has been shown by Ahmadi et al. (2013) that testing whether a quartic polynomial is convex is an NP-hard problem. This implies hardness of testing global nonnegativity of a quartic polynomial.

So, the above examples show that computing the global minimum of a polynomial over simple regions such as the simplex, the sphere, the box, the full space, or the boolean box, are computationally hard already when restricting to small degree (at most 4). Another notable application which can be cast as instance of polynomial optimization problem involving only quadratic polynomials, is the optimal power flow problem in energy (see, e.g., Zohrizadeh et al. (2020)). Polynomial optimization has a broad modeling power and permits to capture problems from various areas, such as probablity, mathematical finance, control, game theory, for which we refer to the exposition by Lasserre (2009).

1.3 Nonnegative Polynomials and Sums of Squares

A polynomial $f \in \mathbb{R}[\mathbf{x}]_{2d}$ is called a *sum of squares* (abbreviated as *sos*) if it can be written as a sum of squares of other polynomials, i.e., $f = \sum_{j=1}^{k} q_j^2$ for some $q_j \in \mathbb{R}[\mathbf{x}]$ and $k \geq 1$. Then, each q_j has degree at most d and can be assumed to be homogeneous if f is homogeneous. We let Σ denote the set of sums of squares of polynomials and set $\Sigma_{2d} = \Sigma \cap \mathbb{R}[\mathbf{x}]_{2d}$. We may use the notation $\Sigma[\mathbf{x}]$ to stress which variables are used.

Let $\mathcal{P}$ denote the set of polynomials f that are globally nonnegative (i.e., $f(\mathbf{x}) \geq 0$ for all $\mathbf{x} \in \mathbb{R}^n$), and $\mathcal{P}(\mathbf{X})$ the set of polynomials that are nonnegative over a given set $\mathbf{X} \subseteq \mathbb{R}^n$.

Clearly, every sum of squares of polynomials is globally nonnegative, i.e., the inclusion $\Sigma \subseteq \mathcal{P}$ holds. As is well-known, this inclusion is in general strict. By a result of Hilbert (1888) we know that any globally nonnegative n-variate polynomial with degree $2d$ is a sum of squares of polynomials only in the following three cases: $n = 1$ (univariate), $d = 1$ (quadratic), and the exceptional case ($n = 2, d = 2$) (i.e., quartic in two variables).

A nonnegative polynomial that is not sos
The first explicit example of a nonnegative polynomial that is not sos was found by Motzkin in 1967, it is bivariate with degree 6 and reads

$$f(x, y) = x^4y^2 + x^2y^4 - 3x^2y^2 + 1. \tag{3}$$

It is depicted in Fig. 1a. To see global nonnegativity one may use the arithmetic-geometric mean inequality, and use 'brute force' to show f is not sos (write f as a sum of squares of polynomials and compare coefficients). Also the Robinson polynomial

$$f(x, y, z) = x^6 + y^6 + z^6 - x^2y^2(x^2 + y^2) - x^2z^2(x^2 + z^2) - y^2z^2(y^2 + z^2) + 3x^2y^2z^2 \tag{4}$$

is globally nonnegative but not sos. We refer to Reznick (1996) and Powers (2025) (and references therein) for a nice historic discussion and many more examples.

At the 1900 International Congress of Mathematicians in Paris Hilbert asked whether every nonnegative polynomial can be written as a sum of squares of *rational* functions (known as *Hilbert's seventeenth problem*). This question was settled in the affirmative by Artin (1927), a result that started the flourishing field of real

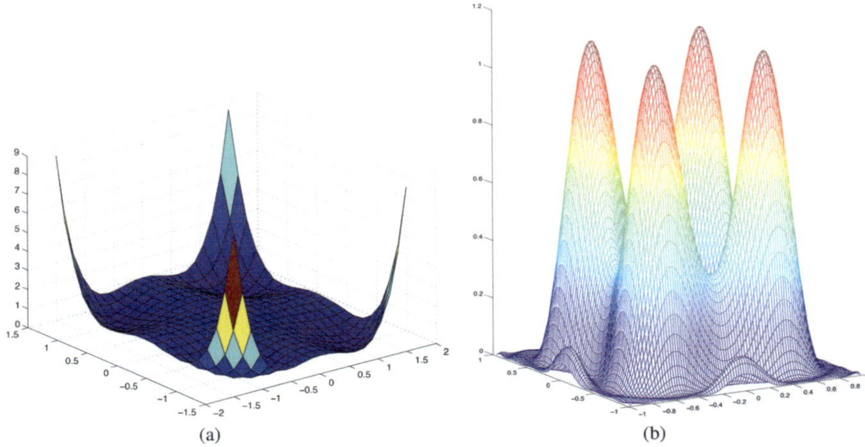

Fig. 1 **a** On the left: the Motzkin polynomial (3); **b** on the right: the optimal degree 16 sum-of-squares polynomial σ in program (14)

algebraic geometry (see, e.g., Prestel and Delzell (2001) and Marshall (2008)). A landmark result is the Positivstellensatz[2] by Krivine (1964) and Stengle (1996) that a.o. characterizes the polynomials that are nonnegative on a semialgebraic set $\mathbf{X}$ as in (2). To cite this result we need the notion of *preordering* $\mathcal{T}(\mathbf{g})$ generated by the polynomials $\mathbf{g} = \{g_1, \ldots, g_m\}$ entering the algebraic description of the set $\mathbf{X}$:

$$\mathcal{T}(\mathbf{g}) = \Big\{ \sum_{J \subseteq [m]} \sigma_J \prod_{j \in J} g_j : \sigma_J \in \Sigma \text{ for } J \subseteq [m] \Big\} \tag{5}$$

(setting $g_\emptyset = 1$). Clearly, $\mathcal{T}(\mathbf{g}) \subseteq \mathcal{P}(\mathbf{X})$. Krivine-Stengle show that a polynomial $f \in \mathbb{R}[x]$ is nonnegative on $\mathbf{X}$ if and only if $p_1 f = f^{2k} + p_2$ for some polynomials $p_1, p_2 \in \mathcal{T}(\mathbf{g})$ and some integer $k \in \mathbb{N}$. This leads to a sos-type decomposition of f as $f = f^{2k}/p_1 + p_2/p_1$, thus 'with a denominator'.

Simpler sos-type decompositions have been shown later under more restrictive assumptions, typically assuming *strict positivity* of f. In particular, Reznick (1995) shows a sharper result for *homogeneous* polynomials: If f is homogeneous and satisfies $f(\mathbf{x}) > 0$ for all $\mathbf{x} \in \mathbb{R}^n \setminus \{0\}$, then there exists an integer $r \geq 0$ such that $(\sum_{i=1}^n x_i^2)^r f(\mathbf{x}) \in \Sigma$.

When the set $\mathbf{X}$ is compact, Schmüdgen (1991) shows membership in the preordering under strict positivity:

[2] The terminology of *Positivtsellensatz* refers to Hilbert's celebrated Nullstellenzatz that characterizes the polynomials vanishing at a given complex variety.

Schmüdgen's Positivstellensatz
If $\mathbf{X}$ is compact and f is strictly positive on $\mathbf{X}$, then $f \in \mathcal{T}(\mathbf{g})$.

Under the Archimedean condition, Putinar (1993) shows membership in the *quadratic module* $Q(\mathbf{g})$, which is defined as

$$Q(\mathbf{g}) = \Big\{ \sum_{j=0}^{m} \sigma_j g_j : \sigma_j \in \Sigma \text{ for } j \in \{0, 1, \dots, m\} \Big\} \tag{6}$$

(setting $g_0 = 1$). So, $Q(\mathbf{g}) \subseteq \mathcal{T}(\mathbf{g}) \subseteq \mathcal{P}(\mathbf{X})$. The advantage of the quadratic module $Q(\mathbf{g})$ over the preordering $\mathcal{T}(\mathbf{g})$ is that it involves less terms: $m + 1$ sos polynomials for $Q(\mathbf{g})$, instead of 2^m for $\mathcal{T}(\mathbf{g})$.

The Archimedean condition asks that $R - \sum_{i=1}^{n} x_i^2 \in Q(\mathbf{g})$ for some $R > 0$ and can be seen as an algebraic certificate of compactness. It is a property of the algebraic description of $\mathbf{X}$ rather than $\mathbf{X}$ itself. It is, however, easy to satisfy, simply by adding the inequality of a ball containing $\mathbf{X}$ to its description. The Archimedean condition holds for most sets $\mathbf{X}$ of interest in applications such as the sphere, box, simplex, etc.

Putinar's Positivstellensatz
If $Q(\mathbf{g})$ is Archimedean and f is strictly positive on $\mathbf{X}$, then $f \in Q(\mathbf{g})$.

So, sums of squares belong to a classical topic in real algebraic geometry, with a rich history going back to early work by Hilbert. It is only recently that their relevance to optimization has been fully appreciated, starting with ground works by Shor (1987), Nesterov (2000), Lasserre (2001), and Parrilo (2000, 2003). A key ingredient for this link to optimization is the fact that sums of squares can be modeled using semidefinite programs, which makes them amenable to numerical algorithms. Next, we explain how to model sos polynomials using semidefinite programming, and thereafter how to use sos polynomials to define hierarchies of (lower and upper) bounds for the original polynomial optimization problem (1).

1.4 Sums of Squares and Semidefinite Optimization

We begin with a quick recap on semidefinite optimization, and refer, e.g., to de Klerk (2002) for a detailed treatment. A semidefinite program (in primal form) reads

$$p^* = \sup\{\langle C, X\rangle : \langle A_j, X\rangle = b_j \ (j \in [m]),\ X \in \mathcal{S}_+^N\}. \tag{7}$$

Here, $C, A_1, \dots, A_m \in \mathcal{S}^N$ are symmetric $N \times N$ matrices and $b \in \mathbb{R}^m$–these are the data of the problem– and $X \in \mathcal{S}_+^N$ is the matrix variable, required to be symmetric and positive semidefinite (also written as $X \succeq 0$). So, the program (7) is a linear program over the cone $\mathcal{S}^N_+$ of positive semidefinite matrices. Its dual program reads

$$d^* = \inf\Big\{\sum_{j=1}^m b_j y_j : \sum_{j=1}^m y_j A_j - C \in \mathcal{S}^N_+\Big\}. \tag{8}$$

Weak duality holds: $p^* \le d^*$, and strong duality $p^* = d^*$ holds under some Slater-type conditions. Semidefinite programs contain linear programs as a special case (when all data matrices C, A_j are diagonal). The crucial fact is that semidefinite programs can be solved efficiently up to any given precision, under some assumptions (like knowing a small ball inside the feasible region of (7) and a ball enclosing it).

As we now see, sums of squares can be modeled with semidefinite programs. As a warm-up observe that a quadratic form $f = x^T M x$ is a sum of squares precisely when the matrix M is positive semidefinite. Consider now a polynomial $f = \sum_{\alpha\in\mathbb{N}^n_{2d}} f_\alpha x^\alpha$ of even degree $2d$ for which we wish to decide whether $f \in \Sigma$. Assume $f \in \Sigma$, i.e., $f = q_1^2 + \cdots + q_k^2$ for some polynomials q_j. Then, each $q_j \in \mathbb{R}[\mathbf{x}]_d$ can be written in the monomial basis as $q_j = [\mathbf{x}]_d^{\mathsf{T}} \mathbf{a}_j$, where $\mathbf{a}_j = ((\mathbf{a}_j)_\beta)_{\beta\in\mathbb{N}^n_d}$ is the vector of coefficients of the polynomial q_j in the monomial basis $[\mathbf{x}]_d$. Thus, we obtain

$$f = \sum_{j=1}^k q_j^2 = \sum_{j=1}^m [\mathbf{x}]_d^{\mathsf{T}} \mathbf{a}_j \mathbf{a}_j^{\mathsf{T}} [\mathbf{x}]_d = [\mathbf{x}]_d \Big(\sum_{j=1}^k \mathbf{a}_j \mathbf{a}_j^{\mathsf{T}}\Big)[\mathbf{x}]_d = [\mathbf{x}]_d^{\mathsf{T}} Q [\mathbf{x}]_d,$$

setting $Q := \sum_{j=1}^k \mathbf{a}_j \mathbf{a}_j^{\mathsf{T}}$. By construction the matrix Q is positive semidefinite and, by equating coefficients at both sides of the above polynomial identity, we arrive at the following characterization.

Modeling sums of squares with semidefinite programs
A polynomial $f = \sum_{\alpha\in\mathbb{N}^n_{2d}} f_\alpha x^\alpha \in \mathbb{R}[\mathbf{x}]_{2d}$ is a sum of squares of polynomials if and only if the following semidefinite program has a feasible solution:

$$Q \in \mathcal{S}^{N_d}_+, \qquad \sum_{\beta,\gamma\in\mathbb{N}^n_d : \beta+\gamma=\alpha} Q_{\beta,\gamma} = f_\alpha \ \text{ for all } \alpha \in \mathbb{N}^n_{2d}, \tag{9}$$

where the matrix variable Q is indexed by $\mathbb{N}^n_d$ and $N_d = |\mathbb{N}^n_d| = \binom{n+d}{d}$.

Note that the choice of the monomial basis is not important in the above derivation, it would work mutatis mutandis using any other polynomial basis of the polynomial space (this fact will be used in Sect. 2.2).

Now that we know how to express sos polynomials using semidefinite programs we can also express polynomials in $Q(\mathbf{g})_{2r}$, the quadratic module truncated at a given degree $2r$, defined by

$$\mathcal{Q}(\mathbf{g})_{2r} = \Big\{ \sum_{j=0}^{m} \sigma_j g_j : \sigma_j \in \Sigma, \deg(\sigma_j g_j) \le 2r \text{ for } j = 0, 1, \dots, m \Big\}, \tag{10}$$

or in the truncated preordering $\mathcal{T}(\mathbf{g})_{2r}$ defined by

$$\mathcal{T}(\mathbf{g})_{2r} = \Big\{ \sum_{J \subseteq [m]} \sigma_J \prod_{j \iota J} g_j : \sigma_J \in \Sigma, \deg\Big(\sigma_J \prod_{j \in J} g_j\Big) \le 2r \text{ for } J \subseteq [m] \Big\}. \tag{11}$$

For this, set $d_j = \lceil \deg(g_j)/2 \rceil$, so that polynomials in $\mathcal{Q}(\mathbf{g})_{2r}$ can be modeled as $\sum_{j=0}^{m} g_j [\mathbf{x}]_{r-d_j}^{\mathsf{T}} Q_j [\mathbf{x}]_{r-d_j}$ for some positive semidefinite matrices $Q_j \in \mathcal{S}^{N_{d_j}}$ (for $j = 0, , \dots, m$). In other words, membership in the truncated quadratic module can be modeled as a semidefinite program. The same holds of course for membership in the truncated preordering.

We now have all tools in hands to define the hierarchies of upper bounds and lower bounds for the polynomial optimization problem (1).

1.5 Upper Bounds

To define upper bounds on $f_{\min}$ the starting point is to observe that problem (1) can be reformulated as a linear optimization problem over the set $\mathcal{M}(\mathbf{X})$ of positive Borel measures supported within the set $\mathbf{X}$:

$$f_{\min} = \min \Big\{ \int_{\mathbf{X}} f(\mathbf{x}) d\nu(\mathbf{x}) : \nu \in \mathcal{M}(\mathbf{X}), \int_{\mathbf{X}} d\nu(\mathbf{x}) = 1 \Big\}. \tag{12}$$

The argument is simple: On the one hand, $\int_{\mathbf{X}} f d\nu \ge f_{\min}$ since ν is a probability measure on $\mathbf{X}$. On the other hand, the Dirac delta $\nu = \delta_{\mathbf{x}^*}$ at a global minimizer $\mathbf{x}^*$ of f in $\mathbf{X}$ provides a feasible solution to (12) with value $f(\mathbf{x}^*) = f_{\min}$.

Let is now fix a (reference) measure $\mu \in \mathcal{M}(\mathbf{X})$ whose support is equal to $\mathbf{X}$. The next step is that one can restrict the optimization in (12) to the measures ν that have a sos density with respect to this given measure μ:

$$f_{\min} = \min \Big\{ \int_{\mathbf{X}} f(\mathbf{x}) \sigma(\mathbf{x}) d\mu(\mathbf{x}) : \sigma \in \Sigma, \int_{\mathbf{X}} \sigma(\mathbf{x}) d\mu(\mathbf{x}) = 1 \Big\} \tag{13}$$

(Lasserre (2010)). Intuitively, this relies on the fact that the Dirac delta $\delta_{\mathbf{x}^*}$ can be well approximated by sos polynomials. The upper bounds on $f_{\min}$ are then obtained by restricting the optimization to sos polynomials of a given degree.

The upper bounds on $f_{\min}$
For any $r \in \mathbb{N}$, define the parameter

$$\mathrm{ub}(f, \mathbf{X}, \mu)_r = \min\left\{\int_{\mathbf{X}} f(\mathbf{x})\sigma(\mathbf{x})d\mu(\mathbf{x}) : \sigma \in \Sigma_{2r},\ \int_{\mathbf{X}} \sigma(\mathbf{x})d\mu(\mathbf{x}) = 1\right\}. \tag{14}$$

We have $f_{\min} \le \mathrm{ub}(f, \mathbf{X}, \mu)_{r+1} \le \mathrm{ub}(f, \mathbf{X}, \mu)_r$.

As an illustration, we show in Fig. 1a the Motzkin polynomial and in (b) the optimal sos density σ for program (14) at order $2r = 16$ (which approximates well the sum of Dirac functions at the four minimizers $(\pm 1, \pm 1)$).

So, the parameter (14) depends on the choice of the reference measure μ on $\mathbf{X}$ and it can be computed via a semidefinite program (in fact, as an eigenvalue problem as we see in Sect. 2.2). In view of (13), the bounds $\mathrm{ub}(f, \mathbf{X}, \mu)_r$ converge asymptotically to $f_{\min}$ as $r \to \infty$. We return to these bounds in Sect. 2, where we will discuss in detail the convergence rate of the error range $\mathrm{ub}(f, \mathbf{X}, \mu)_r - f_{\min}$ as the relaxation order r grows.

For now, we summarize in Table 1 the known results about this convergence rate for various classes of compact sets X (and reference measures μ). As we see there, one can show a convergence rate in $O(1/r^2)$ (up to a $\log^2(r)$ factor) for a large class of compact sets. For instance, one can show a rate in $O(\log^2(r)/r^2)$ for any convex body $\mathbf{X}$, and in $O(1/r^2)$ when $\mathbf{X}$ is 'round', which means that there are tangent inner and outer balls at any point on its boundary. The convergence rate in $O(1/r^2)$ is essentially optimal, as can be seen for the case $\mathbf{X} = [-1, 1]$ (due to the explicit link to extremal roots of orthogonal polynomials, see relation (23) in Sect. 2.3), or for the case $\mathbf{X} = \mathbb{S}^{n-1}$ (due to a link to cubature rules, see Sect. 1.7).

1.6 Lower Bounds

To define lower bounds on $f_{\min}$ the starting point is to observe that the minimum value taken by the polynomial f over the set $\mathbf{X}$ is equal to the largest scalar λ for which the polynomial $f - \lambda$ is nonnegative over $\mathbf{X}$:

$$f_{\min} = \sup\{\lambda : \lambda \in \mathbb{R},\ f(\mathbf{x}) - \lambda \ge 0 \text{ for all } \mathbf{x} \in \mathbf{X}\}. \tag{15}$$

Now, we obtain bounds by replacing the positivity condition by membership in either the truncated quadratic module $\mathcal{Q}(\mathbf{g})_{2r}$ or preordering $\mathcal{T}(\mathbf{g})_{2r}$ (see (10), (11)).

Table 1 Overview of known results on the asymptotic error $\text{ub}(f, \mathbf{X}, \mu)_r - f_{\min}$ of Lasserre's hierarchy of upper bounds

$\mathbf{X}$ (compact)	Error	Measure μ	References
Geometric assumption	$O(1/\sqrt{r})$	Lebesgue	de Klerk et al. (2017)
Convex body	$O(1/r)$	Lebesgue	de Klerk and Laurent (2018)
Semialgebraic with dense interior, convex body	$O(\log^2(r)/r^2)$	Lebesgue	Slot and Laurent (2021)
$\mathbb{S}^{n-1}$	$O(1/r)$	Uniform	Doherty and Wehner (2013)
$\mathbb{S}^{n-1}$	$O(1/r^2)$	Uniform	de Klerk and Laurent (2022)
$[-1, 1]^n$	$O(1/r^2)$	$\prod_i (1 - \mathbf{x}_i)^\lambda d\mathbf{x}$ ($\lambda = -\frac{1}{2}$)	de Klerk and Laurent (2020)
$[-1, 1]^n$	$O(1/r^2)$	$\prod_i (1 - \mathbf{x}_i)^\lambda d\mathbf{x}$ ($\lambda \geq -\frac{1}{2}$)	Slot and Laurent (2022d)
'Round' convex body	$O(1/r^2)$	Lebesgue	Slot and Laurent (2022d)
B^n	$O(1/r^2)$	$(1 - \Vert\mathbf{x}\Vert^2)^\lambda d\mathbf{x}$ ($\lambda \geq 0$)	Slot and Laurent (2022d)
Δ^n	$O(1/r^2)$	Lebesgue	Slot and Laurent (2022d)

The lower bounds on $f_{\min}$
For any integer $r \geq \deg(f)/2$, define the parameters

$$\text{lb}(f, Q(\mathbf{g}))_r = \sup\{\lambda : \lambda \in \mathbb{R},\ f - \lambda \in Q(\mathbf{g})_{2r}\}, \tag{16}$$

$$\text{lb}(f, \mathcal{T}(\mathbf{g}))_r = \sup\{\lambda : \lambda \in \mathbb{R},\ f - \lambda \in \mathcal{T}(\mathbf{g})_{2r}\}. \tag{17}$$

By definition, we have $\text{lb}(f, Q(\mathbf{g}))_r \leq \text{lb}(f, \mathcal{T}(\mathbf{g}))_r$ for all r. Furthermore, we have $\text{lb}(f, Q(\mathbf{g}))_r \leq \text{lb}(f, Q(\mathbf{g}))_{r+1} \leq f_{\min}$, and $\text{lb}(f, \mathcal{T}(\mathbf{g}))_r \leq \text{lb}(f, \mathcal{T}(\mathbf{g}))_{r+1} \leq f_{\min}$.

As an application of the earlier mentioned Positivstellensätze of Putinar and Schmüdgen, if $Q(\mathbf{g})$ is Archimedean (resp., $\mathbf{X}$ is compact), then the bounds $\text{lb}(f, Q(\mathbf{g}))_r$ (resp., $\text{lb}(f, \mathcal{T}(\mathbf{g}))_r$) converge to $f_{\min}$ as $r \to \infty$ (Lasserre (2001)). We return to these bounds in Sect. 3, where we will discuss their asymptotic convergence rates as the relaxation order r grows. We summarize the main known results in Table 2 below. As we see, the results depend on the algebraic structure of the semialgebraic set $\mathbf{X}$.

Table 2 Overview of known results on the asymptotic error $f_{\min} - \text{lb}(f, \mathbf{X})_r$ of Lasserre's hierarchies of lower bounds. (* The result on the sphere $\mathbb{S}^{n-1}$ is shown in Fang and Fawzi (2021) only for homogeneous polynomials f, but it can be extended to general polynomials Blomenhofer et al. (2025))

$\mathbf{X}$ (compact)	Error	Certificate	References
Archimedean	$O(1/\log(r)^c)$	$\mathcal{Q}(\mathbf{g})$	Nie and Schweighofer (2007)
Archimedean	$O(1/r^c)$	$\mathcal{Q}(\mathbf{g})$	Baldi and Mourrain (2023), and Baldi et al. (2025)
General	$O(1/r^c)$	$\mathcal{T}(\mathbf{g})$	Schweighofer (2004)
$[-1, 1]^n$	$O(1/r)$	$\mathcal{Q}(\mathbf{g})$	Baldi and Slot (2024)
$\mathbb{S}^{n-1}$	$O(1/r^2)^\star$	$\mathcal{Q}(\mathbf{g})$ $(= \mathcal{T}(\mathbf{g}))$	Fang and Fawzi (2021)
B^n	$O(1/r^2)$	$\mathcal{Q}(\mathbf{g})$ $(= \mathcal{T}(\mathbf{g}))$	Slot (2022c)
Δ^n	$O(1/r^2)$	$\mathcal{T}(\mathbf{g})$	Slot (2022c)
$[-1, 1]^n$	$O(1/r^2)$	$\mathcal{T}(\mathbf{g})$	Laurent and Slot (2023)

1.7 Link to Cubature Rules

A natural approach to design upper bounds on the minimum value of f over $\mathbf{X}$ is by minimizing f over a well-chosen finite set of points $\mathbf{X}_r \subseteq \mathbf{X}$. For example, for a set $\mathbf{X} \subseteq [0, 1]^n$, one may take for $\mathbf{X}_r$ the set of rational points in $\mathbf{X}$ that have a given denominator $r \geq 1$. This clearly gives a hierarchy of upper bounds that converge asymptotically to $f_{\min}$ when the denominator r grows. For the simplex $\mathbf{X} = \Delta^n$, we have $|\mathbf{X}_r| = \binom{n+r}{r}$, thus polynomial in n for any fixed r. In fact, the parameters $f_{\min,\mathbf{X}_r}$ lead to a polynomial-time approximation scheme with convergence rate $\min_{\mathbf{x}\in\mathbf{X}_r} f(\mathbf{x}) - f_{\min} = O(1/r)$ (de Klerk et al. (2006)). However, for the hypercube $\mathbf{X} = [0, 1]^n$, we have $|\mathbf{X}_r| = (r+1)^n$, which is thus *exponential* in the dimension n. We refer to Martinez et al. (2019) for related work for well-chosen finite meshes on $\mathbf{X}$. Let us mention the following simple link to cubature rules.

Let $\mu \in \mathcal{M}(\mathbf{X})$ with support $\mathbf{X}$. Assume that $\mathbf{X}_r = \{x^{(i)} : i \in [N]\} \subseteq \mathbf{X}$, together with positive weights $w = (w_i)_{i\in[N]}$, provides a positive cubature rule for $(\mathbf{X}, \mu)$ that is exact at degree $\deg(f) + 2r$. That is, for any polynomial p with $\deg(p) \leq \deg(f) + 2r$ we have $\int_{\mathbf{X}} p d\mu = \sum_{i=1}^N w_i p(x^{(i)})$. Then, one easily sees that

$$\text{ub}(f, \mathbf{X}, \mu)_r \geq \min_{\mathbf{x}\in\mathbf{X}_r} f(\mathbf{x}) = f_{\min,\mathbf{X}_r} \geq f_{\min,\mathbf{X}}.$$

Sometimes, this permits to show tightness of the convergence rate for the upper bounds $\text{ub}(f, \mathbf{X}, \mu)_r$ from information about cubature rules. This is done, e.g., for the sphere $\mathbf{X} = \mathbb{S}^{n-1}$ (equipped with the uniform Haar measure) by de Klerk and Laurent (2022), who show $\text{ub}(f, \mathbf{X}, \mu)_r - f_{\min} = \Omega(1/r^2)$ for linear f.

2 Performance Analysis of the Upper Bounds

Recall the set up: $\mathbf{X} \subseteq \mathbb{R}^n$ is a compact set and μ is a reference measure supported by $\mathbf{X}$. Consider the bilinear form $\langle \cdot, \cdot \rangle_\mu$ on $\mathbb{R}[\mathbf{x}]$ induced by μ, defined by

$$\langle p, q \rangle_\mu = \int_{\mathbf{X}} p(\mathbf{x})q(\mathbf{x})d\mu(\mathbf{x}) \ \text{ for } p, q \in \mathbb{R}[\mathbf{x}]. \tag{18}$$

If $\mathbf{X}$ has a nonempty interior, then this defines an inner product on $\mathbb{R}[\mathbf{x}]$. Otherwise, if $\mathbf{X}$ has an empty interior (e.g., $\mathbf{X} = \mathbb{S}^{n-1}$ is the unit sphere), then this defines an inner product on the space $\mathcal{R}(\mathbf{X})$ of polynomials on $\mathbf{X}$. Let $\mathcal{B}_\mu = \{P_\alpha : \alpha \in \mathbb{N}^n\}$ be an orthonormal basis of $\mathcal{R}(\mathbf{X})$ with respect to this inner product, with the property that the set

$$\mathcal{B}_{\mu,d} = \{P_\alpha : \alpha \in \mathbb{N}^n_d\}, \ \text{ sometimes also denoted as the vector } [P_\alpha]_d, \tag{19}$$

is a basis of $\mathcal{R}(\mathbf{X})_d$ (the set of polynomials that agree on $\mathbf{X}$ with a degree d polynomial) for each $d \in \mathbb{N}$.

Our objective here is to give a (rough) overview of the main ideas used to show the results in Table 1, where rates in $O(1/r^2)$ and in $O(\log^2(r)/r^2)$ are presented.

The starting point for showing a $O(r^2)$ rate is establishing an eigenvalue reformulation for the parameter $\text{ub}(f, \mathbf{X}, \mu)_r$ (Sect. 2.2) and a link to extremal roots of orthogonal polynomials in the univariate case (Sect. 2.3). In addition, one uses some simple 'tricks' to reduce the analysis to (at most) quadratic polynomials and to simpler sets and measures (Sect. 2.1).

To show the (slightly weaker) rate $O(\log^2(r)/r^2)$ for much more general sets $\mathbf{X}$ one again follows a univariate strategy, now reducing the search in problem (14) to sos polynomials $\sigma(\mathbf{x}) = s(f(\mathbf{x}))$, where s is a *univariate* sos polynomial (see Sect. 2.4).

2.1 Some Useful Tricks for the Analysis

We group here some simple facts, useful for the analysis of the upper bounds.

Lemma 1 (de Klerk et al. (2017)) *Let $f, g \in \mathbb{R}[\mathbf{x}]$ satisfy (i) $f(\mathbf{x}) \leq g(\mathbf{x})$ for all $\mathbf{x} \in \mathbf{X}$, and (ii) f and g take the same minimum value on $\mathbf{X}$, i.e., $f_{min,\mathbf{X}} = g_{min,\mathbf{X}}$. Then, for any $r \in \mathbb{N}$, we have $ub(f, \mathbf{X}, \mu)_r - f_{min,\mathbf{X}} \leq ub(g, \mathbf{X}, \mu)_r - g_{min,\mathbf{X}}$.*

Using Taylor's expansion, one can see that any polynomial f admits an upper estimator g on $\mathbf{X}$ that satisfies the assumptions (i), (ii) of Lemma 1 and that is (linear or) quadratic.

The next lemma permits to reduce the analysis for a pair $(\mathbf{X}, w)$, where $\mathbf{X}$ is equipped with the absolutely continuous measure $w(\mathbf{x})d\mathbf{x}$, to a possibly simpler pair $(\widehat{\mathbf{X}}, \widehat{w})$, where $\mathbf{X} \subseteq \widehat{\mathbf{X}}$ and $w \leq \widehat{w}$ 'look the same' around a minimizer.

Lemma 2 (Slot and Laurent (2022d)) *Let $\mathbf{X} \subseteq \widehat{\mathbf{X}}$ be compact sets, where $\mathbf{X}$ (resp., $\widehat{\mathbf{X}}$) is equipped with an absolutely continuous measure $w(\mathbf{x})d\mathbf{x}$ (resp., $\widehat{w}(\mathbf{x})d\mathbf{x}$). Let $x^* \in \mathbf{X}$ be a minimizer of f in $\mathbf{X}$. Assume the following conditions hold:*

(i) $\mathbf{X}, \widehat{\mathbf{X}}$ *are 'locally similar' at* $\mathbf{x}^*$: $\mathbf{X} \cap B^n(\mathbf{x}^*, \epsilon) = \widehat{\mathbf{X}} \cap B^n(\mathbf{x}^*, \epsilon)$ *for some* $\epsilon > 0$ *(with $B^n(\mathbf{x}^*, \epsilon)$ the ball centered at $\mathbf{x}^*$ with radius ϵ).*
(ii) $w(\mathbf{x}) \leq \widehat{w}(\mathbf{x})$ *for all* $\mathbf{x} \in int(\mathbf{X})$.
(iii) w, $\widehat{w}$ *are 'locally comparable' at* $\mathbf{x}^*$: $C \cdot \widehat{w}(\mathbf{x}) \leq w(\mathbf{x})$ *for* $\mathbf{x} \in int(K) \cap B^n(\mathbf{x}^*, \epsilon)$ *for some* $\epsilon > 0$ *and* $C > 0$.

Then, there exists a (linear or) quadratic polynomial g such that $f(\mathbf{x}) \leq g(\mathbf{x})$ for all $\mathbf{x} \in \widehat{\mathbf{X}}$, $f_{min,\mathbf{X}} = g_{min,\mathbf{X}}$, and $ub(f, \mathbf{X}, w)_r - f_{min,\mathbf{X}} \leq \frac{2}{C}(ub(g, \widehat{\mathbf{X}}, \widehat{w})_r - g_{min,\widehat{\mathbf{X}}})$.

The up-shot of these two lemmas can be summarized as follows.

Recipe for analyzing the upper bounds
It suffices to analyze the convergence rate of the range $\text{ub}(f, \mathbf{X}, \mu)_r - f_{\min}$ for *(linear or) quadratic* f and 'simple' sets $\mathbf{X}$ (like the box $[-1, 1]^n$ or the ball B^n).

2.2 *Reformulation as an Eigenvalue Problem*

We begin with giving an eigenvalue reformulation for the parameter $\text{ub}(f, \mathbf{X}, \mu)_r$ from (14). For this, we express the polynomial $\sigma \in \Sigma_{2r}$ entering the definition of the parameter $\text{ub}(f, \mathbf{X}, \mu)_r$ using the orthonormal basis $\mathcal{B}_{\mu,r}$, ordered as the vector $[P_\alpha]_r$ (as in (19)). So, we can write $\sigma = \langle Q, [P_\alpha]_r[P_\alpha]_r^\top \rangle$, where Q is the matrix variable (indexed by $\mathcal{B}_{\mu,r}$). Then, we have $\int_\mathbf{X} f\sigma d\mu = \langle Q, M_{\mu,r}(f)\rangle$ and $\int_\mathbf{X} \sigma d\mu = \text{Tr}(Q)$, after defining the matrix

$$M_{\mu,r}(f) = \left(\int_\mathbf{X} f P_\alpha P_\beta d\mu\right)_{\alpha,\beta\in\mathbb{N}^n_r}.$$

In this way we arrive at the following *eigenvalue reformulation*:

$$\text{ub}(f, \mathbf{X}, \mu)_r = \min\{\langle M_{\mu,r}(f), Q\rangle : Q \succeq 0,\ \ \text{Tr}(Q) = 1\} = \lambda_{\min}(M_{\mu,r}(f)), \tag{20}$$

showing that the parameter $\text{ub}(f, \mathbf{X}, \mu)_r$ is equal to the smallest eigenvalue of the matrix $M_{\mu,r}(f)$. Some remarks are in order here: this computation relies on the matrix $M_{\mu,r}(f)$, which in turn requires to be able to integrate a polynomial on the set $\mathbf{X}$ w.r.t. the measure μ. Thus, for practical computation one needs to restrict to some relatively easy sets equipped with well-understood measures.

Estimating this smallest eigenvalue remains a difficult problem in general. However, there is a situation where it is very well-understood: in the univariate case for the interval $\mathbf{X} = [-1, 1]$ equipped with a 'nice' measure. We first consider this case, which will form the basis for understanding the general multivariate case.

2.3 *Univariate Case: Links to Roots of Orthogonal Polynomials*

Here, we consider the univariate case $n = 1$ and the interval $\mathbf{X} = [-1, 1]$ equipped with a measure μ supported on $[-1, 1]$. Consider as above the orthonormal basis $\mathcal{B}_\mu = \{P_k : k \geq 0\}$ of $\mathbb{R}[x]$ w.r.t. the inner product $\langle \cdot, \cdot \rangle_\mu$. Then, the polynomials P_k satisfy the well-known *3-term recurrence*: there exist scalars a_k, b_k $(k \geq 0)$ such that

$$x P_k = a_{k-1} P_{k-1} + b_k P_k + a_k P_{k+1} \quad \text{for } k \geq 0 \quad (\text{setting } a_{-1} = 0). \tag{21}$$

Observe that for the polynomial $f = x$, the matrix $M_{\mu,r}(x) = \left(\int_{-1}^{1} x P_i P_j d\mu \right)_{i,j=0}^{r}$ (also known as the *Jacobi matrix*) is tri-diagonal and its eigenvalues are the roots of the degree $r + 1$ orthogonal polynomial P_{r+1}. In particular,

$$\lambda_{\min}(M_{\mu,r}(x)) = \text{ smallest root of the orthogonal polynomial } P_{r+1} \text{ (w.r.t. } \mu). \tag{22}$$

We refer, e.g., to Dunkl and Xu (2014) for background on orthogonal polynomials.

From now on, we will assume that the reference measure μ is of *Jacobi-type*.

Jacobi-type measures on $[-1, 1]$
The Jacobi-type measure is $d\mu(x) = (1 - x)^\lambda (1 + x)^{\lambda'} dx$, where $\lambda, \lambda' > -1$; the associated orthogonal polynomials are known as the Jacobi polynomials. The case when $\lambda = \lambda' = -1/2$ is of special interest; then, the 3-term recurrence (21) reads $x P_k = (P_{k-1} + P_{k+1})/2$ and the associated orthogonal polynomials are the Chebychev polynomials.
The reason for restricting to Jacobi-type measures is that the behaviour of the extremal roots of their associated orthogonal polynomials is well-understood: The smallest root of P_r is $-1 + \Theta(1/r^2)$ (Dimitrov and Nikolov (2010), and Driver and Jordaan (2012)).

As observed earlier, it suffices to analyze the upper bounds for *linear* or *quadratic* f.

For *linear* $f = x$, since the smallest root of P_{r+1} is $-1 + \Theta(1/r^2)$, by combining with relations (20) and (22), we directly obtain the error analysis

$$\mathrm{ub}(x, [-1, 1], \mu)_r - f_{\min} = \lambda_{\min}(M_{\mu,r}(x)) + 1 = \Theta\Big(\frac{1}{r^2}\Big) \tag{23}$$

(from de Klerk and Laurent (2020)).

We now consider the case when f is *quadratic*. We distinguish two cases, depending whether f has a minimizer in the boundary or in the interior of $[-1, 1]$. In the former case, f admits a linear upper estimator; then we can apply Lemma 1 and use (23) to conclude that $\mathrm{ub}(f, [-1, 1], \mu)_r - f_{\min} = O(1/r^2)$.

So, assume now $f = x^2 + cx$ with $a \in (-2, 2)$, so that f attains its minimum value in $(-1, 1)$. Then, one needs to estimate the smallest eigenvalue of the matrix $M_{\mu,r}(f)$, a difficult task in general. Indeed, $M_{\mu,r}(f)$ is a 5-diagonal matrix, with its entries depending on the parameters a_k, b_k in the 3-term recurrence (21) and the parameter c in the definition of f. Estimating this smallest eigenvalue is, however, easier in the Chebyshev case.

In the case when the measure μ is of Chebyshev-type, i.e., $d\mu = (1 - x^2)^{-1/2}dx$, the matrix $M_{\mu,r}(f)$ is 'almost' a circulant matrix: After modifying its first two rows and columns it can be made a circulant matrix, whose eigenvalues can be explicitly computed. Combining with an interlacing argument, one can show again that $\mathrm{ub}(f, [-1, 1], \mu)_r - f_{\min} = O(1/r^2)$ (de Klerk and Laurent (2020)). Combining this with using Lemma 2 one can show the same result for a general measure μ with weight $(1 - x^2)^{\lambda}$ when $\lambda \geq -1/2$ (Slot and Laurent (2022d)).

In summary, we have $\mathrm{ub}(f, [-1, 1], \mu)_r - f_{\min} = O(1/r^2)$ for any polynomial f, in the case when $[-1, 1]$ is equipped with a measure μ with Jacobi weight $(1 - x^2)^{\lambda}$ and $\lambda \geq -1/2$. This error estimate in $O(1/r^2)$ extends then to the box $[-1, 1]^n$, equipped with a product of such measures.

Similarly, to establish the rate $O(1/r^2)$ for the other sets in Table 1, one uses the result (23) for the univariate case, combined with the results of Lemmas 1 and 2 (see Slot and Laurent (2022d) for the ball, simplex, and round convex bodies), and possibly some 'integration trick' (see de Klerk and Laurent (2022) for the sphere).

2.4 Another Analysis Technique Using Needle Polynomials

At this point there remains to explain how to show the (slightly weaker) rate $O(\log^2(r)/r^2)$ presented in Table 1 for general compact sets $\mathbf{X}$ like semialgebraic sets (with dense interior) and convex bodies. For this the key idea is again to follow a univariate approach, as explained by Slot and Laurent (2021). Instead of searching over all (multivariate) sos polynomials $\sigma \in \mathbb{R}[\mathbf{x}]$ in program (14), the idea is to restrict the search to *univariate* sos polynomials $s \in \mathbb{R}[x]$ and then set $\sigma(\mathbf{x}) = s(f(\mathbf{x}))$. In this way one arrives at the following (weaker) bounds (introduced by Lasserre (2021)):

$$\begin{aligned} \mathrm{ub}_\#(f, \mathbf{X}, \mu)_r &= \min\Big\{ \int_{\mathbf{X}} f(\mathbf{x}) s(f(\mathbf{x})) d\mu(\mathbf{x}) : s \in \mathbb{R}[x]_{2r} \text{ sos}, \int_{\mathbf{X}} s(f(\mathbf{x})) d\mu(\mathbf{x}) = 1 \Big\} \\ &= \min\Big\{ \int_{f(\mathbf{X})} x s(x) d f_\# \mu(x) : s \in \mathbb{R}[x]_{2r} \text{ sos}, \int_{f(\mathbf{X})} s(x) d f_\# \mu(x) = 1 \Big\}, \end{aligned} \tag{24}$$

where $f_\# \mu$ denotes the univariate measure obtained by taking the push-forward of μ by f. Then, we have $\mathrm{ub}(f, \mathbf{X}, \mu)_{rd} \leq \mathrm{ub}_\#(f, \mathbf{X}, \mu)_r$ if $d = \deg(f)$.

In view of relations (20) and (22), the analysis of the bounds $\mathrm{ub}_\#(f, \mathbf{X}, \mu)_r$ relies on the smallest roots of the orthogonal polynomials w.r.t. the push-forward measure $f_\# \mu$. However, these orthogonal polynomials are not well understood in general, so another approach is needed. Up to translation one may assume that $f_{\min} = 0$. Then, the idea is to find a univariate sos polynomial s that approximates well the Dirac delta at the origin. For this, one can employ the so-called *needle polynomials* (from Kroó (2015)) that are widely used in the literature of approximation theory. We refer to Slot and Laurent (2021) for the technical details and extension to compact sets satisfying a suitable geometric assumption.

3 Performance Analysis of the Lower Bounds

We turn now to the lower bounds. Let $\mathbf{X} = \{\mathbf{x} \in \mathbb{R}^n : g_1(\mathbf{x}) \geq 0, \ldots, g_m(\mathbf{x}) \geq 0\}$ be as in (2). We assume that the quadratic module $Q(\mathbf{g})$ is Archimedean. Recall that this implies that, for any f, the bounds $\mathrm{lb}(f, Q(\mathbf{g}))_r$ and $\mathrm{lb}(f, \mathcal{T}(\mathbf{g}))_r$ converge to $f_{\min}$ as $r \to \infty$ by Putinar's and Schmüdgen's Positivstellensätze, respectively. In this section, we outline the main ideas used to obtain the convergence rates presented in Table 2. For ease of writing, we will use the letter C to refer to either $Q(\mathbf{g})$ or $\mathcal{T}(\mathbf{g})$. A useful observation is the following.

Observation 1 *Let $f \in \mathbb{R}[\mathbf{x}]$. Then, for any $\epsilon > 0$ and $r \in \mathbb{N}$, we have*

$$f_{min} - lb(f, C)_r \leq \epsilon \iff f - f_{min} + \epsilon \in C_{2r}.$$

Thus, in order to prove convergence rates for Lasserre's hierarchies of lower bounds, it suffices to find low-degree sos-representations of $f - f_{\min} + \epsilon$. In what follows, we discuss two methods of obtaining such representations. We give particular attention to the so-called *polynomial kernel method* in Sect. 3.1 below, as it reveals an interesting connection between the analysis of the *lower* bounds and the *upper* bounds discussed in Sect. 2.

Throughout this section, we set $d = \deg(f)$, which should be thought of as being fixed, while the relaxation order r will grow.

3.1 Analysis via the Polynomial Kernel Method

The *polynomial kernel method* (PKM) has been successful for proving strong convergence rates for optimization over certain distinguished sets $\mathbf{X}$, including $\mathbf{X} = \mathbb{S}^{n-1}$, B^n, Δ^n, $[-1, 1]^n$. It was initially described[3] in the special case $\mathbf{X} = \mathbb{S}^{n-1}$ by Fang and Fawzi (2021), who use it to show a convergence rate in $O(1/r^2)$ in that setting. It was later applied by Laurent and Slot (2023) and Slot (2022c), to prove rates in $O(1/r^2)$ for optimization over $[-1, 1]^n$ and Δ^n, B^n, respectively. The latter work by Slot (2022c) is the first to describe the technique in full generality, and we follow its exposition here.

Suppose we were able to construct a linear operator $\mathbf{K} : \mathbb{R}[\mathbf{x}] \to \mathbb{R}[\mathbf{x}]$ with the following three properties:

$$\mathbf{K}(1) = 1, \tag{P1}$$

$$\mathbf{K}p \in C_{2r} \quad \text{for all } p \in \mathcal{P}(\mathbf{X}), \tag{P2}$$

$$\max_{\mathbf{x}\in\mathbf{X}} |\mathbf{K}^{-1} f(\mathbf{x}) - f(\mathbf{x})| \leq \epsilon. \tag{P3}$$

Then, we claim that $(f - f_{\min}) + \epsilon \in C_{2r}$ (which shows that $f_{\min} - \operatorname{lb}(f, C)_r \leq \epsilon$ by Observation 1). Indeed, by (P3), we have $\mathbf{K}^{-1} f(\mathbf{x}) \geq f(\mathbf{x}) - \epsilon \geq f_{\min} - \epsilon$ on $\mathbf{X}$. Using (P1), we obtain $\mathbf{K}^{-1}(f - f_{\min} + \epsilon) = \mathbf{K}^{-1} f - f_{\min} + \epsilon \in \mathcal{P}(\mathbf{X})$. By (P2), we may then conclude that

$$f - f_{\min} + \epsilon = \mathbf{K}\big[\mathbf{K}^{-1}\big(f - f_{\min} + \epsilon\big)\big] \in C_{2r}.$$

3.1.1 Constructing Linear Operators

It remains to construct operators $\mathbf{K}$ that enjoy these special properties. For this, we rely on the theory of (polynomial) *reproducing kernels*. Let μ be a (sufficiently nice) measure supported on $\mathbf{X}$. Then, any polynomial $\mathcal{K} \in \mathbb{R}[\mathbf{x}, \mathbf{y}]$ induces a linear operator $\mathbf{K}$ on $\mathbb{R}[\mathbf{x}]$ via convolution:

$$\mathbf{K}p(\mathbf{x}) := \int_{\mathbf{X}} \mathcal{K}(\mathbf{x}, \mathbf{y}) p(\mathbf{y}) d\mu(\mathbf{y}) \quad (p \in \mathbb{R}[\mathbf{x}]). \tag{25}$$

[3] Earlier, weaker analyses of the lower bounds on $\mathbb{S}^{n-1}$ due to Reznick (1995) and Doherty and Wehner (2013) already relied on the PKM *implicitly*.

The goal is to choose $\mathcal{K}$ in such a way that $\mathbf{K}$ satisfies (P1)-(P3). For (P2), it suffices to choose $\mathcal{K}$ so that the polynomial $\mathbf{x} \mapsto \mathcal{K}(\mathbf{x}, \mathbf{y})$ lies in C_{2r} for all *fixed* $\mathbf{y} \in \mathbf{X}$. This is a consequence of Tchakaloff's Theorem, which allows us to write the integral (25) as (positively weighted) finite sum over a cubature rule (see Laurent and Slot (2023)). To establish (P1) and (P3), it turns out that it is enough to control the *eigenvalues* of $\mathbf{K}$. The eigenvalues of $\mathbf{K}$ can be related to the polynomial $\mathcal{K}$ directly if we make the assumption that it is of the form

$$\mathcal{K}(\mathbf{x}, \mathbf{y}) = \sum_{\|\alpha\|_1 \leq 2r} \lambda_\alpha P_\alpha(\mathbf{x}) P_\alpha(\mathbf{y}) \quad (\lambda_\alpha \in \mathbb{R}), \tag{26}$$

where $\{P_\alpha : \alpha \in \mathbb{N}^n\}$ is an orthonormal basis of $\mathbb{R}[\mathbf{x}]$ w.r.t. μ, ordered so that $\deg(P_\alpha) = \|\alpha\|_1$ for all α. Indeed, by orthonormality, the eigenvalues of $\mathbf{K}$ are then given by the coefficients λ_α in (26). If we set $\lambda_\alpha = 1$ for all α, the resulting polynomial is called the *reproducing kernel* for $\mathbb{R}[\mathbf{X}]_{2r}$ w.r.t. μ. Its associated operator $\mathbf{K}$ acts as the identity on $\mathbb{R}[\mathbf{x}]_{2r}$. It thus satisfies (P1) and (P3) (for $\epsilon = 0$). However, it does not have an sos-representation in general, and so it does not satisfy (P2). The idea is to *perturb* the reproducing kernel slightly, choosing $\lambda_\alpha \approx 1$ (for all $|\alpha| \leq d$), in such a way that all three properties hold. The following lemma makes the relation between the λ_α and properties (P1), (P3) precise.

Lemma 3 *Let $\mathcal{K}$ be as in* (26)*, with $\lambda_0 = 1$ and $\lambda_\alpha \in (1/2, 1]$ for all $\|\alpha\|_1 \leq d$. Then, its associated linear operator* $\mathbf{K}$ *satisfies* (P1)*, and*

$$\max_{\mathbf{x} \in \mathbf{X}} |\mathbf{K}^{-1} f(\mathbf{x}) - f(\mathbf{x})| \leq \gamma \cdot \sum_{|\alpha| \leq d} (1 - \lambda_\alpha) \cdot (f_{\max} - f_{\min}).$$

for some constant $\gamma > 0$ depending only on $\mathbf{X}$ *and d (but not on f).*

The inequality in Lemma 3 above can be improved in certain special cases (such as $\mathbf{X} = \mathbb{S}^{n-1}$). In particular, the constant γ can be bounded nontrivially, but this is beyond the scope of this survey.

3.1.2 A Connection to the Upper Bounds

In general, it is not obvious how to choose the λ_α such that (P2) holds. However, for the aforementioned special choices of $\mathbf{X}$, one can rely on special structure of the reproducing kernel to make this problem tractable. For instance, in the original application of the PKM by Fang and Fawzi (2021) for the sphere $\mathbb{S}^{n-1}$, the classical *Funk-Hecke formula* tells us that, for any $k \in \mathbb{N}$,

$$\sum_{\|\alpha\|_1 = k} P_\alpha(\mathbf{x}) P_\alpha(\mathbf{y}) = \mathcal{G}_k(\mathbf{x} \cdot \mathbf{y}) \quad (\mathbf{x}, \mathbf{y} \in \mathbb{S}^{n-1}),$$

where $\mathcal{G}_k$ is the *Gegenbauer polynomial* of degree k. These are the (univariate) orthogonal polynomials on the interval $[-1, 1]$ w.r.t to weight $w(t) = (1-t^2)^{\frac{n-3}{2}}$. Assuming that the λ_α in (26) depend only on $|\alpha|$, the polynomial $\mathcal{K}$ then equals

$$\mathcal{K}(\mathbf{x}, \mathbf{y}) = \sum_{k=0}^{2r} \lambda_k \mathcal{G}_k(\mathbf{x} \cdot \mathbf{y}).$$

Write $q(t) := \sum_{k=0}^{2r} \lambda_k \mathcal{G}_k(t)$. Note that if $q \in \Sigma[t]_{2r}$ is a (univariate) sum of squares, then $\mathcal{K}(\mathbf{x}, \mathbf{y})$ lies in the quadratic module of the sphere for fixed $\mathbf{y} \in \mathbb{S}^{n-1}$, and so $\mathbf{K}$ will satisfy (P2). In light of Lemma 3, we thus wish to select the λ_k so that $\lambda_0 = 1$ (P1); $q \in \Sigma[t]_{2r}$ (P2); and $\sum_{k=1}^{d}(1-\lambda_k)$ is as small as possible (P3). Remarkably, the optimal selection of the λ_k corresponds to a particular instance of Lasserre's *upper* bounds. This correspondence allows us to transport the analysis of these bounds in Sect. 2 to our present setting. Indeed, by orthogonality, and after choosing the right normalization of the Gegenbauer polynomials $\mathcal{G}_k$, we have that $\lambda_k = \int_{-1}^{1} \mathcal{G}_k(t)q(t)w(t)dt$. Now set $g(t) = d - \sum_{k=0}^{d} \mathcal{G}_k(t)$. Then, we have that

$$\int_{-1}^{1} g(t)q(t)w(t)dt = \sum_{k=1}^{d}(1-\lambda_k).$$

Thus, choosing the λ_k optimally reduces to solving the optimization problem

$$\text{opt} := \inf_{q \in \Sigma[t]_{2r}} \Bigg\{ \underbrace{\int_{-1}^{1} \sum_{k=1}^{d} g(t)q(t)w(t)dt}_{\sum_{k=1}^{d}(1-\lambda_k)} : \underbrace{\int_{-1}^{1} q(t)w(t)dt}_{\lambda_0} = 1 \Bigg\}.$$

We recognize this as the program (14) that defines the upper bound $\text{ub}(g, [-1, 1])_r$ for the minimization of g on $[-1, 1]$. As $g_{\min} = 0$ (attained at $t = 1$), we may conclude that opt $= O(1/r^2)$, and a convergence rate of the *lower* bounds on $\mathbb{S}^{n-1}$ of the same order follows. We refer to Fang and Fawzi (2021) (and the exposition in Slot (2022b)) for details.

For $\mathbf{X} \in \{B^n, \Delta^n\}$, one has (more complicated) analogs of the Funk-Hecke formula. This allows one to establish a correspondence between upper and lower bounds similar to the above. For $\mathbf{X} = [-1, 1]^n$, there is no such formula. However, the reproducing kernel has a product structure in that case, which can be exploited to obtain representations of $\mathcal{K}$ in the truncated preordering. We refer to Slot (2022c), Slot (2022b) and Laurent and Slot (2023) for details.

Recipe for analyzing the lower bounds (I): Polynomial kernel method Convergence rates for Lasserre's hierarchy of lower bounds may be obtained from sum-of-squares representations of (perturbed) reproducing kernels on $\mathbf{X}$. For distinguished semialgebraic sets $\mathbf{X}$, such representations can be found via the hierarchy of *upper* bounds.

3.2 Analysis via Algebro-Geometric Reduction

We now discuss a technique employed first by Schweighofer (2004), Nie and Schweighofer (2007), Baldi and Mourrain (2023) and Baldi et al. (2025), to prove convergence rates for Lasserre's lower bounds on general (compact, Archimedean) semialgebraic sets $\mathbf{X} = \mathrm{Se}(\mathbf{g})$ (as in (2)). With respect to the polynomial kernel method, this technique yields weaker guarantees, but it does not rely on special structure of the set $\mathbf{X}$. We follow the exposition of Baldi and Mourrain (2023).

Suppose that, for a given polynomial f, we wish to show that $f - f_{\min} + \epsilon \in \mathcal{Q}(\mathbf{g})_{2r}$ for some small $\epsilon > 0$. The idea is to embed the (potentially complicated) set $\mathbf{X}$ in a simple semialgebraic set $\mathbf{Y} = \mathrm{Se}(\mathbf{h}) \supseteq \mathrm{Se}(\mathbf{g}) = \mathbf{X}$. For instance, $\mathbf{Y}$ might be a (scaled) ball, simplex or box. Following Baldi and Mourrain (2023), let us fix $\mathrm{Se}(\mathbf{h}) = [-1, 1]^n$ for concreteness. Using specialized results on the convergence of Lasserre's hiearchy on $[-1, 1]^n$ (e.g., those obtained via the PKM), we know that, for some $\epsilon = O(1/r^2)$,

$$f - f_{\min,\mathbf{Y}} + \epsilon \in \mathcal{T}(\mathbf{h})_{2r}. \tag{27}$$

This fact is not immediately useful to us, for two reasons. First, we wish to obtain a representation in $\mathcal{Q}(\mathbf{g})$, not $\mathcal{T}(\mathbf{h})$. Second, the minimum $f_{\min,\mathbf{Y}}$ of f on $\mathbf{Y}$ might be (much) smaller than $f_{\min} = f_{\min,\mathbf{X}}$. The first issue is resolved by proving an inclusion $\mathcal{T}(\mathbf{h})_{2r} \subseteq \mathcal{Q}(\mathbf{g})_{2r+\ell}$. Note that the existence of such an inclusion is not surprising, as $\mathbf{Y} \supseteq \mathbf{X}$, and thus $\mathcal{P}(\mathbf{X}) \supseteq \mathcal{P}(\mathbf{Y})$. Nonetheless, careful arguments are needed to control the additional degree term ℓ, but we do not discuss these here. The second issue is more serious. It is resolved as follows: Suppose we were able to construct a regularizing polynomial $q \in \mathcal{Q}(\mathbf{g})_{\ell'}$ with the property that $f_{\min,\mathbf{X}} \leq (f - q)_{\min,\mathbf{Y}}$. Then, setting $\tilde{f} = f - q$, we could apply (27) to $\tilde{f}$, and use the inclusion of $\mathcal{T}(\mathbf{h})$ in $\mathcal{Q}(\mathbf{g})$ to obtain, for some $\tilde{\epsilon} = O(1/r^2)$,

$$\tilde{f} - \tilde{f}_{\min,\mathbf{Y}} + \tilde{\epsilon} \in \mathcal{T}(\mathbf{h})_{2r} \subseteq \mathcal{Q}(\mathbf{g})_{2r+\ell}.$$

Using the identity $f - f_{\min,\mathbf{X}} + \tilde{\epsilon} = (\tilde{f} - \tilde{f}_{\min,\mathbf{Y}} + \tilde{\epsilon}) + q + \tilde{f}_{\min,\mathbf{Y}} - f_{\min,\mathbf{X}}$, combined with $\tilde{f}_{\min,\mathbf{Y}} \geq f_{\min,\mathbf{X}}$ and $q \in \mathcal{Q}(\mathbf{g})_{\ell'}$, this gives us the representation

$$f - f_{\min,\mathbf{X}} + \tilde{\epsilon} \in \mathcal{Q}(\mathbf{g})_{\max\{2r+\ell,\, \ell'\}}.$$

Note that $\tilde{\epsilon}$ depends on the degree of q, which may be substantially larger than d. Therefore, $\tilde{\epsilon} \gg \epsilon$ in general. In addition to the extra degree terms ℓ, ℓ', this is what explains the weaker rates obtained from this technique compared to the PKM.

3.2.1 Construction of the Regularizing Polynomial

It remains to construct the polynomial $q \in Q(\mathbf{g})$ described above. As mentioned, this construction has a large impact on the quality of the resulting convergence guarantees. In fact, it is arguably the construction of q that sets apart the results of Nie and Schweighofer (2007), Baldi and Mourrain (2023) and Baldi et al. (2025), leading to stronger rates in the latter works. We give a sketch of the construction in Baldi and Mourrain (2023).

Recall that after setting $\tilde{f} = f - q$, we want that $\tilde{f}_{\min,\mathbf{Y}} \geq f_{\min,\mathbf{X}}$. Thus, we wish that $q \ll 0$ on $\mathbf{Y} \backslash \mathbf{X}$, and $q \approx 0$ on $\mathbf{X}$ (as $q \in Q(\mathbf{g})$, we know that $q \geq 0$ on $\mathbf{X}$). Without loss of generality, assume that the polynomials g_j defining $\mathbf{X}$ satisfy $-1 \leq g_j(\mathbf{x}) \leq 1$ for all $\mathbf{x} \in \mathbf{Y}$ and $j \in [m]$. The idea is to first construct a *univariate* sum-of-squares polynomial $h \in \Sigma[t]$, with the property that, for some small $\eta > 0$, we have $h(t) \approx 0$ for $t \in [\eta, 1]$ and $h(x) \approx 1$ for $t \in [-1, -\eta]$. This can be done via Chebyshev approximation, see Baldi and Mourrain (2023, Sect. 2.3). Then, we set $q(\mathbf{x}) = M \cdot \sum_{j=1}^{m} g_j(\mathbf{x}) h(g_j(\mathbf{x}))$, with $M > 0$. Note that $q \in Q(\mathbf{g})$ by definition. Note further that, for $\mathbf{x} \notin \mathbf{X}$ sufficiently far from the boundary $\mathrm{bd}(\mathbf{X})$, at least one $g_j(\mathbf{x}) < -\eta$, whence $g_i(\mathbf{x}) h(g_i(\mathbf{x})) \ll 0$, whereas for $\mathbf{x} \in \mathbf{X}$ sufficiently far from $\mathrm{bd}(\mathbf{X})$, all $g_j(\mathbf{X}) \in [\eta, 1]$, whence $q(\mathbf{x}) \approx 0$. It remains to analyze the situation for $\mathbf{x}$ close to $\mathrm{bd}(\mathbf{X})$, where (some of the) constraints $g_j(\mathbf{x})$ are close to 0. There, a careful comparison of the gradients ∇f of f and ∇g_j of the constraints g_j is required to finish the argument. This comparison involves the so-called Łojasiewicz-constant of $\mathbf{g}$, which appears in the exponent of the final convergence rate. See Baldi and Mourrain (2023) for details.

Recipe for analyzing the lower bounds (II): Algebro-geometric reduction
Convergence rates for Lasserre's hierarchy of lower bounds on general semi-algebraic sets may be derived from rates on special sets (such as the simplex or hypercube), through delicate algebraic and geometric arguments.

3.3 Extensions

We end this section with some recent extensions of the techniques described above.

3.3.1 Algebro-Geometric Reduction for the Hypercube

Baldi and Slot (2024) use a variation of the reduction technique described above to prove convergence rates in $O(1/r)$ for the Putinar-type lower bounds on $[-1, 1]^n$, based on the quadratic module (recall that Laurent and Slot (2023) only show rates for the stronger Schmüdgen-type bounds, based on the preordering). This extension shows that the algebro-geometric method can also be useful for analyses on specific, rather than general sets $\mathbf{X}$. The main difference w.r.t. the above is that the 'simple' set $\mathbf{Y} \supseteq [-1, 1]^n$ used in Baldi and Slot (2024) depends on f, r (while it was static before). In fact, somewhat remarkably, $\mathbf{Y} = \text{Se}(\mathbf{h}) = [1 - \eta, 1 + \eta]^n$ is itself just a scaled hypercube, with $\eta = \eta(f, r) > 0$. The use of a dynamic $\mathbf{Y}$ eliminates the need for the regularizing polynomial q, permitting to prove stronger rates than those in Baldi and Mourrain (2023) and Baldi et al. (2025) (which would give a rate in $O(1/\sqrt[10]{r})$ in this setting). On the other hand, proving an inclusion $\mathcal{T}(\mathbf{h}) \subseteq Q(\mathbf{g})$ with proper degree bounds is more complicated. See Baldi and Slot (2024) for details.

3.3.2 Sparse Polynomial Optimization

As we have seen in Sect. 1.4, the semidefinite programs used to model Lasserre's hierarchies in n variables at degree r involve matrices of size $N_r = \binom{n+r}{r}$. These SDPs are thus intractable already for moderately large values of n. To address this issue, variants of Lasserre's hierarchy of lower bounds that exploit *sparsity* of the underlying polynomial optimization problem have been proposed in the literature, see Magron and Wang (2023) for an overview. A recent work by Korda et al. (2024) extends both of the techniques discussed in Sect. 3 to this setting, yielding a performance analysis for the sparse bounds. Furthermore, the convergence rates achieved there are stronger than those of Sect. 3 for sufficiently sparse POPs (relative to the size of the resulting SDPs).

3.3.3 Generalized Moment Problems

Polynomial optimization is a special case of the *generalized problem of moments* (GPM), which asks to minimize a linear function over the cone of positive Borel measures on $\mathbb{R}^n$ under some linear conditions on the moments. Lasserre's approximation hierarchies naturally extend to the GPM, see Lasserre (2009) and de Klerk and Laurent (2019). Convergence rates for these extended hierarchies have been established, especially in the context of dynamical systems, optimal control and volume estimation, by Korda et al. (2017), Korda and Henrion (2018), Schlosser et al. (2024a) and Schlosser and Tacchi (2024b). The analyses in the latter two recent works rely directly on (a combination of) existing convergence results for polynomial optimization discussed above.

4 Discussion

Tightness of the Performance Analysis

As we have seen, the convergence rates for the hierarchy of *upper* bounds presented in Sect. 2 are essentially tight (up to log-factors). For the *lower* bounds, the situation is much less clear. The literature on 'worst-case' examples for the lower bounds is mostly *qualitative* in nature, see, e.g., Scheiderer (2005) and Powers and Scheiderer (2001). On the quantitative side, Baldi and Slot (2024) recently showed that the Putinar-type lower bounds on $\mathbf{X} = [-1, 1]^n$ converge at a rate no better than $\Omega(1/r^8)$. Note that this negative result is still rather far away from the best-known positive result in that setting, which is in $O(1/r)$. It is an interesting research direction to prove negative results on the convergence of the lower bounds that either match the positive results more closely; or apply to more general $\mathbf{X}$; or apply to the Schmüdgen-type bounds.

Exponential Convergence Under Local Optimality Conditions

A common feature of the convergence rates in Table 2 is that they are *sub*exponential, in particular no better than $O(1/r^c)$ for some constant $c > 0$ depending on the set $\mathbf{X}$ and the type of certificate. In light of the negative results mentioned above, these are likely 'optimal' (up to improving the exponent c). On the other hand, Bach and Rudi (2023) show an *exponential* convergence rate for the Schmüdgen-type bounds on $\mathbf{X} = [-1, 1]^n$, which holds under an additional assumption on the objective function f. Roughly speaking, f should have a (strictly) positive definite Hessian at its global minimizer $\mathbf{x}^* \in [-1, 1]^n$, see Bach and Rudi (2023) for details.[4] Under this assumption, $f - f_{\min}$ can be written as a sum of squares of *smooth* functions (not necessarily polynomials). In turn, these smooth functions can be approximated by polynomials via (truncated) expansion in the Fourier basis, leading to a sum-of-squares representation of $f - f_{\min}$. The exponential convergence rate for Lasserre's lower bounds then follows from the fact that the Fourier coefficients of a smooth function decay exponentially quickly. As the approach of Bach and Rudi relies primarily on Fourier analysis, it seem likely that it could be extended to other distinguished $\mathbf{X}$, such as $\mathbf{X} = \mathbb{S}^{n-1}$. More unclear is whether it could also be used to prove guarantees for general semialgebraic sets $\mathbf{X}$, which we believe is an interesting open question.

---

[4] Note that this assumption holds *generically*. Under a similar assumption, Nie (2014) showed earlier that the hierarchy of lower bounds has finite convergence for general $\mathbf{X} = \mathrm{Se}(\mathbf{g})$, i.e., $f - f_{\min}$ belongs to $\mathcal{Q}(\mathbf{g})_r$ for some $r \in \mathbb{N}$. However, his result gives no quantitative information on r.

Optimization Over Finite Semialgebraic Sets

In this survey, we have focused on the setting where $\mathbf{X} \subseteq \mathbb{R}^n$ is an infinite set. Sum-of-squares hierarchies have been extensively studied for optimization over finite semialgebraic $\mathbf{X}$ as well, particularly for (subsets of) the boolean hypercube: $\mathbf{X} \subseteq \{-1, 1\}^n$ (or $\mathbf{X} \subseteq \{0, 1\}^n$). A key difference in the finite setting is that the hierarchies always converge in a finite number of steps (under a minor condition on the description of $\mathbf{X}$), see Nie (2013). In fact, they converge in $\Theta(n)$ steps for a semialgebraic subset of the boolean cube, see Laurent (2023) and Fawzi et al. (2016). For this reason, asymptotic analysis as $r \to \infty$ does not make sense. Rather, one often fixes the level $r \in \mathbb{N}$ of the hierarchy, and lets the number of variables n tend to infinity. There is a large literature on this regime in the theoretical computer science community, see, e.g., Barak and Steurer (2014) and references therein. One could also consider a 'hybrid' regime, where $r = c \cdot n$ for some constant $c > 0$ and $n, r \to \infty$ simultaneously. In this regime, the polynomial kernel method (see Sect. 3.1) yields error guarantees for the upper and lower bounds on $\mathbf{X} = \{-1, 1\}^n$, see Slot and Laurent (2022a) for details.

References

Ahmadi, A.A., Olshevsky, A., Parrilo, P.A., Tsitsiklis, J.N.: NP-hardness of deciding convexity of quartic polynomials and related problems. Math. Program. **137**(1–2), 453–476 (2013)

Artin, E.: Über die Zerlegung definiter Funktionen in Quadrate. Abhandlungen aus dem Mathematischen Seminar der Universität Hamburg **5**, (1927)

Bach, F., Rudi, A.: Exponential convergence of sum-of-squares hierarchies for trigonometric polynomials. SIAM J. Optim. **33**(3), 2137–2159 (2023)

Baldi, L., Mourrain, B.: On moment approximation and the effective Putinar's Positivstellensatz. Math. Program. **200**, 71–103 (2023)

Baldi, L., Slot, L.: Degree bounds for Putinar's Positivstellensatz on the hypercube. SIAM J. Appl. Algebr. Geom. **8**(1), 1–25 (2024)

Baldi, L., Mourrain, B., Parusinski, A.: On Łojasiewicz inequalities and the effective Putinar's Positivstellensatz. J. Algebra **662**, 741–767 (2025)

Barak, B., Steurer, D.: Sum-of-squares proofs and the quest toward optimal algorithms. In: Proceedings of the International Congress of Mathematicians, Seoul, vol. 4, pp. 509–533 (2014)

Blekherman, G., Parrilo, P.A., Thomas, R.R.: Semidefinite Optimization and Convex Algebraic Geometry. MOS-SIAM Series on Optimization, vol. 13 (2012)

Blomenhofer, A.T., Laurent, M., Slot, L.: In preparation

de Klerk, E., Laurent, M.: A survey of semidefinite programming approaches to the generalized problem of moments and their error analysis. In: Araujo, C., Benkart, G., Praeger, C., Tanbay, B. (eds.) World Women in Mathematics 2018. Association for Women in Mathematics Series, vol. 20. Springer, Cham (2019)

de Klerk, E.: Aspects of Semidefinite Programming–Interior Point Algorithms and Selected Applications. Kluwer (2002)

de Klerk, E., Laurent, M.: Comparison of Lasserre's measure-based bounds for polynomial optimization to bounds obtained by simulated annealing. Math. Oper. Res. **43**, 1317–1325 (2018)

de Klerk, E., Laurent, M.: Convergence analysis of a Lasserre hierarchy of upper bounds for polynomial minimization on the sphere. Math. Program. **193**, 665–685 (2022) de Klerk, E., Pasechnik, D.: Approximation of the stability number of a graph via copositive programming. SIAM J. Optim. **12**, 875–892 (2002)

Dimitrov, D.K., Nikolov, G.P.: Sharp bounds for the extreme zeros of classical orthogonal polynomials. J. Approx. Theory **162**, 1793–1804 (2010)

Doherty, C., Wehner, S.: Convergence of SDP hierarchies for polynomial optimization on the hypersphere (2013. arXiv:1210.5048

Driver, K., Jordaan, K.: Bounds for extreme zeros of some classical orthogonal polynomials. J. Approx. Theory **164**, 1200–1204 (2012)

Dunkl, C.F., Xu, Y.: Orthogonal Polynomials of Several Variables. Encyclopedia of Mathematics and its Applications. Cambridge University Press (2014)

Fang, K., Fawzi, H.: The sum-of-squares hierarchy on the sphere, and applications in quantum information theory. Math. Program. Ser. A **190**, 331–360 (2021)

Fawzi, H., Saunderson, J., Parrilo, P.A.: Sparse sums of squares on Abelian groups and improved semidefinite liftings. Math. Program. **160**, 149–191 (2016)

Garey, M.R., Johnson, D.S.: Computers and Intractability. A Guide to the Theory of NP-Completeness. Freeman, San Francisco (1979)

Henrion, D., Korda, M., Lasserre, J.B.: The Moment-SOS Hierarchy—Lectures in Probability, Statistics. Computational Geometry, Control and Nonlinear PDEs. World Scientific (2020)

Hilbert, D.: Über die Darstellung definiter Formen als Summe von Formenquadraten. Math. Ann. **32**, 342–350 (1888)

Klerk, E., Laurent, M.: Worst-case examples for Lasserre's measure-based hierarchy for polynomial optimization on the hypercube. Math. Oper. Res. **45**(1), 86–98 (2020)

Klerk, E., Laurent, M., Parrilo, P.: A PTAS for the minimization of polynomials of fixed degree over the simplex. Theor. Comput. Sci. **361**(2–3), 210–225 (2006)

Klerk, E., Laurent, M., Sun, Z.: Convergence analysis for Lasserre's measure-based hierarchy of upper bounds for polynomial optimization. Math. Program. Ser. A **162**(1), 363–392 (2017)

Korda, M., Magron, V., Rios-Zertuche, R.: Convergence rates for sums-of-squares hierarchies with correlative sparsity. Math. Program. (2024). https://doi.org/10.1007/s10107-024-02071-6

Korda, M., Henrion, D.: Convergence rates of moment-sum-of-squares hierarchies for volume approximation of semialgebraic sets. Optim. Lett. **12**(3), 435–442 (2018)

Korda, M., Henrion, D., Jones, C.N.: Convergence rates of moment-sum-of squares hierarchies for optimal control problems. Syst. & Control Lett. **100**, 1–5 (2017)

Krivine, J.-L.: Anneaux préordonnés. Journal d' Analyse Mathématique **12**, 307–326 (1964)

Kroó, A.: Multivariate needle polynomials with application to norming sets and cubature formulas. Acta Math. Hung. **147**(1), 46–72 (2015)

Lasserre, J.B.: Moments, Positive Polynomials and their Applications. World Scientific, Singapore (2009)

Lasserre, J.B.: Global optimization with polynomials and the problem of moments. SIAM J. Optim. **11**(3), 796–817 (2001)

Lasserre, J.B.: A new look at nonnegativity on closed sets and polynomial optimization. SIAM J. Optim. **21**(3), 864–885 (2010)

Lasserre, J.B.: Introduction to Polynomial and Semi-Algebraic Optimization. Cambridge University Press, Cambridge, UK (2015)

Lasserre, J.B.: Connecting optimization with spectral analysis of tri-diagonal Hankel matrices. Math. Program. **190**, 795–809 (2021)

Laurent, M.: Sums of squares, moment matrices and optimization over polynomials. In: Putinar, M., Sullivant, S. (eds.) Emerging Applications of Algebraic Geometry. IMA Volumes in Mathematics and its Applications, vol. 149, pp. 157–270. Springer (2009)

Laurent, M.: A comparison of the Sherali-Adams, Lovász-Schrijver and Lasserre relaxations for 0–1 programming. Math. Oper. Res. **28**(3), 470–496 (2023)

Laurent, M., Slot, L.: An effective version of Schmüdgen's Positivstellensatz for the hypercube. Optim. Lett. **17**, 515–530 (2023)

Magron, V., Wang, J.: Sparse Polynomial Optimization. Series on Optimization and its Applications. World Scientific (Europe) (2023)

Marshall, M.: Positive Polynomials and Sums of Squares. Mathematical Surveys and Monographs, vol. 146. American Mathematical Society, Providence, Rhode Island (2008)

Martinez, A., Piazzon, F., Sommariva, A., Vianello, M.: Quadrature-based polynomial optimization. Optim. Lett. **14**(5), 1027–1036 (2019)

Nesterov, Y.: Squared functional systems and optimization problems. In: Frenk, J.B.G., Roos, C., Terlaky, T., Zhang, S. (eds.) High Performance Optimization. Applied Optimization, vol. 33, pp. 405–440. Springer, Boston, Massachussets (2000)

Nie, J.: Moment and Polynomial Optimization. MOS-SIAM Series on Optimization (2023)

Nie, J.: Polynomial optimization with real varieties. SIAM J. Optim. **23**(3), 1634–1646 (2013)

Nie, J.: Optimality conditions and finite convergence of Lasserre's hierarchy. Math. Program. **146**(1), 97–121 (2014)

Nie, J., Schweighofer, M.: On the complexity of Putinar's Positivstellensatz. J. Complex. **23**(1), 135–150 (2007)

Parrilo, P.A.: Semidefinite programming relaxations for semialgebraic problems. Math. Program. Ser. B **96**, 293–320 (2003)

Parrilo, P.A.: Structured semidefinite programs and semialgebraic geometry methods in robustness and optimization. Ph.D. thesis, California Institute of Technology (2000)

Pólya, G.: Uber positive Darstellung von Polynomen, Vierteljahresschrift der Naturforschenden Gesellschaft in Zürich **73**, 14–145 (1928). Reprinted. In: Collected Papers, vol. 2, pp. 309–313. MIT Press, Cambridge (1974)

Powers, V.: Certificates of Positivity for Real Polynomials: Theory, Practice, and Applications. Developments in Mathematics, vol. 69 (2021)

Powers, V., Scheiderer, C.: The moment problem for non-compact semialgebraic sets. Adv. Geom. **1**(1), 71–88 (2001)

Prestel, A., Delzell, C.N.: Positive Polynomials—From Hilbert's 17th Problem to Real Algebra. Springer Monographs in Mathematics. Springer, Berlin (2001)

Putinar, M.: Positive polynomials on compact semi-algebraic sets. Indiana Univ. Math. J. **42**, 969–984 (1993)

Reznick, B.: Some concrete aspects of Hilbert's 17th problem. In: Real Algebraic Geometry and Ordered Structures (Baton Rouge, Louisiana, 1996; Delzell, C.N., Madden, J.J. (eds.)) Contemporary Mathematics, vol. 253, pp. 251–272. American Mathematical Society, Providence, Rhode Island (2000)

Reznick, B.: Uniform denominators in Hilbert's seventeenth problem. Math. Z. **220**, 75–97 (1995)

Roos, C., Terlay, T., Vial, J.-P.: Interior Point Methods for Linear Optimization. Springer (2005)

Scheiderer, C.: Non-existence of degree bounds for weighted sums of squares representations. J. Complex. **21**(6), 823–844 (2005)

Schlosser, C., Tacchi, M., Lazarev, A.: Convergence rates for the moment-SoS hierarchy (2024). arXiv:2402.00436

Schlosser, C., Tacchi, M.: Specialized effective Positivstellensätze for improved convergence rates of the moment-SOS hierarchy (2024). arXiv:2403.05934

Schmüdgen, K.: The K-moment problem for compact semi-algebraic sets. Math. Ann. **289**(2), 203–206 (1991)

Schweighofer, M.: On the complexity of Schmüdgen's Positivstellensatz. J. Complex. **20**(4), 529–543 (2004)

Shor, N.Z.: An approach to obtaining global extremums in polynomial mathematical programming problems. Kibernetika **5**, 102–106 (1987)

Slot, L., Laurent, M.: Sum-of-squares hierarchies for binary polynomial optimization. Math. Program. **197**, 621–660 (2022a); Singh, M., Williamson, D.P. (eds.) Integer Programming and Combinatorial Optimization (IPCO 2021). Lecture Notes in Computer Science, vol. 12707, pp. 43–57. Springer, Cham

Slot, L.: Asymptotic analysis of semidefinite bounds for polynomial optimization and independent sets in geometric hypergraphs. Ph.D. thesis, Tilburg University (2022b)
Slot, L.: Sum-of-squares hierarchies for polynomial optimization and the Christoffel-Darboux kernel. SIAM J. Optim. **32**(4), 2612–2635 (2022c)
Slot, L., Laurent, M.: Near-optimal analysis of Lasserre's univariate measure-based bounds for multivariate polynomial optimization. Math. Program. **188**, 443–460 (2021)
Slot, L., Laurent, M.: Improved convergence analysis of Lasserre's measure-based upper bounds for polynomial minimization on compact sets. Math. Program. **193**, 831–871 (2022d)
Stengle, G.: Complexity estimates for the Schmüdgen Positivstellensatz. J. Complex. **12**, 167–174 (1996)
Szegö, G.: Orthogonal Polynomials. American Mathematical Society Colloquium Publications, vol. 23. American Mathematical Society (1959)
Zohrizadeh, F., Josz, C., Jin, M., Madani, R., Lavaei, J., Sojoudi, S.: A survey on conic relaxations of optimal power flow problem. Eur. J. Oper. Res. **287**, 391–409 (2020)

Open Access This chapter is licensed under the terms of the Creative Commons Attribution 4.0 International License (http://creativecommons.org/licenses/by/4.0/), which permits use, sharing, adaptation, distribution and reproduction in any medium or format, as long as you give appropriate credit to the original author(s) and the source, provide a link to the Creative Commons license and indicate if changes were made.

The images or other third party material in this chapter are included in the chapter's Creative Commons license, unless indicated otherwise in a credit line to the material. If material is not included in the chapter's Creative Commons license and your intended use is not permitted by statutory regulation or exceeds the permitted use, you will need to obtain permission directly from the copyright holder.

Aggregation-Diffusion Equations for Collective Behaviour in the Sciences

Rafael Bailo, José A. Carrillo, and David Gómez-Castro

Abstract This is a survey article based on the content of the plenary lecture given by José A. Carrillo at the ICIAM23 conference in Tokyo. It is devoted to produce a snapshot of the state of the art in the analysis, numerical analysis, simulation, and applications of the vast area of aggregation-diffusion equations. We also discuss the implications in mathematical biology explaining cell sorting in tissue growth as an example of this modelling framework. This modelling strategy is quite successful in other timely applications such as global optimisation, parameter estimation and machine learning.

1 Introduction

A surprisingly large number of complex phenomena can be modelled as systems of point particles that interact under short and long-range forces. This modelling framework, which is classical in the study of many-body systems (e.g. gravitational collapse, electron transport in semi-conductors, or kinetic equations in statistical physics), has recently found applications in other areas of physics, as well as in biology, social sciences, machine learning, and even optimisation, see for instance (Carrillo et al. 2010, 2019g, 2022c, 2018a; Bailo et al. 2024), and is a prolific current field of research. The applications are diverse, and span both the microscopic and macroscopic scales: ion channel transport, chemotaxis, bacterial orientation, cellular adhesion, angiogenesis, animal herding, or human crowd motion are but a

R. Bailo
Department of Mathematics and Computer Science, Eindhoven University of Technology, Eindhoven, The Netherlands
e-mail: r.bailo@tue.nl

R. Bailo · J. A. Carrillo (✉)
Mathematical Institute, University of Oxford, Oxford, UK
e-mail: carrillo@maths.ox.ac.uk

D. Gómez-Castro
Departamento de Matemáticas, Universidad Autónoma de Madrid, Madrid, Spain
e-mail: david.gomezcastro@uam.es

© The Author(s) 2026

S. Oishi et al. (eds.), *Recent Developments in Industrial and Applied Mathematics*, ICIAM2023 Springer Series 1, https://doi.org/10.1007/978-981-95-1446-5_9

few examples. The interaction effects between the "individuals" in each example can be approximated by long-range attractive forces (e.g. ligand binding, electrical interactions, or social preferences) and short-range repulsion effects (due to volume constraints or crowding).

There are several agent-based (microscopic) modelling approaches to describe these phenomena, such as cellular automata or driven Brownian particles, all capable of encoding complex behaviour. However, this behaviour is best understood and rigorously analysed by studying the mean-field limit (or its variants), which represents the limit of the microscopic model as the number of agents becomes large (see Oelschläger 1990a; Bolley et al. 2011; Jabin 2014). As an example, suppose that we study two types of cells, each able to express different surface proteins and ligands (such us cadherins or nectins). Assume, for cells of the first kind, that nuclei are located at positions $\{x_i\}$, $i = 1, \ldots, N$, and, for the second, at $\{y_j\}$, $j = 1, \ldots, M$; let also $N = M$ for simplicity. We consider cells that interact either by attraction at medium distances (e.g. through filopodia, protrusions of the cellular membrane), or by strong repulsion at short distances (due to volume and size constraints); a diagram is given in Fig. 1. If the force exerted by cells of type b onto type a is radial and conservative (with potential W_{ab}^N, where $a, b = 1, 2$), we may pose the elementary agent-based model

$$\dot{x}_i = -\frac{m_1}{N}\sum_{j\neq i}\nabla W_{11}^N(x_i - x_j) - \frac{m_2}{N}\sum_{j}\nabla W_{12}^N(x_i - y_j),$$
$$\dot{y}_i = -\frac{m_2}{N}\sum_{j\neq i}\nabla W_{22}^N(y_i - y_j) - \frac{m_1}{N}\sum_{j}\nabla W_{21}^N(y_i - x_j),$$

where m_1 and m_2 are the typical cellular masses of each cell kind.

Under suitable assumptions, the empirical measures (weighted sums of Dirac deltas supported at each particle's position) associated to the model approximate the macroscopic normalised cell densities in the many-particle limit:

$$\rho_1(t,x) \simeq \frac{m_1}{N}\sum_{i=1}^{N}\delta_{x_i(t)}(x) \quad \text{and} \quad \rho_2(t,x) \simeq \frac{m_2}{N}\sum_{i=1}^{N}\delta_{y_i(t)}(x) \quad \text{as } N \to \infty.$$

When considering the biological context, it is reasonable to suppose that the attraction forces act only within a cut-off radius R, and that the repulsion is purely localised (Calvez and Carrillo 2012); a natural scaling choice for the potentials is therefore $W_{ab}^N \simeq \varepsilon\delta_0 + W_{ab}$ as $N \to \infty$ (Oelschläger 1990a; Carrillo et al. 2019g), where ε relates to the typical volume occupied by each cell (assumed here equal across species and interactions). In this scaling, the limit $N \to \infty$ leads to the mean-field model

$$\begin{cases} \partial_t\rho_1 = \nabla\cdot\Big(\rho_1\nabla\big(\varepsilon(\rho_1+\rho_2) + W_{11}*\rho_1 + W_{12}*\rho_2\big)\Big), \\ \partial_t\rho_2 = \nabla\cdot\Big(\rho_2\nabla\big(\varepsilon(\rho_1+\rho_2) + W_{22}*\rho_2 + W_{21}*\rho_1\big)\Big). \end{cases} \tag{1}$$

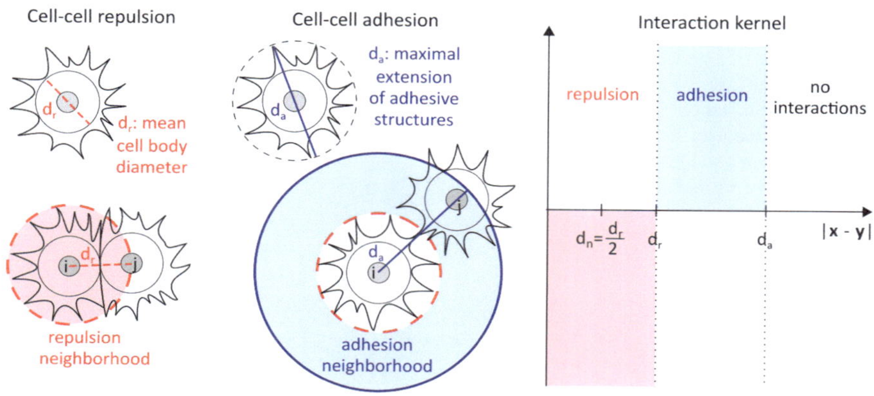

Fig. 1 Interaction forces of an agent-based model for cell adhesion, see Carrillo et al. (2018b), Carrillo et al. (2019g)

Each of the equations in the system above is a particular case of the *aggregation-diffusion equation*

$$\frac{\partial \rho}{\partial t} + \nabla \cdot (\rho u) = 0, \quad u = -\nabla \xi, \quad \xi = U'(\rho) + V + W * \rho, \tag{2}$$

a generic continuum equation for the kinematic evolution of a population density $\rho(t, x)$. The velocity field u drives the evolution of the density through a balance of forces: diffusion, resulting from localised repulsion effects or noise, which is modelled by the convex function $U(\rho)$; drift, due to external forces, described by the potential $V(x)$, and a symmetric interaction force (aggregation, repulsion, or both), given by the potential $W(x)$, with $W(x) = W(-x)$, see Bodnar and Velazquez (2006), Topaz et al. (2006). The diffusion is commonly given by $U(s) = s \log s$ (linear diffusion) or $U(s) = s^m/(m-1)$ for $m > 0$ (non-linear diffusion (Vázquez 2007)). In these cases, the diffusion coefficient is $m\rho^{m-1}$: in the *porous medium* range $m > 1$, diffusion is stronger where the concentration is large; in the *fast diffusion* regime, $m < 1$, where the concentration is small. So-called confining potentials, $V = |x|^p$ for $p > 0$, often appear in Fokker-Planck or McKean-Vlasov equations (Carrillo et al. 2003, 2020b, 2022b).

2 Gradient Flows

The aggregation-diffusion equation (2) can be understood as the gradient flow of the free-energy functional

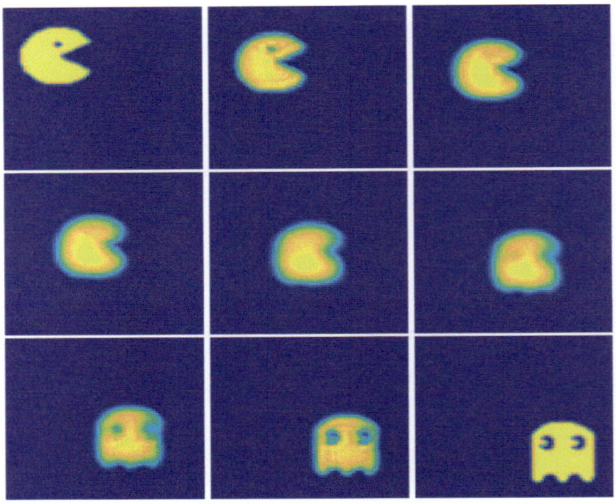

Fig. 2 Numerical approximation of a geodesic curve using the 2-Wasserstein distance between the characteristic sets of Pac-Man and the Ghost (suitably normalised). Video available online

$$\mathcal{F}[\rho] = \int_{\mathbb{R}^d} U(\rho(x))\mathrm{d}x + \int_{\mathbb{R}^d} V(x)\rho(x)\mathrm{d}x + \frac{1}{2}\int_{\mathbb{R}^d}\int_{\mathbb{R}^d} \rho(x)W(x-y)\rho(y)\mathrm{d}y\mathrm{d}x, \tag{3}$$

in a suitable metric space: the 2-Wasserstein space of probability densities over $\mathbb{R}^d$, see Jordan et al. (1998), Villani (2003), Ambrosio et al. (2008). This space arises by equipping the set of probability densities with finite second moment, denoted $\mathcal{P}_2(\mathbb{R}^d)$, with the 2-Wassertein distance, denoted d_2. Said distance, which originated in the theory of optimal transport, has found fruitful applications in the study of mean-field limits because it induces a topology where the evolution of empirical measures in time is continuous, unlike in Sobolev spaces. As a result, d_2 is helpful when quantifying the convergence of particle systems to their mean-field limit. This notion of distance for probabilities may also be used to define transport-based interpolations between densities as geodesic curves in the 2-Wasserstein space; Fig. 2 shows the geodesic interpolation between the normalised characteristics of two sets.

Due to the lack of linear structure in the 2-Wassertein space, the precise construction of solutions to the aggregation-diffusion equation (2) as gradient flows of the free energy functional (3) is quite involved. Roughly, a discrete-in-time sequence of measures is generated from the datum $\rho_0 \in \mathcal{P}_2(\mathbb{R}^d)$ via the so-called *JKO iteration*

$$\rho_{k+1} = \underset{\rho\in\mathcal{P}_2(\mathbb{R}^d)}{\operatorname{argmin}} \left\{ \frac{1}{2\Delta t} d_2^2(\rho, \rho_k) + \mathcal{F}[\rho] \right\},$$

for $k \in \mathbb{N}$, and for any fixed $\Delta t > 0$ (Jordan et al. 1998). A suitable time interpolation leads to a curve of measures $\rho_{\Delta t}$ that converge to the unique solution of (2) as $\Delta t \to 0$, see Jordan et al. (1998), Ambrosio et al. (2008). Solutions constructed by this variational scheme satisfy the fundamental *energy dissipation identity*:

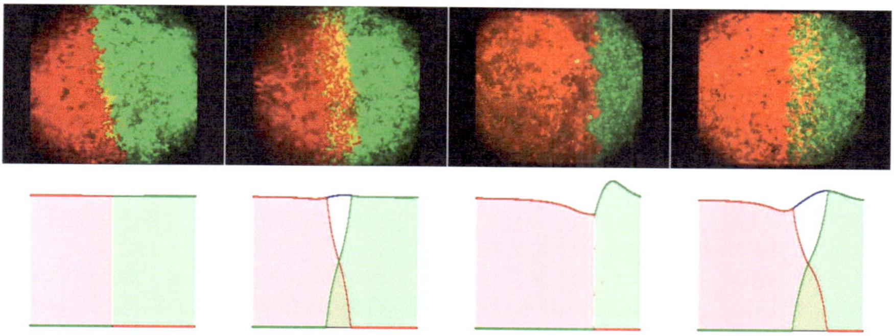

Fig. 3 In vitro experiments from Katsunuma et al. (2016) compared to in silico experiments from Carrillo et al. (2019g). The numerical schemes are based on finite volumes, as in Bailo et al. (2020a, 2023a). Video available online

$$\frac{\mathrm{d}}{\mathrm{d}t}\mathcal{F}[\rho] = \int_{\mathbb{R}^d} \xi\, \partial_t \rho \mathrm{d}x = \int_{\mathbb{R}^d} \xi \nabla \cdot (\rho \nabla \xi)\, \mathrm{d}x = -\int_{\mathbb{R}^d} \rho |\nabla \xi|^2 \mathrm{d}x \leq 0\,, \tag{4}$$

under certain assumptions on the potentials and the internal energy.

If the functional $\mathcal{F}[\rho]$ has a unique global minimiser in $\mathcal{P}_2(\mathbb{R}^d)$, then the solution to (2) converges in time to the minimiser whenever the functional satisfies a suitable notion of convexity (uniform convexity along the geodesics induced by d_2). However, this may not be the case; the purely diffusive setting ($V = W = 0$) leads to asymptotic spreading of $\rho(t)$ in self-similar profiles for large times (Vázquez 2007). In general, the asymptotic balance between aggregation and diffusion is quite subtle, and significant effort has been devoted to its study. Under certain conditions, the existence of radially decreasing, compactly supported, global minimisers of $\mathcal{F}[\rho]$ (unique up to translations) has been established (Cañizo et al. 2015; Carrillo et al. 2019f). In some extreme cases, the minimisation of $\mathcal{F}$ leads to the formation of Dirac deltas (Carrillo et al. 2022a) that accumulate only a fraction of the total mass. When it comes to the dynamical problem, understanding these features is a considerable challenge, see the recent surveys (Carrillo et al. 2019c; Gómez-Castro 2024). Almost-sharp conditions on W have been given in Carrillo et al. (2023) such that $\rho(t)$ behaves like the solution of the purely diffusive problem. Yet, in some settings (Carrillo et al. 2022b), the solution can develop asymptotically the Dirac delta suggested by the minimisation of the free energy.

3 Applications in Mathematical Biology

Mathematical biology has a special place among the applications of aggregation-diffusion systems. Cell population models are particularly relevant, and often use the aggregation-diffusion equation, either in scalar or in system form, to describe the motion of cells as they concentrate or separate from a target. The diffusion terms described above are consistent with population pressure effects, whereby groups of cells naturally spread away from areas of high concentration. The aggregation term has been classically used as a model for chemotaxis, as cells often direct their movement along the increasing gradient of a chemical agent produced either by external sources or by the very cells. The by-now classical Keller-Segel model for chemotaxis (Keller and Segel 1970) is an archetypical example of the aggregation-diffusion equation in mathematical biology.

A new paradigm in the modelling of cell adhesion suggests that ligand binding through filopodia can be captured by short range interaction potentials, as suggested in Fig. 1. An example of such model was developed in Carrillo et al. (2019g): it considers two species of cells, each attracted to itself and to the other in the form described by (1). Despite its apparent simplicity, the model can describe several realistic behaviours. If the all the attraction strengths are equal, all cells simply mix evenly. However, if the relative strengths vary, intricate phenomena arise; for instance, one species of cells might completely envelop the other, forming a "halo". Figure 3 shows the astonishing match between the solutions of the model and the experimental results of Katsunuma et al. (2016). This model demonstrates that the differential adhesion hypothesis introduced by Steinberg (1962) is the driving mechanism for cell sorting (Falcó et al. 2025), modelled here through the asymmetry on the interaction potentials between the two type of cells. Aggregation-diffusion systems (1) with such asymmetry do not typically have a gradient-flow structure, and can develop a wealth of complex phenomena: among others, travelling wave solutions, free boundaries, and mixing or segregation between species; each and every one of them a challenging open problem in the analysis of Partial Differential Equations.

Beyond biology, the applications of these ideas in global optimisation and machine learning are not only timely, but potentially quite impactful. Many researchers are exploring them in order to perform gradient-free optimisation (Pinnau et al. 2017; Carrillo et al. 2022c, 2021b, 2024b; Bailo et al. 2024), gradient-free sampling (Garbuno-Inigo et al. 2020; Korba et al. 2020; Ding and Li 2021; Carrillo and Vaes 2021; Carrillo et al. 2022c), Bayesian parameter estimation (Bongini et al. 2017; Lu et al. 2019, 2022; Gomes et al. 2019; Lang and Lu 2022; Carrillo et al. 2024a), or neural network calibration (Mokrov et al. 2021; Chizat et al. 2022; Fernández-Real and Figalli 2022; Lee et al. 2023).

4 Analytical Results

There is a rich theory of the aggregation equation (2), yet its analysis remains a very active field, producing an array of results each year. This section will provide an overview, referring the reader to the recent surveys (Carrillo et al. 2019c; Gómez-Castro 2024) for detailed explanations and additional references.

4.1 Local Minimisers of the Free Energy

The Euler-Lagrange equation for the minimisation of the free energy (3) in $\mathcal{P}_2(\mathbb{R}^d)$ is

$$\begin{cases} U'(\rho) + V + W * \rho = C & \text{if } \rho > 0, \\ U'(\rho) + V + W * \rho \geq C & \text{in } \mathbb{R}^d. \end{cases}$$

The constant C can sometimes be expressed in terms of the energy of candidates for the minimiser of the free energy. Global minimisers may have a multiply connected support, yet C must be the same on each, which is not the case for local minimisers. If the map $U' : (0, \infty) \to (0, \infty)$ is invertible, the conditions can be equivalently expressed as the fixed point problem

$$\widehat{\rho} = \max\{0, (U')^{-1}(C - V(x) - W * \widehat{\rho})\} \quad \text{in } \mathbb{R}^d. \tag{5}$$

We refer to Balagué et al. (2013b), Carrillo et al. (2019e) for a proof of these conditions, as well as for the Euler-Lagrange conditions with respect to other distances between probability measures. Global minimisers in the purely aggregation equation, $U = 0$, have also been studied a lot in the last years, see Bertozzi et al. (2012), Carrillo et al. (2016, 2017, 2020c), Cañizo et al. (2015), Carrillo and Huang (2017), Frank (2022), Davies et al. (2022, 2023), Carrillo and Shu (2023, 2024) for instance. This is a subtopic of the theory of aggregation-diffusion equations on its own with many interesting open problems such as the dimension of the support of global minimisers in certain ranges (Balagué et al. 2013a, b), see the survey (Carrillo and Vázquez 2015) and the references therein.

4.2 Steady States

Steady states are solutions (in a suitable sense) to the time-independent problem

$$-\nabla \cdot (\rho \nabla (U'(\rho) + V + W * \rho)) = 0.$$

Under mild assumptions, steady states are also characterised by the densities where the free-energy dissipation (4) vanishes; that is,

$$\int_{\mathbb{R}^d} \rho |\nabla (U'(\rho) + V + W * \rho)|^2 \mathrm{d}x = 0.$$

Partitioning the support of the density, $\{x : \rho > 0\}$, as a union of disjoint open sets A_i, the condition is equivalent to finding a density ρ and constants C_i such that

$$U'(\rho) + V + W * \rho = C_i \quad \text{in } A_i.$$

These steady states need not be minimisers of the energy.

In particular cases (degenerate diffusion, $W = 0$, and certain choices of non-convex V, such as a double-well potential), steady states can be created by "glueing" disjoint, compactly supported solutions, leading to

$$\widehat{\rho} = (U')^{-1}(C_i - V(x)), \quad \text{in } A_i.$$

These steady states sometimes have a non-trivial basin of attraction.

4.3 Radial Symmetry Versus Symmetry Breaking

When V and W are both radially symmetric, we may expect the stationary states of (2) to be radially symmetric as well, yet this is not true in general. For instance, if W is attractive, smooth, and compactly supported, V is zero, and the diffusion is degenerate, there is a single "bump" stationary state which is radially symmetric, but multiple copies of the state can be present in a non-interacting way (sufficiently far from each other) to form non-trivial states. In one dimension, the "gluing" example of the previous section is valid if V is a symmetric double-well potential. In general, *symmetry breaking* is a well-documented phenomenon in PDEs (e.g. Dolbeault et al. 2019), and there are many examples that are derived from particle systems (see for instance (Balagué et al. 2013b; Bertozzi et al. 2015)). We highlight (Delgadino et al. 2022), where examples of W with infinitely many radially decreasing steady states are constructed.

Even in the case of globally supported interaction potentials, the radial symmetry of steady states is generally difficult to prove. The global minimisers of the free energy can be proven to be radially symmetric when V and W are radially increasing, through the method of Schwarz rearrangement (Talenti 2016). This method replaces ρ by a radially symmetric non-increasing function $\rho^\star$ for minimizing sequences of the free energy in (3) such that the measure of the level sets is preserved:

$$|\{x : \rho^\star \geq h\}| = |\{x : \rho \geq h\}|, \quad \forall h \in \mathbb{R}.$$

Proving radial symmetry for the steady states of the equation is much more difficult than for the global minimisers of the energy. The idea there is to use continuous Steiner rearrangement (see Brock (2000)), where one continuously "slides" each level set. This was used in Carrillo et al. (2019f) to prove radial symmetry of the steady states for V and W radially increasing, and subsequently generalised for potentials which are more singular at the origin (Carrillo et al. 2018c).

Establishing whether the radially symmetric and non-increasing steady states are unique (up to translations, if $V = 0$) would be useful in order to understand the long-time asymptotics of these models; note, further, that uniqueness of the steady states would also imply uniqueness of the global minimisers for the energy. However, the question of sharp conditions for uniqueness is still open, though important advances have been made under various assumptions on the interaction potentials, see Lieb (1983), Chan et al. (2020), Calvez et al. (2021), Delgadino et al. (2022). The uniqueness of absolutely continuous stationary states is very important in certain applications, such as neural networks models based on aggregation-diffusion equations, where the uniqueness would provide guarantees on the performance of the network (Rotskoff and Vanden-Eijnden 2022; Bach and Chizat 2023).

4.4 Steady States Without Aggregation

When $W = 0$, the condition (5) gives an explicit candidate for the global minimiser, provided a value of C consistent with the mass of the steady state can be found (this may not always be possible). Uniqueness follows from a monotonicity argument.

In the case of fast diffusion, steady states might involve Dirac deltas, and their characterisation may require the lower-semi-continuous extension of the free energy,

$$\widetilde{\mathcal{F}}(\mu) = \inf_{\rho_n \rightharpoonup \mu} \liminf_n \mathcal{F}(\rho_n),$$

where ρ_n are densities. Under a radially symmetric potential V, the global minimiser in $\mathcal{P}(\mathbb{R}^d)$ of $\widetilde{\mathcal{F}}$ sometimes takes the form

$$\widehat{\mu} = (1 - \mathfrak{m})\delta_0 + \widehat{\rho}, \quad \text{where } \mathfrak{m} = \int_{\mathbb{R}^d} \widehat{\rho} \mathrm{d}x, \tag{6}$$

where $\widehat{\rho}$ is non-negative and $\mathfrak{m} \leq 1$ is a density. This is indeed the global attractor in some cases (Carrillo et al. 2022b), where the density spontaneously forms a Dirac delta at the origin, an effect known as *concentration*.

4.5 A Clearer Picture for the Homogeneous Setting

A particularly interesting setting is the so-called *homogeneous setting*:

$$U(\rho) = \frac{\rho^m}{m-1}, \quad W(x) = \chi \frac{|x|^k}{k}, \tag{7}$$

with $V = 0$ for simplicity, and where we formally denote $|x|^0/0 := \log|x|$. The parameter $\chi > 0$ represents the relative strength of attraction over diffusion in the system. A direct scaling analysis (for $k \neq 0$) leads to

$$\mathcal{F}(\rho_\lambda) = \lambda^{d(m-1)} \frac{1}{m-1} \int_{\mathbb{R}^d} \rho_1^m(x) \, \mathrm{d}x + \lambda^{-k} \frac{\chi}{2k} \int_{\mathbb{R}^d} \int_{\mathbb{R}^d} |x-y|^k \rho_1(x) \rho_1(y) \, \mathrm{d}x \, \mathrm{d}y,$$

where $\rho_\lambda(x) := \lambda^d \rho_1(\lambda x)$, for all $\lambda > 0$, and where ρ_1 is a fixed density. This reveals the critical exponent $m_c = \frac{d-k}{d}$ (the *fair competition range*), where the free energy is homogeneous. The range $m > m_c$ is known as the *diffusion-dominated regime* (where spreading is more energetically favourable than concentration) and $m \in (0, m_c)$ is the *aggregation-dominated regime*, (where the opposite is true). The homogeneous setting has been the subject of significant work, and is relatively well-understood. We summarise the results:

Theorem 9.1 (Homogeneous Interaction Potential (Carrillo et al. 2019e, d; Calvez et al. 2017))

1. (Degenerate diffusion – Diffusion dominated) *If $m > 1$ and $k \in ((1-m)d, 0)$, then the energy*

$$\widetilde{\mathcal{F}}(\rho) = \begin{cases} \mathcal{F}(\rho) & \text{if } \rho \in L^m(\mathbb{R}^d) \\ +\infty & \text{if } \rho \in \mathcal{P}(\mathbb{R}^d) \setminus L^m(\mathbb{R}^d) \end{cases}$$

has a global minimiser for all values of the parameter $\chi > 0$.

2. (Degenerate diffusion—Fair competition) *If $m > 1$ and $k = (1-m)d$, then there is a dichotomy: the energy $\widetilde{\mathcal{F}}(\rho)$ has global minimisers if and only if χ coincides with a critical parameter χ_c. If $0 < \chi < \chi_c$, the free energy functional has zero infimum, but it is not achieved. If $\chi > \chi_c$, the free energy functional is not bounded below.*
3. (Fast diffusion—Aggregation dominated) *If $m \le \frac{d}{d+k}$ and $k > 0$, then the free energy is not bounded below in $\mathcal{P}(\mathbb{R}^d) \cap L^\infty(\mathbb{R}^d)$.*
4. (Fast diffusion—Diffusion dominated) *If $\frac{d}{d+k} < m < 1$ and $k > 0$, then the free energy is bounded below in $\mathcal{P}(\mathbb{R}^d) \cap L^\infty(\mathbb{R}^d)$ and there are radially decreasing global minimisers of a suitably extended free energy functional (allowing for a possible partial mass concentration at the origin).*

The proofs of these results hinge drastically on the sharp constants of related functional inequalities.

4.6 Degenerate Diffusion

The case of degenerate diffusions is handled with the classical Hardy-Littlewood-Sobolev (HLS) inequality; to be precise, with the following variant obtained by interpolation. Let $k \in (-d, 0)$, and $m \in (1, 2)$ such that $d(m-1) + k = 0$. Then there exists a positive and finite optimal constant C_* such that, for $f \in L^1(\mathbb{R}^d) \cap L^m(\mathbb{R}^d)$, we have

$$\left| \iint_{\mathbb{R}^d \times \mathbb{R}^d} f(x) |x-y|^k f(y) \mathrm{d}x \mathrm{d}y \right| \le C_* ||f||_1^{\frac{d+k}{d}} ||f||_m^m. \tag{8}$$

This inequality permits the comparison of competing terms in the free energy $\mathcal{F}[\rho]$ in (3), where $V = 0$ and U and W are given by (7), since the internal energy is positive for $m > 1$, and the interaction energy is negative. By defining the critical interaction strength χ_c as the biggest χ such that $\mathcal{F}[\rho] \geq 0$ for all ρ, one can establish the following:

1. (*Critical case* $\chi = \chi_c$) The variant of the HLS inequality is equivalent to $\mathcal{F}[\rho] \geq 0$. Stationary states are equivalent to optimisers of the variant of the HLS inequality. Moreover, for the critical case, all of them are global minimisers of the free energy. Furthermore, there exists $\rho^* \in \mathcal{P}(\mathbb{R}^d) \cap L^m(\mathbb{R}^d)$, non-increasing, radially symmetric, and with $||\rho_*||_m = 1$ such that $\mathcal{F}[\rho_*] = 0$; all steady states are translations and dilations of ρ_*. These are compactly supported, bounded, and integrable enough for the singular convolution to make sense in the full range. Moreover, at the boundary, they have limited regularity, like the Barenblatt profiles of the porous medium equation (Vázquez 2007).
2. (*Subcritical case* $0 < \chi < \chi_c$, $k < 0$) There are stationary states in rescaled coordinates, which lead to self-similar solutions in original coordinates.
3. (*Supercritical case* $\chi > \chi_c$, $k < 0$) The free energy $\mathcal{F}[\rho]$ is not bounded below. Moreover, global solutions that decay fast at infinity do not exist; this is shown by a contradiction argument, as for the classical Keller-Segel model (Dolbeault and Perthame 2004; Blanchet et al. 2006a).

In short, the variant of the HLS inequality (8) for the degenerate diffusion range $m > 1$ and the fair competition regimes $m = 1 - \frac{k}{d}$ leads to the same dichotomy as for the classical Keller-Segel model (Dolbeault and Perthame 2004; Blanchet et al. 2006b) (which uses the logarithmic Hardy-Littlewood-Sobolev inequality (Carlen and Loss 1992)). For more details, we refer for more details to Dolbeault and Perthame (2004), Blanchet et al. (2006a, 2009, 2008, 2012), Calvez et al. (2017), Carrillo et al. (2018c).

4.7 Fast Diffusion

In Carrillo et al. (2019d), the authors introduce a reversed HLS (RHLS) inequality to deal with the diffusion-dominated regime:

$$\iint_{\mathbb{R}^d \times \mathbb{R}^d} f(x)|x-y|^k f(y)\mathrm{d}x\mathrm{d}y \geq C\left(\int_{\mathbb{R}^d} f(x)\mathrm{d}x\right)^{\alpha}\left(\int_{\mathbb{R}^d} f(x)^m \mathrm{d}x\right)^{\frac{2-\alpha}{m}}. \quad (9)$$

for $f \geq 0$. Again, the RHLS inequality serves to control the negative part of the free energy $\mathcal{F}[\rho]$ in (3), where $V = 0$ and U and W are given by (7). To be precise, we can control the negative part (the internal energy in the fast diffusion range), by the positive part (the interaction energy, which now is a sort of non-local moment).

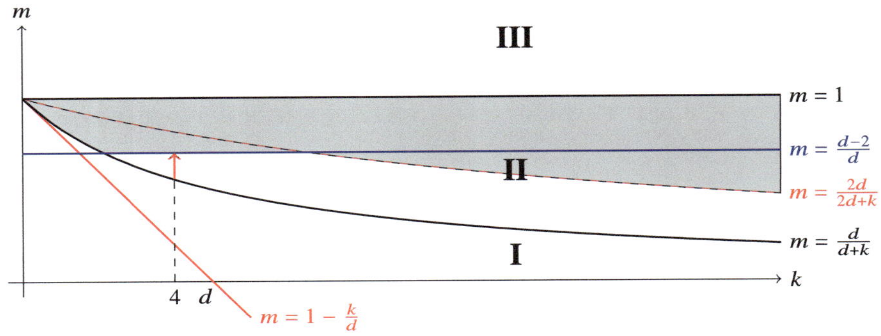

Fig. 4 Sketch of the main results known for the fast diffusion range. The dark grey region corresponds to the area in which global minimisers of the free energy are integrable. The arrowed red line is the region for which we know the exact value of m separating integrable from partially mass-concentrated global minimisers. The remaining white region in Zone II corresponds to the set of parameters in which the separation of integrable and partial mass concentrated global minimisers is not known yet

Notice that the roles of the competing mechanisms in the free energy functional has swapped with respect to the degenerate diffusion case. The interesting fact for $m < 1$ is that this can only happen for $m > \frac{d}{d+k}$, as stated in Theorem 9.1. Moreover, the homogeneity line $m = 1 - \frac{k}{d}$ does not play any role, since it lies below the curve $m = \frac{d}{d+k}$ (and is tangent to it at $m = 1$); thus, it lies in the region where the free energy is not bounded below. We refer to Fig. 4 for a sketch of the main results.

We denote by $L^1 \cap L^m(\mathbb{R}^d)$ the set of integrable functions on $\mathbb{R}^d$ such that ρ^m is an integrable function (recall $0 < m < 1$ in this section, so this is not a Banach space). The main results can be summarised as:

- Zone I, $m \leq \frac{d}{d+k}$: The free energy in $L^1 \cap L^m(\mathbb{R}^d)$ is not bounded below up to $m = \frac{d}{d+k}$. The dilation scaling relation, $m = 1 - \frac{k}{d}$, does not play a role.
- Zone II, $\frac{d}{d+k} < m < 1$: The reversed HLS inequality (9) holds.
  1. The minimiser is attained in $L^1 \cap L^m(\mathbb{R}^d)$ in the dark grey region. It is a positive and radially decreasing density (up to translations).
  2. In the rest of this zone, the minimiser is attained by suitably extending the free energy functional to the set of probability measures. It consists of a radially decreasing function in $\mathrm{L}^1 \cap \mathrm{L}^m(\mathbb{R}^d)$ with **possibly** a Dirac Delta at the origin (modulo translations). We call this a possible partial mass concentration steady state (6).
  3. The black-red dashed line corresponds to the only case of the RHLS inequality known previous to Carrillo et al. (2019d), see Dou and Zhu (2015), Ngô and Nguyen (2017).

4. The arrow-red line is the area where partial mass concentration happens if m is below an explicit sharp value of the exponent m as proven in Carrillo et al. (2022a).

- Zone III, $m \geq 1$: The minimiser is attained and it is a radially symmetric, decreasing, and compactly supported density, modulo translations.

Part of these results can be found in Carrillo et al. (2019d). They also appeared as a preprint (Carrillo and Delgadino 2018) written in terms of (2) and its free energy functional (3), instead of functional inequalities. The authors also give suitable Euler-Lagrange equations for $\widehat{\mu}$, which are more involved than those in Sect. 4.1 due to the presence of the Dirac delta. It is worth remarking that $k = 2$ is trivial, since $W * \rho$ is explicit for radial functions. The case $k = 4$ was studied in full detail in Carrillo et al. (2022a). There, the authors characterise precisely the dimensions d and indices m for the global minimiser to have a Dirac delta with partial mass at the origin, i.e., (6) with $\mathfrak{m} = \|\widehat{\rho}\|_{L^1} < 1$.

There are a number of interesting open problems in this direction. Finding the general relation between m and k for which there is concentration in the white region of zone II is a very interesting open question. Studying the evolution problem associated to the gradient flow of these free energy functionals is also a difficult problem. Characterising the conditions on the initial data leading to a partial mass concentration at the origin, either asymptotically, or in finite time, eventually converging to the final global minimiser of the energy with partial mass concentrated at the origin, is an even more challenging open problem.

4.8 *Bifurcation of Steady States*

The existence of multiple solutions of the fixed-point problem (5) is particularly interesting as we increase the influence of W, for instance, by including a multiplicative parameter $\chi > 0$, as in $W(x) = \chi \frac{|x|^k}{k}$. A simple example is provided by the case of linear diffusion $U = \rho \log \rho$ and $V = 0$; there, we can re-state the problem as finding all solutions of the non-linear integral equation:

$$\widehat{\rho}(x) = \frac{e^{-W*\widehat{\rho}(x)}}{\displaystyle\int_{\mathbb{R}^d} e^{-W*\widehat{\rho}(y)} \mathrm{d}y}, \quad \text{for all } x \in \mathbb{R}^d.$$

Barbaro et al. (2016), Carrillo et al. (2020b) are two examples of works dealing with the bifurcation branches for (2): the first one, on the whole space (reducible to a one dimensional bifurcation problem), and the second one, with periodic conditions (by using Crandall-Rabinowitz theory). The main results in this direction show the

the characterisation of bifurcation branches and the multiplicity of solutions for this family of equations is a complex and intricate process. The bifurcations from constant solutions in the periodic case or from symmetric solutions in the whole space have been reported for both linear and non-linear diffusions (Barbaro et al. 2016; Barré et al. 2018; Carrillo et al. 2020a; Carrillo and Gvalani 2021).

In the periodic setting an almost full characterization of all bifurcation points from constants steady states can be given in terms of Fourier modes of the interaction potential W, see Carrillo et al. (2020b), Carrillo and Gvalani (2021). The classical case is the Kuramoto model for synchronization, which leads to a single branch of bumped solutions at a well-know critical value (Kuramoto 1975; Strogatz and Mirollo 1991; Acebrón et al. 2005; Carrillo et al. 2019a). The stability analysis of these stationary solutions for (2) is only known in very special situations, and a full clarification of the basin of attractions or stable/unstable manifolds of bifurcation branches is still out of reach. Some of these aspects need a full exploration with suitable numerical methods, see Bailo et al. (2020a).

To conclude, we discuss another interesting aspect of this problem: the continuity of the behaviour of the system when a bifurcation takes place. Could it happen, in the periodic case, that there are branches of global minimisers different from the constant solution for values of the parameter χ before the first bifurcation branches at parameter χ_*? This is known as a *discontinuous phase transition*. On the other hand, the classical Kuramoto model is the archetypical example of continuous phase transition, in which a branch of stable equilibria appears at a critical value in a continuous way emanating from the constant steady state. Sufficient conditions for continuous and discontinuous phase transitions to occur, both in the linear and non-linear diffusion cases, have been given in Carrillo et al. (2020b), Carrillo and Gvalani (2021).

5 Numerical Schemes

The aggregation-diffusion equation (2) can be solved numerically in myriad ways, such as mesh-based schemes (Cancès et al. 2023; Jüngel et al. 2024), particle methods (Craig and Bertozzi 2015; Campos Pinto et al. 2018; Carrillo et al. 2019b), and methods based in the JKO iteration (Carrillo et al. 2021a). In some cases, classical numerical methods for the Fokker-Planck equation such as the Scharfetter-Gummel scheme have been generalised to (2) (Schlichting and Seis 2021; Hraivoronska et al. 2023). Other approaches treat the equation as a conservation law in order to establish structure-preserving properties (Sun et al. 2018; Almeida et al. 2019; Cancès et al. 2020) (as was done in Bessemoulin-Chatard and Filbet (2012) for non-linear diffusions).

An example of the latter is Carrillo et al. (2015), which introduces the finite-volume spatial semi-discretisation (presented in one dimension for simplicity)

$$\frac{\mathrm{d}\rho_i}{\mathrm{d}t} + \frac{F_{i+\frac{1}{2}} - F_{i-\frac{1}{2}}}{\Delta x} = 0, \qquad F_{i+\frac{1}{2}} = \rho_i (u_{i+\frac{1}{2}})^+ + \rho_{i+1} (u_{i+\frac{1}{2}})^-, \tag{10a}$$

$$u_{i+\frac{1}{2}} = -\frac{\xi_{i+1} - \xi_i}{\Delta x}, \qquad \xi_i = H'(\rho_i) + V_i + (W * \rho)_i, \tag{10b}$$

where $(s)^+ := \max\{0, s\}$ and $(s)^- := \min\{0, s\}$, and where the index i represents the value at a point x_i in a discretised domain. The terms V_i and $(W * \rho)_i$ are suitable approximations of the potential $V(x)$ and the convolution term $(W * \rho)(x)$. A fully discrete scheme was given in Bailo et al. (2020a, b):

$$\frac{\rho_i^{n+1} - \rho_i^n}{\Delta t} + \frac{F_{i+\frac{1}{2}}^{n+1} - F_{i-\frac{1}{2}}^{n+1}}{\Delta x} = 0, \quad F_{i+\frac{1}{2}}^{n+1} = \rho_i^{n+1} (u_{i+\frac{1}{2}}^{n+1})^+ + \rho_{i+1}^{n+1} (u_{i+\frac{1}{2}}^{n+1})^-, \tag{11a}$$

$$u_{i+\frac{1}{2}}^{n+1} = -\frac{\xi_{i+1}^{n+1} - \xi_i^{n+1}}{\Delta x}, \qquad \xi_i^{n+1} = H'(\rho_i^{n+1}) + V_i + \left(W * \rho^{**}\right)_i, \tag{11b}$$

where $\rho_i^{**} = (\rho_i^n + \rho_i^{n+1})/2$, and where the index n indicates the value at time $t^n = n\Delta t$. Schemes (10) and (11) capture the (respectively) semi-discrete and fully discrete analogues of three crucial properties of the aggregation-diffusion equation (2): the conservation of mass, the non-negativity of solutions, and the dissipation of the free energy.

The conservation of mass and non-negativity properties of the continuous problem can be rigorously established by studying (2) as a conservation law, but they are intuitively suggested by the derivation of the equation as a mean-field limit: the equation models a density of individuals (which cannot be negative), and their motion in space, rather than their birth or death (thus the number of individuals is constant in time). Because these principles are so intrinsic to the model, it is desirable to preserve them when conducting simulations, which is the case in these numerical schemes.

Proposition 9.2 (Conservation of mass and non-negativity) *Let $\rho_i(t)$ be a solution to (10) with $\rho_i(0) \geq 0$ and $\Delta x \sum_i \rho_i(0) = M$. Then, $\rho_i(t) \geq 0$ and $\Delta x \sum_i \rho_i(t) = M$ for all $t > 0$.*

Similarly, let ρ_i^n be a solution to (11) with $\rho_i^0 \geq 0$ and $\sum_i \rho_i^0 = M$. Then, $\rho_i^n \geq 0$ and $\Delta x \sum_i \rho_i^n = M$ for all $n > 0$.

The dissipation of the free energy in the aggregation-diffusion equation was already discussed in Sect. 2. While solutions to (2) can be rigorously constructed as gradient flows of the energy (3), the milder energy dissipation identity (4) can be shown formally by directly computing the time derivative of the free energy along solutions to the equation assuming suitable boundary conditions. The last term in the identity reveals that the non-negativity of the solution ρ is in fact a crucial ingredient of the energy dissipation. In part because said non-negativity is preserved by the numerical schemes above, discrete energy dissipation *inequalities* can be established:

Theorem 9.3 (Energy dissipation) *Let $\rho_i(t)$ be a solution to (10) with $\rho_i(0) \geq 0$. Define the* semi-discrete free energy *by*

$$\mathcal{F}_\Delta(t) = \sum_i H(\rho_i(t))\,\Delta x + \sum_i V_i \rho_i(t)\Delta x + \frac{1}{2}\sum_i\sum_k W_{i-k}\rho_i(t)\rho_k(t)\Delta x^2. \tag{12}$$

Then,

$$\frac{\mathrm{d}}{\mathrm{d}t}\mathcal{F}_\Delta(t) \le -\sum_i \min\{\rho_i(t), \rho_{i+1}(t)\}\,|u_{i+\frac{1}{2}}|^2\Delta x \le 0. \tag{13}$$

Similarly, let ρ_i^n be a solution to (11) with $\rho_i^0 \ge 0$. Define the fully-discrete free energy *by*

$$\mathcal{F}_\Delta^n = \sum_i H(\rho_i^n)\,\Delta x + \sum_i V_i \rho_i^n \Delta x + \frac{1}{2}\sum_i\sum_k W_{i-k}\rho_i^n\rho_k^n\Delta x^2. \tag{14}$$

Then,

$$\frac{\mathcal{F}_\Delta^{n+1} - \mathcal{F}_\Delta^n}{\Delta t} \le -\sum_i \min\left\{\rho_i^{n+1}, \rho_{i+1}^{n+1}\right\}|u_{i+\frac{1}{2}}^{n+1}|^2\Delta x \le 0. \tag{15}$$

Recalling the definition of the velocity field $u = -\nabla\xi$ in the aggregation-diffusion equation (2), the inequalities (13) and (15) can be understood as discretisations of the dissipation identity (4).

These properties do, in fact, generalise quite well. Scheme (10) is also given with a second-order spatial discretisation (using flux limiters) and in two dimensions in Carrillo et al. (2015), and can be trivially generalised to higher dimensional settings, always preserving the conservation of mass, non-negativity, and energy dissipation properties. The properties do not survive a trivial discretisation in time (for instance, a fully explicit treatment), and their preservation may require stringent restrictions on the time step Δt. However, the time discretisation (11) preserves the properties unconditionally, and is also generalised to higher dimensions in Bailo et al. (2020a). The latter introduces a new type of dimensional splitting for (11) as well which nevertheless preserves the dissipation Theorem 9.3, even in the presence of general non-local potentials W; this is needed in practice, as the implicit treatment of higher-dimensional problems can be very costly.

Figure 5 reproduces a numerical example from Bailo et al. (2020a): a solution of the aggregation-diffusion equation (2) which exhibits metastability. The particles undergo non-linear diffusion and experience a global but exponentially decaying attraction force. As a result, the density can be organised into compactly supported "bumps", each approximately a steady state of (2), which interact with each other very weakly. These bumps approach each other very slowly over large periods of time, before rapidly merging into a new bump. The slow and fast dynamics are best understood by looking at the free energy and its plateaus. Structure preserving schemes such as (10) are useful to simulate such a scenario because a large time step can be used without introducing stability issues, oscillations around the boundary of the compact support of each bump, or numerical artifacts in the evolution of the free energy.

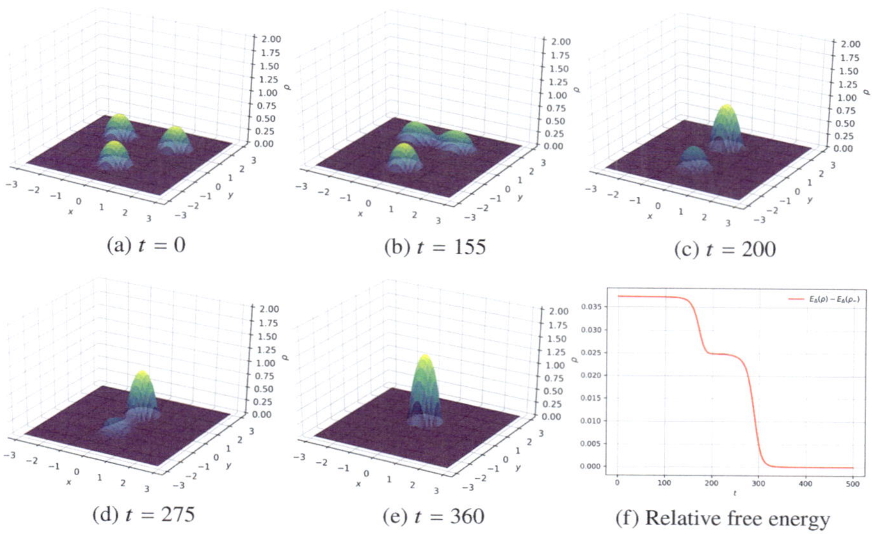

(a) $t = 0$ (b) $t = 155$ (c) $t = 200$

(d) $t = 275$ (e) $t = 360$ (f) Relative free energy

Fig. 5 Metastability in the aggregation-diffusion equation (2) under a force with exponential decay from Bailo et al. (2020a). Long periods of slow motion are interspersed by rapid merging phases until a steady state is reached

The techniques of Carrillo et al. (2015), Bailo et al. (2020a) have also been generalised to equations with non-linear mobilities, as well as systems of aggregation-diffusion equations in Bailo et al. (2023a), always preserving non-negativity and dissipation properties, as well the upper bounds that may be introduced by the mobility terms. Their setting includes the mean-field model (1) discussed above, and preserves the energy dissipation in the setting where the equation itself is a gradient flow. These ideas have also been extended to other types of gradient flows, such as the Cahn-Hilliard equation (Bailo et al. 2023b), a gradient flow in H^{-1}.

Figure 6 adapts a numerical test of Bailo et al. (2023a), which performs cell adhesion experiments similar to those of Fig. 3. The cellular model (1) is solved with initially separate populations: just as in Carrillo et al. (2019g), the experimental behaviour described in Katsunuma et al. (2016) can be reproduced by adjusting the adhesion parameters between the cells. Full mixing (not shown in the figure) can be achieved by making all adhesion strengths equal. Figure 6a, b show scenarios where the populations remain strongly cohesive, but also attract each other via weaker adhesion; depending on the specific strengths, a clean or a diffusive boundary between the populations can be achieved. Figure 6c shows a scenario where one population completely envelops the other, forming the halo described in Sect. 3. The same scenario with a non-linear mobility function is shown in Fig. 6d; the additional term models volume constraint effects, where the combined cell density cannot exceed a threshold. This limit forces cells to move around the areas of highest density (such

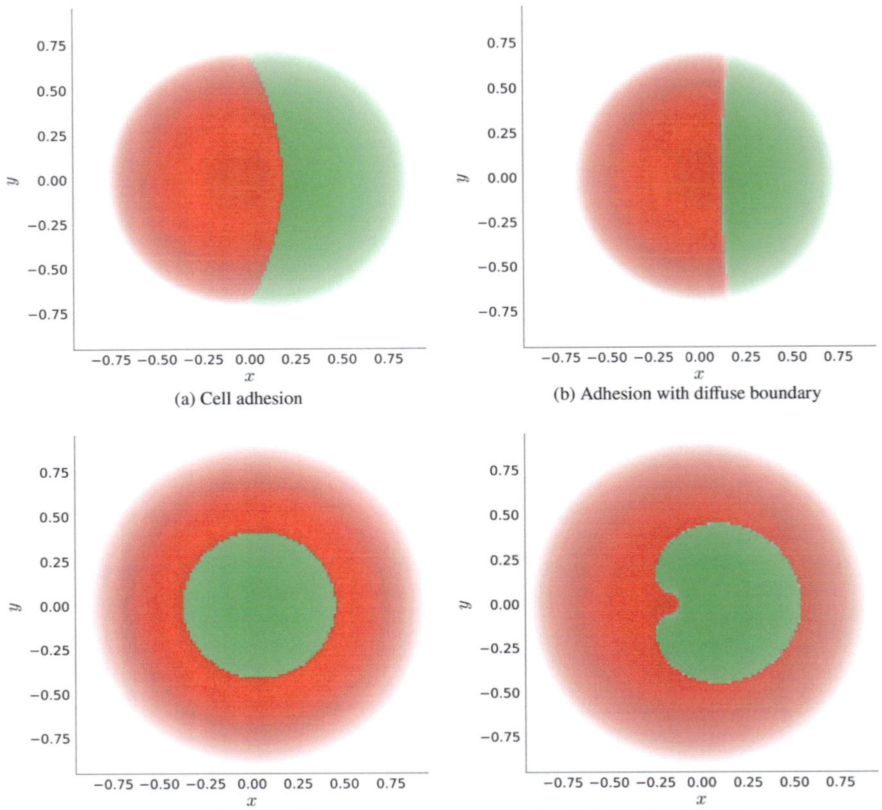

(a) Cell adhesion

(b) Adhesion with diffuse boundary

(c) Cellular engulfment

(d) Engulfment limited by volume constraints

Fig. 6 Cell aggregates mediated by cellular adhesion in model (1) under different parameters, coloured by population, adapted from Bailo et al. (2023a)

as the point where the two populations initially come into contact, which forms a pinch in the figure), leading to asymmetric behaviour.

6 Conclusions

Our goal in this exposition has been to showcase some of the most interesting aspects in the study of aggregation-diffusion equations and systems developed over the past 30 years. The field continues to grow, and while many questions have received answers in recent years, many remain open. To recognise our limited understanding, we remark that long-time asymptotics, bifurcation branches, stable and unstable manifolds, and basins of attraction are only understood for a handful of these models.

The applications of aggregation-diffusion equations are vast: mathematical biology, data science, machine learning, and optimisation, to name but a few. The opportunities for applied mathematicians to continue exploring this fascinating topic are boundless.

Acknowledgements The authors were supported by the Advanced Grant Nonlocal-CPD (Nonlocal PDEs for Complex Particle Dynamics: Phase Transitions, Patterns and Synchronization) of the European Research Council Executive Agency (ERC) under the European Union's Horizon 2020 research and innovation programme (grant agreement No. 883363). JAC is partially supported by the "Maria de Maeztu" Excellence Unit IMAG, reference CEX2020-001105-M, funded by MCIN/AEI/10.13039/501100011033/. DGC is support by RYC2022-037317-I and PID2023-151120NA-I00 from the Spanish government MCIN/AEI/10.13039/501100011033/FEDER, UE.

References

Acebrón, J.A., Bonilla, L.L., Vicente, C.J.P., Ritort, F., Spigler, R.: The Kuramoto model: a simple paradigm for synchronization phenomena. Rev. Mod. Phys. **77**(1), 137 (2005)

Almeida, L., Bubba, F., Perthame, B., Pouchol, C.: Energy and implicit discretization of the Fokker-Planck and Keller-Segel type equations. Netw. Heterog.s Med. **14**(1), 23–41 (2019). https://doi.org/10.3934/nhm.2019002

Ambrosio, L., Gigli, N., Savaré, G.: Gradient Flows in Metric Spaces and in the Space of Probability Measures. Lectures in Mathematics ETH Zürich. Birkhäuser Verlag, Basel (2008)

Bach, F., Chizat, L.: Gradient descent on infinitely wide neural networks: global convergence and generalization. In: ICM—International Congress of Mathematicians. Vol. 7. Sections 15–20, pp. 5398–5419. EMS Press, Berlin (2023)

Bailo, R., Barbaro, A., Gomes, S.N., Riedl, K., Roith, T., Totzeck, C., Vaes, U.: CBX: Python and Julia packages for consensus-based interacting particle methods. J. Open Source Softw. **9**(98), 6611 (2024). https://doi.org/10.21105/joss.06611

Bailo, R., Carrillo, J.A., Hu, J.: Fully discrete positivity-preserving and energy-dissipating schemes for aggregation-diffusion equations with a gradient-flow structure. Commun. Math. Sci. **18**(5), 1259–1303 (2020a). https://doi.org/10.4310/cms.2020.v18.n5.a5

Bailo, R., Carrillo, J.A., Hu, J.: Bound-preserving finite-volume schemes for systems of continuity equations with saturation. SIAM J. Appl. Math. **83**(3), 1315–1339 (2023a). https://doi.org/10.1137/22m1488703

Bailo, R., Carrillo, J.A., Kalliadasis, S., Perez, S.P.: Unconditional bound-preserving and energy-dissipating finite-volume schemes for the Cahn-Hilliard equation. Commun. Comput. Phys. **34**(3), 713–748 (2023b). https://doi.org/10.4208/cicp.oa-2023-0049

Bailo, R., Carrillo, J.A., Murakawa, H., Schmidtchen, M.: Convergence of a fully discrete and energy-dissipating finite-volume scheme for aggregation-diffusion equations. Math. Models Method Appl. Sci. **30**(13), 2487–2522 (2020b). https://doi.org/10.1142/s0218202520500487

Balagué, D., Carrillo, J., Laurent, T., Raoul, G.: Nonlocal interactions by repulsive-attractive potentials: radial ins/stability. Phys. D. **260**, 5–25 (2013a)

Balagué, D., Carrillo, J.A., Laurent, T., Raoul, G.: Dimensionality of local minimizers of the interaction energy. Arch. Ration. Mech. Anal. **209**(3), 1055–1088 (2013b). https://doi.org/10.1007/s00205-013-0644-6

Barbaro, A.B.T., Cañizo, J.A., Carrillo, J.A., Degond, P.: Phase transitions in a kinetic flocking model of Cucker-Smale type. Multiscale Model. Simul. **14**(3), 1063–1088 (2016). https://doi.org/10.1137/15M1043637

Barré, J., Carrillo, J.A., Degond, P., Peurichard, D., Zatorska, E.: Particle interactions mediated by dynamical networks: assessment of macroscopic descriptions. J. Nonlinear Sci. **28**(1), 235–268 (2018). https://doi.org/10.1007/s00332-017-9408-z

Bertozzi, A.L., Kolokolnikov, T., Sun, H., Uminsky, D., Brecht, J.V.: Ring patterns and their bifurcations in a nonlocal model of biological swarms. Commun. Math. Sci. **13**(4), 955–985 (2015)

Bertozzi, A.L., Laurent, T., Léger, F.: Aggregation and spreading via the Newtonian potential: the dynamics of patch solutions. Math. Models Methods Appl. Sci. **22**, 1140005, 39 (2012). https://doi.org/10.1142/S0218202511400057

Bessemoulin-Chatard, M., Filbet, F.: A finite volume scheme for nonlinear degenerate parabolic equations. SIAM J. Sci. Comput. **34**(5), B559–B583 (2012). https://doi.org/10.1137/110853807

Blanchet, A., Carlen, E.A., Carrillo, J.A.: Functional inequalities, thick tails and asymptotics for the critical mass Patlak-Keller-Segel model. J. Funct. Anal. **262**(5), 2142–2230 (2012). https://doi.org/10.1016/j.jfa.2011.12.012

Blanchet, A., Carrillo, J.A., Laurençot, P.: Critical mass for a Patlak-Keller-Segel model with degenerate diffusion in higher dimensions. Calc. Var. Partial. Differ. Equ. **35**(2), 133–168 (2009)

Blanchet, A., Carrillo, J.A., Masmoudi, N.: Infinite time aggregation for the critical Patlak-Keller-Segel model in $\mathbb{R}^2$. Comm. Pure Appl. Math. **61**(10), 1449–1481 (2008). https://doi.org/10.1002/cpa.20225

Blanchet, A., Dolbeault, J., Perthame, B.: Two-dimensional Keller-Segel model: optimal critical mass and qualitative properties of the solutions. Electron. J. Differ. Eq. (2006a)

Blanchet, A., Dolbeault, J., Perthame, B.: Two-dimensional Keller-Segel model: optimal critical mass and qualitative properties of the solutions. Electron. J. Diff. Eq. **2006**, 1–33 (2006b)

Bodnar, M., Velazquez, J.J.L.: An integro-differential equation arising as a limit of individual cell-based models. J. Differ. Eq. **222**(2), 341–380 (2006). https://doi.org/10.1016/j.jde.2005.07.025

Bolley, F., Cañizo, J.A., Carrillo, J.A.: Stochastic mean-field limit: non-Lipschitz forces and swarming. Math. Models Methods Appl. Sci. **21**(11), 2179–2210 (2011). https://doi.org/10.1142/S0218202511005702

Bongini, M., Fornasier, M., Hansen, M., Maggioni, M.: Inferring interaction rules from observations of evolutive systems I: the variational approach. Math. Models Methods Appl. Sci. **27**(5), 909–951 (2017). https://doi.org/10.1142/S0218202517500208

Brock, F.: Continuous rearrangement and symmetry of solutions of elliptic problems. Proc. Indian Acad. Sci. Math. Sci. **110**(2), 157–204 (2000). https://doi.org/10.1007/BF02829490

Cañizo, J.A., Carrillo, J.A., Patacchini, F.S.: Existence of compactly supported global minimisers for the interaction energy. Arch. Ration. Mech. Anal. **217**(3), 1197–1217 (2015). https://doi.org/10.1007/s00205-015-0852-3

Calvez, V., Carrillo, J.A.: Refined asymptotics for the subcritical Keller-Segel system and related functional inequalities. Proc. Amer. Math. Soc. **140**(10), 3515–3530 (2012). https://doi.org/10.1090/S0002-9939-2012-11306-1

Calvez, V., Carrillo, J.A., Hoffmann, F.: Equilibria of homogeneous functionals in the fair-competition regime. Nonlinear Anal. **159**, 85–128 (2017). https://doi.org/10.1016/j.na.2017.03.008

Calvez, V., Carrillo, J.A., Hoffmann, F.: Uniqueness of stationary states for singular Keller-Segel type models. Nonlinear Anal. **205**, Paper No. 112222, 24 (2021). https://doi.org/10.1016/j.na.2020.112222

Campos Pinto, M., Carrillo, J.A., Charles, F., Choi, Y.P.: Convergence of a linearly transformed particle method for aggregation equations. Numer. Math. **139**(4), 743–793 (2018). https://doi.org/10.1007/s00211-018-0958-2

Cancès, C., Chainais-Hillairet, C., Fuhrmann, J., Gaudeul, B.: A numerical-analysis-focused comparison of several finite volume schemes for a unipolar degenerate drift-diffusion model. IMA J. Numer. Anal. **41**(1), 271–314 (2020). https://doi.org/10.1093/imanum/draa002

Cancès, C., Herda, M., Massimini, A.: Finite Volumes for a Generalized Poisson-Nernst-Planck System with Cross-Diffusion and Size Exclusion, pp. 57–73. Springer Nature Switzerland (2023). https://doi.org/10.1007/978-3-031-40864-9_4

Carlen, E.A., Loss, M.: Competing symmetries, the logarithmic HLS inequality and Onofri's inequality on S^n. Geom. Funct. Anal. **2**(1), 90–104 (1992)

Carrillo, J.A., Chen, X., Wang, Q., Wang, Z., Zhang, L.: Phase transitions and bump solutions of the Keller-Segel model with volume exclusion. SIAM J. Appl. Math. **80**(1), 232–261 (2020a). https://doi.org/10.1137/19M125827X

Carrillo, J.A., Chertock, A., Huang, Y.: A finite-volume method for nonlinear nonlocal equations with a gradient flow structure. Commun. Comput. Phys. **17**(1), 233–258 (2015). https://doi.org/10.4208/cicp.160214.010814a

Carrillo, J.A., Choi, Y.P., Pareschi, L.: Structure preserving schemes for the continuum Kuramoto model: phase transitions. J. Comput. Phys. **376**, 365–389 (2019a). https://doi.org/10.1016/j.jcp.2018.09.049

Carrillo, J.A., Choi, Y.P., Totzeck, C., Tse, O.: An analytical framework for consensus-based global optimization method. Math. Models Methods Appl. Sci. **28**(6), 1037–1066 (2018a). https://doi.org/10.1142/S0218202518500276

Carrillo, J.A., Colombi, A., Scianna, M.: Adhesion and volume constraints via nonlocal interactions determine cell organisation and migration profiles. J. Theoret. Biol. **445**, 75–91 (2018b)

Carrillo, J.A., Craig, K., Patacchini, F.S.: A blob method for diffusion. Calc. Var. Partial Dif. **58**(53) (2019b). https://doi.org/10.1007/s00526-019-1486-3

Carrillo, J.A., Craig, K., Wang, L., Wei, C.: Primal dual methods for Wasserstein gradient flows. Found. Comput. Math. **22**(2), 389–443 (2021a). https://doi.org/10.1007/s10208-021-09503-1

Carrillo, J.A., Craig, K., Yao, Y.: Aggregation-diffusion equations: dynamics, asymptotics, and singular limits. In: Active particles, Vol. 2, Model. Simul. Sci. Eng. Technol., pp. 65–108. Birkhäuser/Springer, Cham (2019c)

Carrillo, J.A., Delgadino, M.G.: Free energies and the reversed HLS inequality (2018). https://doi.org/10.48550/arxiv.1803.06232

Carrillo, J.A., Delgadino, M.G., Dolbeault, J., Frank, R.L., Hoffmann, F.: Reverse Hardy-Littlewood-Sobolev inequalities. J. Math. Pures Appl. **132**, 133–165 (2019d). https://doi.org/10.1016/j.matpur.2019.09.001

Carrillo, J.A., Delgadino, M.G., Frank, R.L., Lewin, M.: Fast diffusion leads to partial mass concentration in Keller-Segel type stationary solutions. Math. Models Methods Appl. Sci. **32**(4), 831–850 (2022a). https://doi.org/10.1142/S021820252250018X

Carrillo, J.A., Delgadino, M.G., Mellet, A.: Regularity of local minimizers of the interaction energy via obstacle problems. Comm. Math. Phys. **343**(3), 747–781 (2016). https://doi.org/10.1007/s00220-016-2598-7

Carrillo, J.A., Delgadino, M.G., Patacchini, F.S.: Existence of ground states for aggregation-diffusion equations. Anal. Appl. (Singap.) **17**(3), 393–423 (2019e). https://doi.org/10.1142/S0219530518500276

Carrillo, J.A., Estrada-Rodriguez, G., Mikolas, L., Tang, S.: Sparse identification of nonlocal interaction kernels in nonlinear gradient flow equations via partial inversion (2024a)

Carrillo, J.A., Figalli, A., Patacchini, F.S.: Geometry of minimizers for the interaction energy with mildly repulsive potentials. Ann. Inst. H. Poincaré C Anal. Non Linéaire **34**(5), 1299–1308 (2017). https://doi.org/10.1016/j.anihpc.2016.10.004

Carrillo, J.A., Fornasier, M., Toscani, G., Vecil, F.: Particle, kinetic, and hydrodynamic models of swarming. In: Mathematical Modelling of Collective Behavior in Socio-Economic and Life Sciences, pp. 297–336 (2010)

Carrillo, J.A., Gómez-Castro, D., Vázquez, J.L.: Infinite-time concentration in aggregation-diffusion equations with a given potential. J. Math. Pures Appl. **9**(157), 346–398 (2022b). https://doi.org/10.1016/j.matpur.2021.11.002

Carrillo, J.A., Gómez-Castro, D., Yao, Y., Zeng, C.: Asymptotic simplification of aggregation-diffusion equations towards the heat kernel. Arch. Ration. Mech. Anal. **247**(1), Paper No. 11, 45 (2023). https://doi.org/10.1007/s00205-022-01838-5

Carrillo, J.A., Gvalani, R.S.: Phase transitions for nonlinear nonlocal aggregation-diffusion equations. Comm. Math. Phys. **382**(1), 485–545 (2021). https://doi.org/10.1007/s00220-021-03977-4

Carrillo, J.A., Gvalani, R.S., Pavliotis, G.A., Schlichting, A.: Long-time behaviour and phase transitions for the Mckean-Vlasov equation on the Torus. Arch. Ration. Mech. Anal. **235**(1), 635–690 (2020b). https://doi.org/10.1007/s00205-019-01430-4

Carrillo, J.A., Hittmeir, S., Volzone, B., Yao, Y.: Nonlinear aggregation-diffusion equations: radial symmetry and long time asymptotics. Inventiones Math. **218**(3), 889–977 (2019f). https://doi.org/10.1007/s00222-019-00898-x

Carrillo, J.A., Hoffmann, F., Mainini, E., Volzone, B.: Ground states in the diffusion-dominated regime. Calc. Var. Partial. Differ. Equ. **57**(5), 127 (2018c)

Carrillo, J.A., Hoffmann, F., Stuart, A.M., Vaes, U.: Consensus-based sampling. Stud. Appl. Math. **148**(3), 1069–1140 (2022c). https://doi.org/10.1111/sapm.12470

Carrillo, J.A., Huang, Y.: Explicit equilibrium solutions for the aggregation equation with power-law potentials. Kinet. Relat. Models **10**(1), 171–192 (2017). https://doi.org/10.3934/krm.2017007

Carrillo, J.A., Jin, S., Li, L., Zhu, Y.: A consensus-based global optimization method for high dimensional machine learning problems. ESAIM Control Optim. Calc. Var. **27**, Paper No. S5, 22 (2021b). https://doi.org/10.1051/cocv/2020046

Carrillo, J.A., Jin, S., Zhang, H., Zhu, Y.: An interacting particle consensus method for constrained global optimization (2024b)

Carrillo, J.A., Mateu, J., Mora, M.G., Rondi, L., Scardia, L., Verdera, J.: The ellipse law: Kirchhoff meets dislocations. Comm. Math. Phys. **373**(2), 507–524 (2020c). https://doi.org/10.1007/s00220-019-03368-w

Carrillo, J.A., McCann, R.J., Villani, C.: Kinetic equilibration rates for granular media and related equations: entropy dissipation and mass transportation estimates. Rev. Mat. Iberoam. **19**(3), 971–1018 (2003). https://doi.org/10.4171/RMI/376

Carrillo, J.A., Murakawa, H., Sato, M., Togashi, H., Trush, O.: A population dynamics model of cell-cell adhesion incorporating population pressure and density saturation. J. Theoret. Biol. **474**, 14–24 (2019g)

Carrillo, J.A., Shu, R.: From radial symmetry to fractal behavior of aggregation equilibria for repulsive-attractive potentials. Calc. Var. Partial Differ. Equ. **62**(1), Paper No. 28, 61 (2023). https://doi.org/10.1007/s00526-022-02368-4

Carrillo, J.A., Shu, R.: Global minimizers of a large class of anisotropic attractive-repulsive interaction energies in 2D. Comm. Pure Appl. Math. **77**(2), 1353–1404 (2024)

Carrillo, J.A., Vaes, U.: Wasserstein stability estimates for covariance-preconditioned Fokker-Planck equations. Nonlinearity **34**(4), 2275–2295 (2021). https://doi.org/10.1088/1361-6544/abbe62

Carrillo, J.A., Vázquez, J.L.: Some free boundary problems involving non-local diffusion and aggregation. Philos. Trans. Roy. Soc. A **373**(2050), 20140275, 16 (2015). https://doi.org/10.1098/rsta.2014.0275

Chan, H., González, M.d.M., Huang, Y., Mainini, E., Volzone, B.: Uniqueness of entire ground states for the fractional plasma problem. Calc. Var. Partial Differ. Equ. **59**(6), Paper No. 195, 42 (2020). https://doi.org/10.1007/s00526-020-01845-y

Chizat, L., Colombo, M., Fernández-Real, X., Figalli, A.: Infinite-width limit of deep linear neural networks. Commun. Pure Appl. Math. (2022)

Craig, K., Bertozzi, A.L.: A blob method for the aggregation equation. Math. Comp. **85**(300), 1681–1717 (2015). https://doi.org/10.1090/mcom3033

Davies, C., Lim, T., McCann, R.J.: Classifying minimum energy states for interacting particles: spherical shells. SIAM J. Appl. Math. **82**(4), 1520–1536 (2022). https://doi.org/10.1137/21M1455309

Davies, C., Lim, T., McCann, R.J.: Classifying minimum energy states for interacting particles: regular simplices. Comm. Math. Phys. **399**(2), 577–598 (2023). https://doi.org/10.1007/s00220-022-04564-x

Delgadino, M.G., Yan, X., Yao, Y.: Uniqueness and nonuniqueness of steady states of aggregation—diffusion equations. Comm. Pure Appl. Math. **75**(1), 3–59 (2022). https://doi.org/10.1002/cpa.21950

Ding, Z., Li, Q.: Ensemble Kalman sampler: mean-field limit and convergence analysis. SIAM J. Math. Anal. **53**(2), 1546–1578 (2021). https://doi.org/10.1137/20M1339507

Dolbeault, J., Esteban, M.J., Loss, M., Esteban, M.: Symmetry and symmetry breaking: rigidity and flows in elliptic PDEs. In: Proceedings of the International Congress of Mathematicians (ICM 2018), pp. 2261–2285. World Scientific (2019). https://doi.org/10.1142/9789813272880_0138

Dolbeault, J., Perthame, B.: Optimal critical mass in the two-dimensional Keller-Segel model in $\mathbb{R}^2$. C. R. Math. Acad. Sci. Paris **339**(9), 611–616 (2004). https://doi.org/10.1016/j.crma.2004.08.011

Dou, J., Zhu, M.: Reversed Hardy-Littewood-Sobolev inequality. Int. Math. Res. Not. IMRN 9696–9726 (2015). https://doi.org/10.1093/imrn/rnu241

Falcó, C., Baker, R.E., Carrillo, J.A.: A local continuum model of cell-cell adhesion. SIAM J. Appl. Math. **0**(0), S17–S42 (0). https://doi.org/10.1137/22M1506079

Fernández-Real, X., Figalli, A.: The continuous formulation of shallow neural networks as Wasserstein-type gradient flows. In: Analysis at Large: Dedicated to the Life and Work of Jean Bourgain, pp. 29–57. Springer (2022)

Frank, R.L.: Minimizers for a one-dimensional interaction energy. Nonlinear Anal. **216**, Paper No. 112691, 10 (2022). https://doi.org/10.1016/j.na.2021.112691

Garbuno-Inigo, A., Hoffmann, F., Li, W., Stuart, A.M.: Interacting Langevin diffusions: gradient structure and ensemble Kalman sampler. SIAM J. Appl. Dyn. Syst. **19**(1), 412–441 (2020). https://doi.org/10.1137/19M1251655

Gomes, S.N., Stuart, A.M., Wolfram, M.T.: Parameter estimation for macroscopic pedestrian dynamics models from microscopic data. SIAM J. Appl. Math. **79**(4), 1475–1500 (2019). https://doi.org/10.1137/18M1215980

Gómez-Castro, D.: Beginner's guide to aggregation-diffusion equations. SeMA J. (2024). https://doi.org/10.1007/s40324-024-00350-y

Hraivoronska, A., Schlichting, A., Tse, O.: Variational convergence of the Scharfetter-Gummel scheme to the aggregation-diffusion equation and vanishing diffusion limit (2023). arXiv:2306.02226

Jabin, P.E.: A review of the mean field limits for Vlasov equations. Kinet. Relat. Models **7**(4), 661–711 (2014). https://doi.org/10.3934/krm.2014.7.661

Jordan, R., Kinderlehrer, D., Otto, F.: The variational formulation of the Fokker-Planck equation. SIAM J. Math. Anal. **29**(1), 1–17 (1998). https://doi.org/10.1137/S0036141096303359

Jüngel, A., Portisch, S., Zurek, A.: A convergent finite-volume scheme for nonlocal cross-diffusion systems for multi-species populations. ESAIM: Math. Model. Numer. Anal. **58**(2), 759–792 (2024). https://doi.org/10.1051/m2an/2024016

Katsunuma, S., Honda, H., Shinoda, T., Ishimoto, Y., Miyata, T., Kiyonari, H., Abe, T., Nibu, K.i., Takai, Y., Togashi, H.: Synergistic action of nectins and cadherins generates the mosaic cellular pattern of the olfactory epithelium. J. Cell Biol. **212**(5), 561–575 (2016)

Keller, E., Segel, L.: Initiation of slide mold aggregation viewed as an instability. J. Theoret. Biol. **26** (1970)

Korba, A., Salim, A., Arbel, M., Luise, G., Gretton, A.: A non-asymptotic analysis for stein variational gradient descent. Adv. Neural. Inf. Process. Syst. **33**, 4672–4682 (2020)

Kuramoto, Y.: International symposium on mathematical problems in theoretical physics. Lect. Notes Phys. **30**, 420 (1975)

Lang, Q., Lu, F.: Learning interaction kernels in mean-field equations of first-order systems of interacting particles. SIAM J. Sci. Comput. **44**(1), A260–A285 (2022). https://doi.org/10.1137/20M1377072

Lee, W., Wang, L., Li, W.: Deep JKO: time-implicit particle methods for general nonlinear gradient flows (2023)

Lieb, E.H.: Sharp constants in the Hardy-Littlewood-Sobolev and related inequalities. Ann. Math. **118**(2), 349–374 (1983). https://doi.org/10.2307/2007032

Lu, F., Maggioni, M., Tang, S.: Learning interaction kernels in stochastic systems of interacting particles from multiple trajectories. Found. Comput. Math. **22**(4), 1013–1067 (2022). https://doi.org/10.1007/s10208-021-09521-z

Lu, F., Zhong, M., Tang, S., Maggioni, M.: Nonparametric inference of interaction laws in systems of agents from trajectory data. Proc. Natl. Acad. Sci. U.S.A. **116**(29), 14424–14433 (2019). https://doi.org/10.1073/pnas.1822012116

Mokrov, P., Korotin, A., Li, L., Genevay, A., Solomon, J.M., Burnaev, E.: Large-scale Wasserstein gradient flows. Adv. Neural. Inf. Process. Syst. **34**, 15243–15256 (2021)

Ngô, Q.A., Nguyen, V.H.: Sharp reversed Hardy-Littlewood-Sobolev inequality on R^n. Israel J. Math. **220**(1), 189–223 (2017). https://doi.org/10.1007/s11856-017-1515-x

Oelschläger, K.: Large systems of interacting particles and the porous medium equation. J. Differ. Equ. **88**(2), 294–346 (1990). https://doi.org/10.1016/0022-0396(90)90101-T

Pinnau, R., Totzeck, C., Tse, O., Martin, S.: A consensus-based model for global optimization and its mean-field limit. Math. Models Methods Appl. Sci. **27**(1), 183–204 (2017). https://doi.org/10.1142/S0218202517400061

Rotskoff, G.M., Vanden-Eijnden, E.: Trainability and accuracy of artificial neural networks: an interacting particle system approach. Comm. Pure Appl. Math. **75**(9), 1889–1935 (2022)

Schlichting, A., Seis, C.: The Scharfetter-Gummel scheme for aggregation-diffusion equations. IMA J. Numer. Anal. **42**(3), 2361–2402 (2021). https://doi.org/10.1093/imanum/drab039

Steinberg, M.S.: On the mechanism of tissue reconstruction by dissociated cells, I. Population kinetics, differential adhesiveness, and the absence of directed migration. Proc. Natl. Acad. Sci. U.S.A. **48**(9), 1577 (1962)

Strogatz, S.H., Mirollo, R.E.: Stability of incoherence in a population of coupled oscillators. J. Statist. Phys. **63**(3–4), 613–635 (1991). https://doi.org/10.1007/BF01029202

Sun, Z., Carrillo, J.A., Shu, C.W.: A discontinuous Galerkin method for nonlinear parabolic equations and gradient flow problems with interaction potentials. J. Comput. Phys. **352**, 76–104 (2018). https://doi.org/10.1016/j.jcp.2017.09.050

Talenti, G.: The art of rearranging. Milan J. Math. **84**(1), 105–157 (2016). https://doi.org/10.1007/s00032-016-0253-6

Topaz, C.M., Bertozzi, A.L., Lewis, M.A.: A nonlocal continuum model for biological aggregation. Bull. Math. Biol. **68**(7), 1601–1623 (2006). https://doi.org/10.1007/s11538-006-9088-6

Vázquez, J.L.: The porous medium equation. In: Mathematical Theory. Oxford Mathematical Monographs. The Clarendon Press, Oxford University Press, Oxford (2007)

Villani, C.: Topics in Optimal Transportation. Graduate Studies in Mathematics, vol. 58. American Mathematical Society, Providence, RI (2003)

Open Access This chapter is licensed under the terms of the Creative Commons Attribution 4.0 International License (http://creativecommons.org/licenses/by/4.0/), which permits use, sharing, adaptation, distribution and reproduction in any medium or format, as long as you give appropriate credit to the original author(s) and the source, provide a link to the Creative Commons license and indicate if changes were made.

The images or other third party material in this chapter are included in the chapter's Creative Commons license, unless indicated otherwise in a credit line to the material. If material is not included in the chapter's Creative Commons license and your intended use is not permitted by statutory regulation or exceeds the permitted use, you will need to obtain permission directly from the copyright holder.

Kinetic Models of Dilute Polymeric Fluids: Existence of Global Weak Solutions and Rigorous Macroscopic Closure

Endre Süli

Abstract We review recent results concerning the existence of large-data global-in-time weak solutions to Navier–Stokes–Fokker–Planck systems featuring in bead-spring-chain models for dilute polymeric fluids and discuss the rigorous macroscopic closure of macro-micro models of this type.

1 Introduction

Mathematical models of non-Newtonian fluids give rise to interesting and challenging analytical problems, which have been the driving force behind a large body of research in the field of analysis of nonlinear partial differential equations over the last two decades, aimed primarily at understanding the well-posedness of various models proposed in the non-Newtonian fluid mechanics and polymer physics literature that are of relevance in scientific, engineering and industrial applications.

Here we shall concentrate on a specific class of non-Newtonian fluids, dilute polymers, the underlying modelling assumption in their mathematical description being that polymer molecules, which are suspended in a (typically) Newtonian (i.e., Navier–Stokes) fluid, the solvent, exhibit no self-interaction and do not interact with each other. The associated mathematical model involves the coupling of the Fokker–Planck equation, describing the evolution of a probability density function whose role is to capture the random motion of polymer molecules at the microscopic scale in the solvent, to the Navier–Stokes system that models the evolution of macroscopic features of the polymeric fluid. We shall focus on questions of existence of global-in-time large-data weak solutions to the resulting macro-micro Navier–Stokes–Fokker–Planck (NSFP) model and the problem of rigorous macroscopic closure of models of this type.

The paper is structured as follows. In the next section we present a brief derivation of the model in the simple case when the bead-spring chain, which is the idealization

E. Süli (✉)
Mathematical Institute, University of Oxford, Oxford, UK
e-mail: endre.suli@maths.ox.ac.uk

© The Author(s) 2026

S. Oishi et al. (eds.), *Recent Developments in Industrial and Applied Mathematics*,
ICIAM2023 Springer Series 1, https://doi.org/10.1007/978-981-95-1446-5_10

of a polymer molecule, consists of just a pair of massless beads connected by a single massless elastic spring, referred to as the *dumbbell model*. We shall further assume that the fluid density is constant and that the viscosity coefficient in the Navier–Stokes equation is independent of both the density and the polymer number density. While there has been significant progress in recent years with the mathematical analysis of compressible Navier–Stokes–Fokker–Planck systems (cf., for example, Barrett and Süli (2016a, b), Feireisl et al. (2016), Süli and Wróblewska-Kamińska (2020) and references therein), for the sake of brevity we shall restrict ourselves here to incompressible flows. In Sect. 3 we shall focus on a more general setting than in Sect. 2 and will consider bead-spring chain models consisting of $K + 1$ massless beads linearly connected by K massless springs; we shall allow the viscosity coefficient to depend on the fluid density and the polymer number density, and the friction coefficient appearing in the hydrodynamic drag force acting on the beads will also be allowed to depend on the density. We shall in particular state the (formal) energy equality, which is at the heart of the proof of the existence of a global-in-time large-data weak solution to the model, given in the recent paper He and Süli (2024). Throughout Sect. 3 we shall be concerned with finitely-extensible nonlinear elastic (or, briefly, FENE-type) spring potentials. In Sect. 4 we shall review recent results from Dębiec and Süli (2023) concerning the existence of a large-data global weak solution to the Hookean model, where the spring force in the dumbbell describing the stretching of the spring is, simply, a linear function of the extension of the spring. This apparent simplification leads to significant complications in the analysis of the resulting Navier–Stokes–Fokker–Planck system, because Kramers' polymeric extra-stress tensor, whose divergence appears on the right-hand side of the Navier–Stokes balance-of-linear-momentum equation, looses weak compactness in L^1. In this case only the existence of a *generalized dissipative solution* can be proved, permitting the appearance of a positive semidefinite tensor-valued defect measure in the extra-stress tensor. In two space dimensions, however, the defect measure can be shown to vanish, and the generalized dissipative solution is then just a standard global-in-time weak solution of the Navier–Stokes–Fokker–Planck system. The proof of this result also implies in a mathematically rigorous sense that, in two space dimensions at least, the classical Oldroyd-B model with stress-diffusion is the macroscopic closure of the Hookean dumbbell model. In three space dimensions, in contrast, the presence of the defect measure can only be ruled out by demanding that the velocity field $\boldsymbol{u}$ has additional regularity over the space-time domain $\Omega \times [0, T]$ (e.g., by supposing that $\partial_t \boldsymbol{u} \in L^1(0, T; L^2(\Omega; \mathbb{R}^3))$ and $\nabla_x \boldsymbol{u} \in L^1(0, T; C(\overline{\Omega}; \mathbb{R}^{3\times 3})))$, in which case the generalized dissipative solution of the coupled Navier–Stokes–Fokker–Planck system, once again, becomes a global-in-time weak solution in the usual sense. The presence of this extra regularity in three space dimensions however cannot be guaranteed globally in time in the framework of large-data weak solutions to the problem. An interesting open question is then whether generalized dissipative solutions (or a subclass of these) satisfy the weak-strong uniqueness property, i.e., whether they necessarily coincide with the strong solution emanating from the same initial data on the (possibly short) time interval on which strong solutions can be shown to exist.

2 Derivation of the Model

There are a number of important books concerned with the statistical physics of polymeric fluids; see, for example, the monographs of Doi and Edwards (1986) or Bird et al. (1987); a brief historical survey of the field is contained at the back of H. C. Öttinger's *Stochastic Processes in Polymeric Fluids* (Öttinger 2012). We draw the reader's attention to the fact that in Bird et al. (1987), which is frequently cited in the mathematical literature concerned with Navier–Stokes–Fokker–Planck systems for the derivation of the Fokker–Planck equation (cf. in particular Renardy (1991) and a number of subsequent publications concerned with the mathematical analysis of Navier–Stokes–Fokker–Planck systems that cite Bird et al. (1987) and Renardy (1991)), the exposition is limited to the case of a homogeneous velocity field, $\boldsymbol{u}$, in the sense that the symmetric velocity gradient $\mathbb{D}(\boldsymbol{u}) := \frac{1}{2}(\nabla_x \boldsymbol{u} + (\nabla_x \boldsymbol{u})^{\mathrm{T}})$ is the same at all points of the flow field (cf. p. 56, Sect. 13.1 of Bird et al. (1987) and, in particular, eq. (13.2-13) on p. 62, where the velocity gradient $\kappa := \nabla_x \boldsymbol{u}$ is explicitly assumed to be independent of the spatial position $\boldsymbol{x}$ in the flow field Ω). The justification to couple the Fokker–Planck equation derived in Bird et al. (1987) to the Navier–Stokes system in the context of mathematical analysis of the existence of weak solutions, where $\mathbb{D}(\boldsymbol{u}) \in L^2(0, T; L^2(\Omega; \mathbb{R}^{d\times d}))$ with $d = 2, 3$, and may be highly irregular as a function of $\boldsymbol{x}$, is therefore questionable. For this reason, following Barrett and Süli (2007) and Fritz et al. (2024), we shall revisit the derivation of the Fokker–Planck equation featuring in the coupled Navier–Stokes–Fokker–Planck system under consideration here in the physically most relevant case of $d = 3$ space dimensions; we shall do so *without* assuming that $\nabla_x \boldsymbol{u}$ is independent of $\boldsymbol{x}$. Significantly, our derivation will reveal that the resulting Fokker–Planck equation contains an additional term, referred to as *centre-of-mass diffusion*, which is absent from the Fokker–Planck equation derived in Bird et al. (1987).

Let $\Omega \subset \mathbb{R}^3$ denote the flow domain and let us denote by $\boldsymbol{u}(\boldsymbol{x}, t)$ the velocity of the fluid under consideration at a point $\boldsymbol{x} \in \Omega$ at time $t \geq 0$. We shall idealize each polymer molecule immersed in the solvent, a viscous incompressible Newtonian fluid, as a collection of $K + 1$ massless beads connected pairwise by a massless elastic spring (cf. Fig. 1), with $\boldsymbol{r}_j$ denoting the position vector of the centroid of the j-th bead with respect to an arbitrary but fixed reference point $O \in \mathbb{R}^3$, and with $\boldsymbol{q}_j := \boldsymbol{r}_{j+1} - \boldsymbol{r}_j$ signifying the so-called *conformation vector* of the j-th spring in the chain for $j \in \{1, \ldots, K\}$. This collection of beads and springs is called the *Rouse chain*, after P. E. Rouse Jr. (1953). In the simplest case, when $K = 1$, the Rouse chain, consisting of two massless beads connected by a single massless spring, is referred to as a *dumbbell*. For the sake of simplicity of the exposition we shall, initially at least, confine ourselves to the dumbbell model; we shall return to the general case in Sect. 3, where we shall suppose that $K \geq 1$.

Let us denote the position vectors with respect to O of the centroids of the two beads of the dumbbell at time t by $\boldsymbol{r}_j(t) \in \mathbb{R}^3$ for $j \in \{1, 2\}$. At time t, the conformation vector pointing from $\boldsymbol{r}_1(t)$ to $\boldsymbol{r}_2(t)$ is $\boldsymbol{q}^{1,2}(t) := \boldsymbol{r}_2(t) - \boldsymbol{r}_1(t)$. We denote the vector pointing in the opposite direction by $\boldsymbol{q}^{2,1}(t) := -\boldsymbol{q}^{1,2}(t)$ and assume that

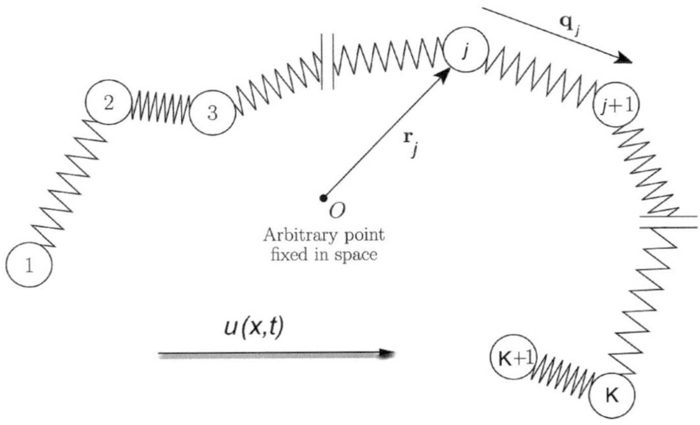

Fig. 1 Rouse chain, consisting of $K+1$ massless beads and K massless elastic springs connecting the beads, where $\boldsymbol{r}_j$ is the position vector of the centroid of the j-th bead with respect to an arbitrary but fixed reference point $O \in \mathbb{R}^3$, and $\boldsymbol{q}_j := \boldsymbol{r}_{j+1} - \boldsymbol{r}_j$ signifies the *conformation vector* of the j-th spring in the chain for $j \in \{1, \ldots, K\}$. The polymer molecule is assumed to be immersed in a Newtonian fluid, which at point $\boldsymbol{x} \in \mathbb{R}^3$ and time $t \geq 0$ has velocity $\boldsymbol{u}(\boldsymbol{x}, t)$

$\boldsymbol{q}^{1,2}(t)$ and $\boldsymbol{q}^{2,1}(t)$ are contained, for all $t \geq 0$, within a given convex open set $D \subset \mathbb{R}^3$ such that $\mathbf{0} \in D$, and $-\boldsymbol{q} \in D$ whenever $\boldsymbol{q} \in D$. Typically, D is assumed to be a bounded open ball centred at the origin $\mathbf{0}$ (not to be confused with O) in $\mathbb{R}^3$ in the case of FENE-type spring models, or it is taken to be the whole of $\mathbb{R}^3$ in the case of Hookean-type spring models, involving springs that can be, theoretically at least, stretched to any positive length.

In the absence of external forces, three kinds of force act on the j-th bead in the dumbbell suspended in the solvent: a drag force (Stokes drag) arising from the motion of the bead through the solvent; an elastic spring force exerted by the spring; and a random force, which is assumed to be Brownian, modelling the random collisions of the bead with the molecules of the surrounding Newtonian solvent. As each of the two beads is assumed to be massless, Newton's second law implies that, at time $t > 0$,

$$\mathbf{0} = \textbf{Drag Force}_j(t) + \textbf{Spring Force}_j(t) + \textbf{Brownian Force}_j(t), \quad j = 1, 2. \tag{1}$$

According to Stokes' law, the hydrodynamic drag force acting on the j-th bead of the dumbbell suspended in the fluid is

$$\textbf{Drag Force}_j = -\zeta\Big(\frac{\mathrm{d}}{\mathrm{d}t}\boldsymbol{r}_j(t) - \boldsymbol{u}(\boldsymbol{r}_j(t), t)\Big), \quad j \in \{1, 2\},$$

because the bead is suspended in a solvent, which at point $\boldsymbol{r}_j(t)$ and time t has velocity $\boldsymbol{u}(\boldsymbol{r}_j(t), t)$, and the relative velocity of the bead is therefore $\frac{\mathrm{d}}{\mathrm{d}t}\boldsymbol{r}_j(t) - \boldsymbol{u}(\boldsymbol{r}_j(t), t)$.

Here $\zeta > 0$ is the friction coefficient, which carries the SI unit [kg/s] and is linear in the dynamic viscosity $\eta > 0$; for a sphere of radius $R > 0$. one has that $\zeta = 6\pi R\eta$.

The elastic spring force $\boldsymbol{F} : D \to \mathbb{R}^3$ of the spring connecting the two beads is

$$\boldsymbol{F}(\boldsymbol{q}) := H\,U'\big(\tfrac{1}{2}|\boldsymbol{q}|^2\big)\boldsymbol{q}, \quad \boldsymbol{q} \in \{\boldsymbol{q}^{1,2}, \boldsymbol{q}^{2,1}\},$$

where U is a nonnegative continuously differentiable potential defined on a bounded interval of the form $[0, b)$ with $b > 0$ in the case of FENE-type models, or on the whole of $[0, \infty)$ in the case of Hookean-type models, and $H > 0$ is a spring constant having the SI unit [kg/s^2]. In the case of the classical Hookean dumbbell model the spring force is linear in $\boldsymbol{q}$ and is given by $\boldsymbol{F}(\boldsymbol{q}) = H\boldsymbol{q}$ with $\boldsymbol{q} \in D = \mathbb{R}^3$ and the corresponding potential is $U(s) = s$ for $s \in [0, \infty)$, while in the case of a classical FENE model proposed by Warner (1972) one has, instead,

$$D = B(\boldsymbol{0}, q_{\max}), \quad \boldsymbol{F}(\boldsymbol{q}) = \frac{H\boldsymbol{q}}{1 - |\boldsymbol{q}|^2/q_{\max}^2}, \quad U(s) = -\frac{q_{\max}^2}{2}\ln\left(1 - \frac{2s}{q_{\max}^2}\right)$$

for $\boldsymbol{q} \in D$ and $s \in [0, \frac{1}{2}q_{\max}^2)$, where $B(\boldsymbol{0}, q_{\max})$ is an open ball in $\mathbb{R}^3$ with radius $q_{\max} > 0$ centred at the origin $\boldsymbol{0} \in \mathbb{R}^3$ and $q_{\max}$ is a strict upper bound on the maximal admissible length to which a dumbbell can be stretched, i.e., $|\boldsymbol{q}| < q_{\max}$.

The Brownian force acting on the j-th bead at time $t > 0$ is denoted by $\boldsymbol{B}_j(t)$ and is formally defined by

$$\boldsymbol{B}_j(t) := \sqrt{2k_B\mu_T\zeta}\,\frac{\mathrm{d}\boldsymbol{W}_j(t)}{\mathrm{d}t}, \quad j \in \{1, 2\},$$

where $k_B = 1.380649 \times 10^{-23}$[kg m^2/(s^{2}K)] is the Boltzmann constant, μ_T denotes the temperature in Kelvins [K], and $\boldsymbol{W}_1(t)$ and $\boldsymbol{W}_2(t)$ are two independent Wiener processes. We define $\boldsymbol{W}(t) := \big(\boldsymbol{W}_1(t)^{\mathrm{T}}, \boldsymbol{W}_2(t)^{\mathrm{T}}\big)^{\mathrm{T}}$ (as is usual, for a column-vector $z \in \mathbb{R}^3$, z^{T} denotes the row-vector obtained by transposition of z). As each of the two Wiener processes is distributed according to $\mathcal{N}(0, t)$, i.e., has normal distribution with mean 0 and variance t, the SI unit of $\mathrm{d}\boldsymbol{W}_j(t)/\mathrm{d}t$ is $[1/\sqrt{\mathrm{s}}]$, $j = 1, 2$.

By introducing the notation

$$\boldsymbol{R}(t) := \begin{pmatrix} \boldsymbol{r}_1(t) \\ \boldsymbol{r}_2(t) \end{pmatrix} \in \mathbb{R}^6, \quad \boldsymbol{b}(\boldsymbol{R}(t), t) := \begin{pmatrix} \boldsymbol{u}(\boldsymbol{r}_1(t), t) + \zeta^{-1}\boldsymbol{F}(\boldsymbol{q}^{1,2}(t)) \\ \boldsymbol{u}(\boldsymbol{r}_2(t), t) + \zeta^{-1}\boldsymbol{F}(\boldsymbol{q}^{2,1}(t)) \end{pmatrix} \in \mathbb{R}^6, \tag{2}$$

and dividing the equality (1) by ζ, we obtain from (1) the following Itô stochastic differential equation

$$\mathrm{d}\boldsymbol{R}(t) = \boldsymbol{b}(\boldsymbol{R}(t), t)\,\mathrm{d}t + \sqrt{\frac{2k_B\mu_T}{\zeta}}\,\mathrm{d}\boldsymbol{W}(t), \quad t > 0,$$

called *Langevin's equation*. Instead of tracking the evolution in time of the random vector $\boldsymbol{R}(t)$, $t > 0$, we shall be interested in the evolution of its probability density function $\tilde{\psi} = \tilde{\psi}(\boldsymbol{r}_1, \boldsymbol{r}_2, t)$. The partial differential equation describing the evolution of $\tilde{\psi}$ is the Fokker–Planck equation (cf., for example, Pavliotis (2014)):

$$\partial_t \tilde{\psi} = \sum_{j=1}^{2} \left(-\operatorname{div}_{\boldsymbol{r}_j} \big(\boldsymbol{b}_j((\boldsymbol{r}_1^{\mathrm{T}}, \boldsymbol{r}_2^{\mathrm{T}})^{\mathrm{T}}, t)\, \tilde{\psi}\big) + \frac{k_B \mu_T}{\zeta} \Delta_{\boldsymbol{r}_j} \tilde{\psi} \right),$$

with $\boldsymbol{r}_1$ and $\boldsymbol{r}_2$ considered to be column-vectors in $\mathbb{R}^3$, while $\boldsymbol{b}_j$, $j \in \{1, 2\}$, are the two components of the vector-function $\boldsymbol{b}$ defined in (2).

Next, we perform the linear change of variables $(\boldsymbol{r}_1, \boldsymbol{r}_2) \mapsto (\frac{1}{2}(\boldsymbol{r}_1 + \boldsymbol{r}_2), \boldsymbol{r}_2 - \boldsymbol{r}_1) =: (\boldsymbol{x}, \boldsymbol{q})$; hence, $\boldsymbol{r}_1 = \boldsymbol{x} - \frac{1}{2}\boldsymbol{q}$ and $\boldsymbol{r}_2 = \boldsymbol{x} + \frac{1}{2}\boldsymbol{q}$. By recalling the definition (2) of the vector-function $\boldsymbol{b}(\cdot, \cdot)$ and setting $\psi(\boldsymbol{x}, \boldsymbol{q}, t) := \tilde{\psi}(\boldsymbol{x} - \frac{1}{2}\boldsymbol{q}, \boldsymbol{x} + \frac{1}{2}\boldsymbol{q}, t)$, we find that

$$\begin{aligned} &\partial_t \psi + \operatorname{div}_{\boldsymbol{x}} \left(\frac{\boldsymbol{u}(\boldsymbol{x} - \frac{1}{2}\boldsymbol{q}, t) + \boldsymbol{u}(\boldsymbol{x} + \frac{1}{2}\boldsymbol{q}, t)}{2} \psi \right) \\ &+ \operatorname{div}_{\boldsymbol{q}} \left(\big(\boldsymbol{u}(\boldsymbol{x} + \tfrac{1}{2}\boldsymbol{q}, t) - \boldsymbol{u}(\boldsymbol{x} - \tfrac{1}{2}\boldsymbol{q}, t)\big)\psi - \frac{2\boldsymbol{F}(\boldsymbol{q})}{\zeta} \psi \right) \\ &= \frac{k_B \mu_T}{2\zeta} \Delta_{\boldsymbol{x}} \psi + \frac{2 k_B \mu_T}{\zeta} \Delta_{\boldsymbol{q}} \psi. \end{aligned} \tag{3}$$

In contrast with the global homogeneity assumption adopted in Bird et al. (1987) mentioned at the beginning of this section, which requires that $\nabla_{\boldsymbol{x}} \boldsymbol{u}$ is independent of $\boldsymbol{x} \in \Omega$, we shall assume instead 'local homogeneity', i.e., that the spatial variation of the velocity field over the microscopic length-scale of a single dumbbell is negligibly small. Consequently, the arithmetic mean $\big(\boldsymbol{u}(\boldsymbol{x} - \frac{1}{2}\boldsymbol{q}, t) + \boldsymbol{u}(\boldsymbol{x} + \frac{1}{2}\boldsymbol{q}, t)\big)/2 = \boldsymbol{u}(\boldsymbol{x}, t) + \mathcal{O}(|\boldsymbol{q}|^2)$, and the second term on the left-hand side of (3) can be therefore approximated by $\operatorname{div}_{\boldsymbol{x}}(\boldsymbol{u}\psi)$ on the grounds that $|\boldsymbol{q}| \ll 1$. In the third term on the left-hand side of (3) we use Taylor expansion of $\boldsymbol{u}$ about the point $\boldsymbol{x}$ to obtain

$$\boldsymbol{u}(\boldsymbol{x} + \tfrac{1}{2}\boldsymbol{q}, t) - \boldsymbol{u}(\boldsymbol{x} - \tfrac{1}{2}\boldsymbol{q}, t) = \nabla_{\boldsymbol{x}} \boldsymbol{u}(\boldsymbol{x}, t)\boldsymbol{q} + \mathcal{O}(|\boldsymbol{q}|^3),$$

and we neglect the $\mathcal{O}(|\boldsymbol{q}|^3)$ term, again on the grounds that $|\boldsymbol{q}| \ll 1$. Thus, up to these, respectively, $\mathcal{O}(|\boldsymbol{q}|^2)$ and $\mathcal{O}(|\boldsymbol{q}|^3)$ approximation errors that we ignore, we end up with the following Fokker–Planck equation:

$$\partial_t \psi + \operatorname{div}_{\boldsymbol{x}}(\boldsymbol{u}\psi) + \operatorname{div}_{\boldsymbol{q}} \big((\nabla_{\boldsymbol{x}} \boldsymbol{u})\boldsymbol{q}\psi\big) = \frac{k_B \mu_T}{2\zeta} \Delta_{\boldsymbol{x}} \psi + \frac{2 k_B \mu_T}{\zeta} \Delta_{\boldsymbol{q}} \psi + \operatorname{div}_{\boldsymbol{q}} \Big(\frac{2\boldsymbol{F}(\boldsymbol{q})}{\zeta} \psi \Big), \tag{4}$$

where $\boldsymbol{u} = \boldsymbol{u}(\boldsymbol{x}, t)$, with $\boldsymbol{x} \in \Omega$ and $t > 0$. The partial differential equation (4) can be initialized by requiring that $\psi(\boldsymbol{x}, \boldsymbol{q}, 0) = \psi_0(\boldsymbol{x}, \boldsymbol{q})$, where $\psi_0 \geq 0$ is a given function such that $\int_{\Omega \times D} \psi_0(\boldsymbol{x}, \boldsymbol{q})\, \mathrm{d}\boldsymbol{q}\, \mathrm{d}\boldsymbol{x} = 1$.

Suppose for simplicity that the boundary $\partial\Omega$ of the macroscopic flow domain $\Omega \subset \mathbb{R}^3$ has no inflow or outflow parts. Suppose further that $T > 0$ and assume that $t \in [0, T]$. The polymeric fluid under consideration is therefore confined to Ω, and its velocity will be supposed to satisfy the no-slip boundary condition $\boldsymbol{u}(\boldsymbol{x}, t) = \boldsymbol{0}$ for all $(\boldsymbol{x}, t) \in \partial\Omega \times (0, T]$. Thus, the total number $N \gg 1$ of polymer molecules contained in Ω remains constant in time. The number density $\varrho_P := \varrho_P(\boldsymbol{x}, t)$, in $[m^{-3}]$, of the polymer molecules contained in Ω is called the *polymer number density* and it is related to N by $N = \int_\Omega \varrho_P(\boldsymbol{x}, t)\,\mathrm{d}\boldsymbol{x}$ for all $t \in [0, T]$. The polymer number density is further related to the probability density function ψ satisfying the Fokker–Planck equation (4) by

$$\varrho_P(\boldsymbol{x}, t) = N \int_D \psi(\boldsymbol{x}, \boldsymbol{q}, t)\,\mathrm{d}\boldsymbol{q}, \quad (\boldsymbol{x}, t) \in \Omega \times [0, T], \tag{5}$$

with the understanding that ψ has the SI unit of $[m^{-6}]$. We shall supplement the Fokker–Planck equation (4) with a no-flux boundary condition on $\partial\Omega \times D \times (0, T]$ and (if D is bounded) on $\Omega \times \partial D \times (0, T]$. By integrating (4) over $\Omega \times D$ and using the divergence theorem, it then follows that $\frac{\mathrm{d}}{\mathrm{d}t} \int_{\Omega \times D} \psi(\boldsymbol{x}, \boldsymbol{q}, t)\,\mathrm{d}\boldsymbol{q}\,\mathrm{d}\boldsymbol{x} = 0$, and therefore

$$\int_{\Omega \times D} \psi(\boldsymbol{x}, \boldsymbol{q}, t)\,\mathrm{d}\boldsymbol{q}\,\mathrm{d}\boldsymbol{x} = \int_{\Omega \times D} \psi(\boldsymbol{x}, \boldsymbol{q}, 0)\,\mathrm{d}\boldsymbol{q}\,\mathrm{d}\boldsymbol{x} = \int_{\Omega \times D} \psi_0(\boldsymbol{x}, \boldsymbol{q})\,\mathrm{d}\boldsymbol{q}\,\mathrm{d}\boldsymbol{x} = 1$$

for all $t \in [0, T]$. By integrating the Fokker–Planck equation (4) over D and recalling the assumed no-flux boundary condition for ψ imposed on $\Omega \times \partial D \times (0, T]$, we deduce that ϱ_P satisfies the following parabolic partial differential equation:

$$\partial_t \varrho_P + \mathrm{div}_{\boldsymbol{x}}(\boldsymbol{u}\varrho_P) = \frac{k_B \mu_T}{2\zeta} \Delta_{\boldsymbol{x}} \varrho_P, \quad (\boldsymbol{x}, t) \in \Omega \times (0, T],$$

subject to the initial condition $\varrho_P(\boldsymbol{x}, 0) = N \int_D \psi_0(\boldsymbol{x}, \boldsymbol{q})\,\mathrm{d}\boldsymbol{q}$ for $\boldsymbol{x} \in \Omega$ and a no-flux boundary condition on $\partial\Omega \times (0, T]$ obtained from the no-flux boundary condition imposed on $\partial\Omega \times D \times (0, T]$ for Eq. (4) by integrating it over $\boldsymbol{q} \in D$.

The presence of the $N \gg 1$ polymer molecules in the Newtonian solvent contributes to the stress tensor appearing in the balance-of-linear-momentum equation in the Navier–Stokes system an additional term, $\mathbb{T} = \mathbb{T}(\psi)$, a symmetric polymeric extra-stress tensor, which we shall define below. In other words, the Cauchy stress tensor featuring in the macroscopic equation is $2\eta\,\mathbb{D}(\boldsymbol{u}) + (-p)\mathbb{I}_3 + \mathbb{T}(\psi)$, where $\eta > 0$ denotes the (constant) dynamic viscosity, $\mathbb{I}_3$ is the 3×3 identity matrix, and p is the pressure. The evolution of the velocity field $\boldsymbol{u}$ and the pressure p is thereby governed by the incompressible Navier–Stokes-like system

$$\begin{aligned} \partial_t(\varrho\boldsymbol{u}) + \mathrm{div}_{\boldsymbol{x}}(\varrho\boldsymbol{u} \otimes \boldsymbol{u}) - \mathrm{div}_{\boldsymbol{x}}(2\eta\,\mathbb{D}(\boldsymbol{u}) - p\mathbb{I}_3) = \varrho\boldsymbol{f} + \mathrm{div}_{\boldsymbol{x}}\,\mathbb{T}(\psi) \quad & \text{in } \Omega \times (0, T], \\ \mathrm{div}_{\boldsymbol{x}}\,\boldsymbol{u} = 0 \quad & \text{in } \Omega \times (0, T], \end{aligned}$$

supplemented with the initial condition $\boldsymbol{u}(\boldsymbol{x},0)=\boldsymbol{u}_0(\boldsymbol{x})$ for $\boldsymbol{x}\in\Omega$, and the homogeneous Dirichlet boundary condition $\boldsymbol{u}=\boldsymbol{0}$ on $\partial\Omega\times(0,T]$.

Here, $\varrho>0$ stands for the macroscopic density, which we shall, at this point at least, assume to be constant in space and time, and $\boldsymbol{f}$ is the density of external body forces (for example, gravitational force). The polymeric extra-stress tensor $\mathbb{T}=\mathbb{T}(\psi)$ is defined by the so-called *Kramers expression*, see, e.g., Bird et al. (1987),

$$\mathbb{T}(\psi):=\varrho_P\big(\mathbb{C}(\psi)-k_B\mu_T\mathbb{I}_3\big),$$

where, as before, ϱ_P is the polymer number density, and k_B and μ_T are the Boltzmann constant and the temperature, respectively. Finally, the 3×3 symmetric tensor $\mathbb{C}(\psi)$ appearing in the expression for $\mathbb{T}(\psi)$ is defined by

$$\mathbb{C}(\psi)(\boldsymbol{x},t):=\frac{\int_D \boldsymbol{F}(\boldsymbol{q})\boldsymbol{q}^{\mathrm{T}}\psi(\boldsymbol{x},\boldsymbol{q},t)\,\mathrm{d}\boldsymbol{q}}{\int_D \psi(\boldsymbol{x},\boldsymbol{q},t)\,\mathrm{d}\boldsymbol{q}},$$

see, e.g., Süli and Yahiaoui (2018). Because the polymer number density ϱ_P is related to ψ by (5), the polymeric extra-stress tensor is simplified to

$$\mathbb{T}(\psi)=N\left\{\int_D \boldsymbol{F}(\boldsymbol{q})\boldsymbol{q}^{\mathrm{T}}\psi\,\mathrm{d}\boldsymbol{q}-k_B\mu_T\int_D\psi\,\mathrm{d}\boldsymbol{q}\;\mathbb{I}_3\right\}.$$

As ψ is assumed to have the SI unit $[m^{-6}]$, it follows from the definition of $\mathbb{C}(\psi)$ that $\mathbb{T}=\mathbb{T}(\psi)$ has SI unit $[kg/(s^2m^2)]$. After dividing the Navier–Stokes momentum equation by the constant macroscopic density ϱ and introducing the *kinematic viscosity* as $\nu:=\eta/\varrho$, we arrive at the following Navier–Stokes-like system:

$$\begin{aligned}
\partial_t\boldsymbol{u}+\mathrm{div}_{\boldsymbol{x}}(\boldsymbol{u}\otimes\boldsymbol{u})-\nu\Delta_x\boldsymbol{u}+\frac{1}{\varrho}\nabla_x p&=\boldsymbol{f}+\frac{1}{\varrho}\,\mathrm{div}_{\boldsymbol{x}}\,\mathbb{T}(\psi) &&\text{in }\Omega\times(0,T],\\
\mathrm{div}_{\boldsymbol{x}}\,\boldsymbol{u}&=0 &&\text{in }\Omega\times(0,T],
\end{aligned}\tag{6}$$

where we have used that $\mathrm{div}_{\boldsymbol{x}}(\nabla_x\boldsymbol{u})=\Delta_x\boldsymbol{u}$ and $\mathrm{div}_{\boldsymbol{x}}((\nabla_x\boldsymbol{u})^{\mathrm{T}})=\nabla_x(\mathrm{div}_{\boldsymbol{x}}\boldsymbol{u})=\boldsymbol{0}$.

Next, we shall perform a nondimensionalization of the Navier–Stokes–Fokker–Planck system. We define the nondimensional quantities $\widehat{\boldsymbol{x}}$, $\widehat{t}$ and $\widehat{\boldsymbol{u}}$ by setting

$$\boldsymbol{x}=L_0\widehat{\boldsymbol{x}},\quad t=T_0\widehat{t},\quad \boldsymbol{u}(\boldsymbol{x},t)=U_0\widehat{\boldsymbol{u}}(\widehat{\boldsymbol{x}},\widehat{t}),$$

where L_0 and T_0 signify the characteristic macroscopic length-scale and time-scale, respectively, and U_0 denotes the characteristic speed of the macroscopic flow. Similarly, we introduce $\widehat{\boldsymbol{q}}$ by letting $\boldsymbol{q}=\ell_0\widehat{\boldsymbol{q}}$, where ℓ_0 is a characteristic microscopic length-scale and we recall that ψ was assumed to have the SI unit $[m^{-6}]$; we then define the dimensionless quantity $\widehat{\psi}$ by

$$\widehat{\psi}(\widehat{\boldsymbol{x}},\widehat{\boldsymbol{q}},\widehat{t}):=(L_0\ell_0)^3\,\psi(\boldsymbol{x},\boldsymbol{q},t).\tag{7}$$

Thus we obtain from Eq. (4) the following partial differential equation:

$$\frac{1}{T_0}\partial_{\hat{t}}\widehat{\psi} + \frac{U_0}{L_0}\,\mathrm{div}_{\widehat{x}}(\widehat{\boldsymbol{u}}(\psi)\widehat{\psi}) + \frac{U_0}{L_0}\,\mathrm{div}_{\widehat{q}}\left(\nabla_{\widehat{x}}(\widehat{\boldsymbol{u}}(\psi))\widehat{\boldsymbol{q}}\,\widehat{\psi}\right) \\ = \frac{k_B\mu_T}{2\zeta L_0^2}\Delta_{\widehat{x}}\,\widehat{\psi} + \frac{2k_B\mu_T}{\zeta\ell_0^2}\Delta_{\widehat{q}}\,\widehat{\psi} + \frac{2\,H}{\zeta}\,\mathrm{div}_{\widehat{q}}(\widehat{\boldsymbol{F}}(\widehat{\boldsymbol{q}})\widehat{\psi}), \tag{8}$$

defined on the nondimensionalized domain $\widehat{\Omega}\times\widehat{D}\times(0,\widehat{T})$, where $\widehat{\Omega} := \{\widehat{\boldsymbol{x}} = \boldsymbol{x}/L_0 : \boldsymbol{x}\in\Omega\}$, $\widehat{D} := \{\widehat{\boldsymbol{q}} = \boldsymbol{q}/L_0 : \boldsymbol{q}\in D\}$ and $\widehat{T} := T/T_0$. Here we have used the notation $\widehat{\boldsymbol{u}}(\psi)$ to indicate that the velocity field is understood to depend on the original (dimensional) probability density function ψ, and will only be expressed as a function depending on the nondimensionalized probability density function $\widehat{\psi}$ after the Navier–Stokes system, coupled to the Fokker–Planck equation through the polymeric extra-stress tensor, has also been nondimensionalized. We set the characteristic macroscopic time-scale to $T_0 := L_0/U_0$. The nondimensionalized force in the case of the classical Hookean dumbbell model is $\widehat{\boldsymbol{F}}(\widehat{\boldsymbol{q}}) := \widehat{\boldsymbol{q}}$ with $\widehat{\boldsymbol{q}}\in\widehat{D} = \mathbb{R}^d$, while for the classical FENE model it is defined by

$$\widehat{\boldsymbol{F}}(\widehat{\boldsymbol{q}}) := \frac{\widehat{\boldsymbol{q}}}{1-|\widehat{\boldsymbol{q}}|^2/\widehat{q}_{\max}^2},\quad \widehat{\boldsymbol{q}}\in B(\boldsymbol{0},\widehat{q}_{\max})\subset\mathbb{R}^d,\quad \widehat{q}_{\max} := q_{\max}/\ell_0.$$

We shall fix the microscopic length-scale ℓ_0 by demanding that the positive prefactors of the last two terms on the right-hand side of (8) are equal; hence, $\ell_0^2 := k_B\mu_T/H$. Next we introduce the Deborah number λ, defined as the ratio of the time $\zeta/(4H)$ it takes for the material to adjust to applied stresses/deformations and the characteristic time-scale, T_0:

$$\lambda := \frac{\zeta/(4H)}{T_0} = \frac{\zeta}{4HT_0} = \frac{\zeta\ell_0^2}{4k_B\mu_T T_0}.$$

Finally, we define the positive nondimensional parameter

$$\varepsilon := \frac{k_B\mu_T}{2\zeta U_0 L_0},$$

whose numerator and denominator both carry the SI unit of $\mathrm{J} = [\mathrm{kg\,m^2/s^2}]$ corresponding to energy. Thus, we obtain from (8) by multiplying it with T_0 the following nondimensional Fokker–Planck equation:

$$\partial_{\hat{t}}\widehat{\psi} + \mathrm{div}_{\widehat{x}}(\widehat{\boldsymbol{u}}(\psi)\widehat{\psi}) + \mathrm{div}_{\widehat{q}}\left(\nabla_{\widehat{x}}(\widehat{\boldsymbol{u}}(\psi))\widehat{\boldsymbol{q}}\,\widehat{\psi}\right) = \varepsilon\Delta_{\widehat{x}}\,\widehat{\psi} + \frac{1}{2\lambda}\,\mathrm{div}_{\widehat{q}}(\nabla_{\widehat{q}}\,\widehat{\psi} + \widehat{\boldsymbol{F}}(\widehat{\boldsymbol{q}})\widehat{\psi}). \tag{9}$$

The first term on the right-hand side of (9) is called the *centre-of-mass diffusion term*. As the ratio of the diffusion coefficients ε and $1/(2\lambda)$ featuring in the equation (9) is equal to $\frac{1}{4}(\ell_0/L_0)^2$, which is $\ll 1$, one may formally argue that the centre-of-mass diffusion term in (9) can be neglected; making this assertion rigorous is however far from obvious. Indeed, rigorous passage to the limit $\varepsilon \to 0_+$ *in the class of weak solutions* to the coupled Navier–Stokes–Fokker–Planck system is a nontrivial open problem. We shall therefore retain this term in the equation and will continue to work with a strictly positive *centre-of-mass diffusion coefficient* ε, as is implied by the derivation of the Fokker–Planck equation (9) performed above; emphatically, $\varepsilon \Delta_{\widehat{x}} \widehat{\psi}$ is *not* an artificially introduced regularization term: rather, it naturally arises in the derivation of the model (see also, El-Kareh and Leal (1989), Bhave et al. (1991), Süli and Yahiaoui (2018), Dostalík et al. (2020) concerning the presence of the centre-of-mass diffusion term in the Fokker–Planck equation).

Next, we perform a nondimensionalization of the Navier–Stokes system (6) by letting $\boldsymbol{x} = L_0\widehat{\boldsymbol{x}}$, $t = T_0\widehat{t}$ and $\boldsymbol{u}(\boldsymbol{x}, t) = U_0\widehat{\boldsymbol{u}}(\widehat{\boldsymbol{x}}, \widehat{t})$, and we multiply the resulting equation by $T_0/U_0 = L_0/U_0^2$; this then leads to the following nondimensionalized system posed on $\widehat{\Omega} \times (0, \widehat{T}]$:

$$\partial_{\widehat{t}}\widehat{\boldsymbol{u}} + \mathrm{div}_{\widehat{x}}(\widehat{\boldsymbol{u}} \otimes \widehat{\boldsymbol{u}}) - \frac{1}{\mathrm{Re}}\Delta_{\widehat{x}}\widehat{\boldsymbol{u}} + \nabla_{\widehat{x}}\widehat{p} = \widehat{\boldsymbol{f}} + \mathrm{div}_{\widehat{x}}\, \widehat{\mathbb{T}}(\widehat{\psi}), \quad \mathrm{div}_{\widehat{x}}\, \widehat{\boldsymbol{u}} = 0,$$

where $\mathrm{Re} := U_0L_0/\nu = \varrho U_0L_0/\eta$ is the Reynolds number and $\widehat{p}$ and $\widehat{\boldsymbol{f}}$ are the scaled pressure and source term, respectively. Furthermore, the dimensionless form of the Kramers tensor is

$$\widehat{\mathbb{T}}(\widehat{\psi})(\widehat{\boldsymbol{x}}, \widehat{t}) := \gamma \left\{ \int_{\widehat{D}} \widehat{\boldsymbol{F}}(\widehat{\boldsymbol{q}})\widehat{\boldsymbol{q}}^{\mathrm{T}}\widehat{\psi}(\widehat{\boldsymbol{x}}, \widehat{\boldsymbol{q}}, \widehat{t})\, \mathrm{d}\widehat{\boldsymbol{q}} - \int_{\widehat{D}} \widehat{\psi}(\widehat{\boldsymbol{x}}, \widehat{\boldsymbol{q}}, \widehat{t})\, \mathrm{d}\widehat{\boldsymbol{q}}\, \mathbb{I}_3 \right\},$$

where $\gamma := k_B\mu_T N/(\varrho U_0^2 L_0^3)$ is a positive dimensionless parameter. For the sake of simplicity, hereafter we omit all $\widehat{\cdot}$ symbols and we arrive at the following NSFP system:

$$\begin{aligned}
\partial_t\boldsymbol{u} + \mathrm{div}_{\boldsymbol{x}}(\boldsymbol{u} \otimes \boldsymbol{u}) - \frac{1}{\mathrm{Re}}\Delta_{\boldsymbol{x}}\boldsymbol{u} + \nabla_{\boldsymbol{x}} p &= \boldsymbol{f} + \mathrm{div}_{\boldsymbol{x}}\mathbb{T}(\psi) && \text{in } \Omega \times (0, T],\\
\mathrm{div}_{\boldsymbol{x}}\boldsymbol{u} &= 0 && \text{in } \Omega \times (0, T],\\
\partial_t\psi + \mathrm{div}_{\boldsymbol{x}}(\boldsymbol{u}\psi) + \mathrm{div}_{\boldsymbol{q}}\big((\nabla_{\boldsymbol{x}}\boldsymbol{u})\boldsymbol{q}\psi\big) & \\
- \tfrac{1}{2\lambda}\,\mathrm{div}_{\boldsymbol{q}}(\nabla_{\boldsymbol{q}}\psi + \boldsymbol{F}(\boldsymbol{q})\psi) &= \varepsilon\Delta_{\boldsymbol{x}}\psi && \text{in } \Omega \times D \times (0, T],
\end{aligned}$$

which we supplement with the following boundary conditions: $\boldsymbol{u} = \boldsymbol{0}$ on $\partial\Omega \times (0, T]$, $\varepsilon\nabla_{\boldsymbol{x}}\psi \cdot \boldsymbol{n}_{\partial\Omega} = 0$ on $\partial\Omega \times D \times (0, T]$, and, if D if bounded, then also

$$\big(\tfrac{1}{2\lambda}(\nabla_{\boldsymbol{q}}\psi + \boldsymbol{F}(\boldsymbol{q})\psi) - (\nabla_{\boldsymbol{x}}\boldsymbol{u})\boldsymbol{q}\psi\big) \cdot \boldsymbol{n}_{\partial D} = 0 \qquad \text{on } \Omega \times \partial D \times (0, T],$$

where $\boldsymbol{n}_{\partial\Omega}$ and $\boldsymbol{n}_{\partial D}$ are the unit outward normal vectors to $\partial\Omega$ and ∂D, respectively. The nondimensional initial data $\boldsymbol{u}_0 = \boldsymbol{u}_0(\boldsymbol{x})$ and $\psi_0 = \psi_0(\boldsymbol{x}, \boldsymbol{q})$ for the velocity field and the probability density function are obtained from their respective dimensional counterparts. We note that the scaling (7) ensures that $\int_{\Omega\times D} \psi_0(\boldsymbol{x}, \boldsymbol{q})\, \mathrm{d}\boldsymbol{x}\, \mathrm{d}\boldsymbol{q} = 1$; this then guarantees, thanks to the no-flux boundary conditions imposed on ψ, that $\int_{\Omega\times D} \psi(\boldsymbol{x}, \boldsymbol{q}, t)\, \mathrm{d}\boldsymbol{x}\, \mathrm{d}\boldsymbol{q} = 1$ for all $t \geq 0$.

3 The FENE Bead-Spring-Chain Model

We shall generalize the model derived in Sect. 2 in two directions: instead of a simple dumbbell model consisting of a pair of massless beads connected by a single massless elastic spring we shall consider a bead-spring-chain model consisting of $K+1$ massless beads, which are linearly connected by K massless springs, as shown in Fig. 1. In addition, we shall assume that the solvent, which occupies a bounded open Lipschitz domain $\Omega \subset \mathbb{R}^d$ where $d \in \{2, 3\}$, with boundary $\partial\Omega$, is an incompressible, viscous, isothermal Newtonian fluid with variable density ϱ and dynamic viscosity $\eta = \eta(\varrho, \varrho_P)$, where ϱ_P is the variable polymer number density. Let $T \in \mathbb{R}_{>0}$ denote the length of the time interval of interest and let $Q := \Omega \times (0, T]$. We consider the following system of partial differential equations over the space-time domain Q:

$$\frac{\partial \varrho}{\partial t} + \mathrm{div}_{\boldsymbol{x}}(\boldsymbol{u}\varrho) = 0, \quad \mathrm{div}_{\boldsymbol{x}}\boldsymbol{u} = 0,$$

$$\frac{\partial(\varrho\boldsymbol{u})}{\partial t} + \mathrm{div}_{\boldsymbol{x}}(\varrho\boldsymbol{u} \otimes \boldsymbol{u}) - \mathrm{div}_{\boldsymbol{x}}(2\eta(\varrho, \varrho_P)\,\mathbb{D}(\boldsymbol{u})) + \nabla_{\boldsymbol{x}} p = \varrho\boldsymbol{f} + \mathrm{div}_{\boldsymbol{x}}\mathbb{T},$$

subject to the initial conditions $\varrho(\cdot, 0) = \varrho_0(\cdot)$ and $(\varrho\boldsymbol{u})(\cdot, 0) = (\varrho_0\boldsymbol{u}_0)(\cdot)$ on Ω, and the no-slip boundary condition $\boldsymbol{u} = \boldsymbol{0}$ on $\partial\Omega \times (0, T]$. In these equations $\varrho : \Omega \times [0, T] \to \mathbb{R}$ denotes the solvent density, $\boldsymbol{u} : \overline{Q} \to \mathbb{R}^d$ is the solvent velocity, $\mathbb{T} : Q \to \mathbb{R}^{d\times d}_{\mathrm{sym}}$ is Kramers' polymeric extra-stress tensor, $\boldsymbol{f} : Q \to \mathbb{R}^d$ is the density of the external body forces, and $p : Q \to \mathbb{R}$ is the pressure.

In a bead-spring chain model for dilute polymers consisting of $K+1$ beads coupled by K elastic springs to represent a polymer chain, the polymeric extra-stress tensor $\mathbb{T}$ is defined by a version of the Kramers expression depending on the probability density function ψ, which in addition to $\boldsymbol{x} \in \Omega$ and $t \in [0, T]$ also depends on the conformation vector $\boldsymbol{q}$ of the chain, where now $\boldsymbol{q} := ((\boldsymbol{q}^1)^{\mathrm{T}}, \ldots, (\boldsymbol{q}^K)^{\mathrm{T}})^{\mathrm{T}} \in \mathbb{R}^{d\times K}$, with the column vector $\boldsymbol{q}^j := (q^j_1, \ldots, q^j_d)^{\mathrm{T}} \in \mathbb{R}^d$ representing the d-component conformation vector of the jth spring in the bead-spring chain. Let $D := D^1 \times \cdots \times D^K \subset \mathbb{R}^{d\times K}$ be the set of all admissible conformation vectors. Typically D^j is the whole space $\mathbb{R}^d$ or a bounded open ball centred at the origin $\boldsymbol{0}$ in $\mathbb{R}^d$, for each $j = 1, \ldots, K$. We focus on finitely extensible nonlinear elastic (FENE-type) models where $D^j := B(\boldsymbol{0}, \sqrt{b_j})$, a ball centred at the origin $\boldsymbol{0}$ in $\mathbb{R}^d$ and of radius $\sqrt{b_j}$, with $b_j > 0$ for each $j \in \{1, \ldots, K\}$. The continuously differentiable function $U^j : [0, b_j/2) \to [0, \infty)$, $j \in \{1, \ldots, K\}$, is the spring potential of the j-th spring, which is assumed to be such that $U^j(0) = 0$, $\lim_{s\to(b_j/2)_-} U^j(s) = +\infty$. When $K = 1$, the model reduces to the dumbbell model discussed in the previous section, with $q_{\max} = \sqrt{b_1}$. Kramers' polymeric extra-stress tensor $\mathbb{T}$ is now defined by the formula:

$$\mathbb{T}(\boldsymbol{x}, t) := \gamma \left(\sum_{j=1}^{K} \int_D \psi(\boldsymbol{x}, \boldsymbol{q}, t) \boldsymbol{q}^j {\boldsymbol{q}^j}^{\mathrm{T}} H\,(U^j)' \left(\frac{1}{2} |\boldsymbol{q}^j|^2 \right) \mathrm{d}\boldsymbol{q} - K \varrho_P(\boldsymbol{x}, t)\, \mathbb{I}_d \right), \tag{10}$$

where γ is a positive constant, $H > 0$ is a spring constant, $\mathbb{I}_d \in \mathbb{R}^{d\times d}$ denotes the $d \times d$ identity matrix, $\mathrm{d}\boldsymbol{q} := \mathrm{d}\boldsymbol{q}^1 \cdots \mathrm{d}\boldsymbol{q}^K$, and the polymer number density at position $\boldsymbol{x} \in \Omega$ at time $t \in [0, T]$ is $\varrho_P(\boldsymbol{x}, t)$. Finally, we shall assume that $\zeta(\varrho) = \zeta_0 z(\varrho)$ is a density-dependent positive function, where $\zeta_0 > 0$ is a characteristic friction coefficient and $z(\cdot)$ is a nondimensional positive function. We define the normalized partial Maxwellian M^j with respect to the variable $\boldsymbol{q}^j$ by

$$M^j(\boldsymbol{q}^j) := \frac{1}{Z^j} \mathrm{e}^{-U^j(\frac{1}{2}|\boldsymbol{q}^j|^2)}, \quad \text{where } Z^j := \int_{D^j} \mathrm{e}^{-U^j(\frac{1}{2}|\boldsymbol{q}^j|^2)}\, \mathrm{d}\boldsymbol{q}^j,$$

where $\mathrm{d}\boldsymbol{q}^j := \mathrm{d}\boldsymbol{q}^j_1 \cdots \mathrm{d}\boldsymbol{q}^j_d$, $j = 1, \dots, K$. The Maxwellian in the model is then

$$M(\boldsymbol{q}) := M^1(\boldsymbol{q}^1) \cdots M^K(\boldsymbol{q}^K), \qquad \boldsymbol{q} := (\boldsymbol{q}^1, \dots, \boldsymbol{q}^K) \in D := D^1 \times \cdots \times D^K.$$

We require the following notational conventions: $\nabla_{\boldsymbol{q}^j} := (\partial/\partial q^j_1, \dots, \partial/\partial q^j_d)^{\mathrm{T}}$, for $j = 1, \dots, K$, and we define $\mathrm{div}_{\boldsymbol{q}^j} := \nabla_{\boldsymbol{q}^j}\cdot$. For a general mapping $\boldsymbol{q} \in D \mapsto \mathbb{B}(\boldsymbol{q}) \in \mathbb{R}^{d\times K}$, we define $\mathrm{div}_{\boldsymbol{q}} \mathbb{B} := \mathrm{div}_{q^1} \boldsymbol{B}^1 + \cdots + \mathrm{div}_{q^K} \boldsymbol{B}^K$, where $\boldsymbol{B}^j$, $j = 1, \dots, K$, denote the d-component column vectors of the matrix $\mathbb{B} = \mathbb{B}(\boldsymbol{q})$, and we consider the $d \times K$-component differential operator $\nabla_{\boldsymbol{q}} := (\nabla_{\boldsymbol{q}^1}, \dots, \nabla_{\boldsymbol{q}^K})$.

The probability density function ψ satisfies the following Fokker–Planck equation:

$$\begin{aligned} &\frac{\partial \psi}{\partial t} + \mathrm{div}_{\boldsymbol{x}}(\boldsymbol{u}\psi) + \sum_{j=1}^{K} \nabla_{\boldsymbol{q}^j} \cdot ((\nabla_{\boldsymbol{x}} \boldsymbol{u}) \boldsymbol{q}^j \psi) \\ &\qquad = \varepsilon \Delta_{\boldsymbol{x}} \left(\frac{\psi}{z(\varrho)} \right) + \frac{1}{2\lambda} \sum_{i=1}^{K} \sum_{j=1}^{K} A_{ij} \nabla_{\boldsymbol{q}^j} \left(M \nabla_{\boldsymbol{q}^i} \left(\frac{\psi}{z(\varrho) M} \right) \right) \end{aligned} \tag{11}$$

on $\Omega \times D \times (0, T]$. Let $\partial \bar{D}^j := D^1 \times \cdots \times D^{j-1} \times \partial D^j \times D^{j+1} \times \cdots \times D^K$. We require that

$$\varepsilon \nabla_x \left(\frac{\psi}{z(\varrho)} \right) \cdot \boldsymbol{n} = 0 \qquad \text{on } \partial\Omega \times D \times (0, T],$$

$$\left[\frac{1}{2\lambda} \sum_{i=1}^{K} A_{ij} M \nabla_{q^i} \left(\frac{\psi}{z(\varrho) M} \right) - (\nabla_x \boldsymbol{u}) \boldsymbol{q}^j \psi \right] \cdot \frac{\boldsymbol{q}^j}{|\boldsymbol{q}^j|} = 0 \qquad \text{on } \Omega \times \partial \bar{D}^j \times (0, T],$$

for $j = 1, \ldots, K$, where $\boldsymbol{n} = \boldsymbol{n}_{\partial\Omega}$ is the unit outward normal vector to $\partial\Omega$, $\boldsymbol{q}^j/|\boldsymbol{q}^j| = \boldsymbol{n}_{\partial D^j}$ is the unit outward normal vector to ∂D^j. We further impose the initial condition $\psi(\boldsymbol{x}, \boldsymbol{q}, 0) = \psi_0(\boldsymbol{x}, \boldsymbol{q})$ on $\Omega \times D$. In (10), the constant $\gamma > 0$ involves the product of the Boltzmann constant k_B and the absolute temperature μ_T. In (11), the centre-of-mass diffusion coefficient $\varepsilon > 0$ is defined as $\varepsilon := (\ell_0/L_0)^2/(4(K+1)\lambda)$ with $\ell_0 := \sqrt{k_B \mu_T / H}$ signifying the characteristic microscopic length-scale and $\lambda := \zeta_0/(4HT_0)$. As in the previous section, the dimensionless positive parameter λ, called the Deborah number, characterizes the elastic relaxation property of the fluid, and L_0 and T_0 are the characteristic macroscopic length-scale and time-scale, respectively. The constant matrix $A = (A_{ij})_{i,j=1}^{K}$, called the *Rouse matrix*, is symmetric and positive definite. We associate with A the linear mapping $\mathbb{A} : \mathbb{R}^{d\times K} \to \mathbb{R}^{d\times K}$ defined, for any $B = (B_i^j)_{i=1,\ldots,d}^{j=1,\ldots,K} \in \mathbb{R}^{d\times K}$, by $(\mathbb{A}(B))_i^j := \sum_{k=1}^{K} B_i^k A_{kj}$. Finally, let

$$\mathcal{F}(s) := s(\log s - 1) + 1 \quad \text{for } s > 0, \qquad \text{with } \mathcal{F}(0) := 1.$$

We shall derive a formal energy identity under the assumption that $\boldsymbol{u}$, ϱ and ψ are sufficiently smooth, ϱ and ψ are nonnegative, and z is positive. By taking the $L^2(\Omega)$ inner product of the balance-of-mass equation with $\frac{1}{2}|\boldsymbol{u}|^2$ and the $L^2(\Omega; \mathbb{R}^d)$ inner product of the balance-of-linear-momentum equation in the Navier–Stokes system with $\boldsymbol{u}$, we deduce by partial integration and noting the homogeneous Dirichlet boundary condition for $\boldsymbol{u}$ and the divergence-free property of $\boldsymbol{u}$ that

$$\begin{aligned} &\frac{\mathrm{d}}{\mathrm{d}t} \int_\Omega \frac{1}{2} \varrho |\boldsymbol{u}|^2 \,\mathrm{d}\boldsymbol{x} + \int_\Omega 2\eta(\varrho, \varrho_P) |\mathbb{D}(\boldsymbol{u})|^2 \,\mathrm{d}\boldsymbol{x} \\ &\quad = \int_\Omega \varrho \boldsymbol{f} \cdot \boldsymbol{u} \,\mathrm{d}\boldsymbol{x} - \gamma \sum_{j=1}^{K} \int_{\Omega\times D} \psi \, (U^j)' \left(\frac{1}{2} |\boldsymbol{q}^j|^2 \right) \boldsymbol{q}^j \boldsymbol{q}^{j^{\mathrm{T}}} : \nabla_x \boldsymbol{u} \,\mathrm{d}\boldsymbol{q} \,\mathrm{d}\boldsymbol{x}. \end{aligned} \tag{12}$$

Next, we multiply the Fokker–Planck equation by $\mathcal{F}'\left(\frac{\psi}{z(\varrho)M}\right)$ and integrate over $\Omega \times D$. Using partial integration and the assumed boundary conditions in the second and third term on the left-hand side and the two terms on the right-hand side of the Fokker–Planck equation, we get

$$\begin{aligned} &\frac{\mathrm{d}}{\mathrm{d}t} \int_{\Omega\times D} M z(\varrho) \mathcal{F}\left(\frac{\psi}{z(\varrho)M} \right) \mathrm{d}\boldsymbol{q} \,\mathrm{d}\boldsymbol{x} - \sum_{j=1}^{K} \int_{\Omega\times D} \psi (U^j)' \left(\frac{1}{2} |\boldsymbol{q}^j|^2 \right) \boldsymbol{q}^j \boldsymbol{q}^{j^{\mathrm{T}}} : \nabla_x \boldsymbol{u} \,\mathrm{d}\boldsymbol{q} \,\mathrm{d}\boldsymbol{x} \\ &\quad + 4\varepsilon \int_{\Omega\times D} M \left| \nabla_x \sqrt{\frac{\psi}{z(\varrho)M}} \right|^2 \mathrm{d}\boldsymbol{q} \,\mathrm{d}\boldsymbol{x} \end{aligned}$$

$$+\frac{2}{\lambda}\int_{\Omega\times D} M\mathbb{A}\left(\nabla_{\boldsymbol{q}}\sqrt{\frac{\psi}{z(\varrho)M}}\right):\left(\nabla_{\boldsymbol{q}}\sqrt{\frac{\psi}{z(\varrho)M}}\right)\mathrm{d}\boldsymbol{q}\,\mathrm{d}\boldsymbol{x}=0.$$

We now multiply the last equality by γ and add the resulting identity to (12) to deduce the desired formal energy equality:

$$\begin{aligned}&\frac{\mathrm{d}}{\mathrm{d}t}\int_{\Omega}\left[\frac{1}{2}\varrho|\boldsymbol{u}|^2+\gamma\int_D Mz(\varrho)\mathcal{F}\left(\frac{\psi}{z(\varrho)M}\right)\mathrm{d}\boldsymbol{q}\right]\mathrm{d}\boldsymbol{x}+\int_{\Omega}2\eta(\varrho,\varrho_P)|\mathbb{D}(\boldsymbol{u})|^2\,\mathrm{d}\boldsymbol{x}\\&+4\gamma\varepsilon\int_{\Omega\times D}M\left|\nabla_{\boldsymbol{x}}\sqrt{\frac{\psi}{z(\varrho)M}}\right|^2\mathrm{d}\boldsymbol{q}\,\mathrm{d}\boldsymbol{x}\\&+\frac{2\gamma}{\lambda}\int_{\Omega\times D}M\mathbb{A}\left(\nabla_{\boldsymbol{q}}\sqrt{\frac{\psi}{z(\varrho)M}}\right):\left(\nabla_{\boldsymbol{q}}\sqrt{\frac{\psi}{z(\varrho)M}}\right)\mathrm{d}\boldsymbol{q}\,\mathrm{d}\boldsymbol{x}=\int_{\Omega}\varrho\boldsymbol{f}\cdot\boldsymbol{u}\,\mathrm{d}\boldsymbol{x}.\end{aligned}$$

This formal energy equality is the starting point of the proof of the existence of a global-in-time large-data weak solution to the model. It proceeds by constructing sequences of approximations to ϱ, $\boldsymbol{u}$ and ψ that satisfy an energy (in)equality, analogous to the one above, and using a compactness argument to pass to the respective limits to infer that the resulting limiting functions represent a weak solution to the problem (cf. Barrett and Süli (2012), He (2023), He and Süli (2024); see also Barrett and Süli (2011) and Bulíček et al. (2013) in the case of constant density, $K \geq 1$ and $\varepsilon > 0$, as well as (Masmoudi 2013), with constant density, $K = 1$, and $\varepsilon = 0$). By using techniques from stochastic analysis, in the case of a planar shear flow and the FENE dumbbell model a local-in-time existence and uniqueness result was proved in Jourdain et al. (2004).

4 The Hookean Dumbbell Model and Its Macroscopic Closure

In the case of the Hookean dumbbell model the polymeric extra-stress tensor $\mathbb{T}$ is again related to the probability density function ψ by *Kramers' expression* for the polymeric extra-stress tensor (with all physical constants scaled to 1):

$$\mathbb{T}(\boldsymbol{x},t):=\int_D\psi(\boldsymbol{x},\boldsymbol{q},t)\,\boldsymbol{q}\boldsymbol{q}^{\mathrm{T}}\,\mathrm{d}\boldsymbol{q}-\left(\int_D\psi(\boldsymbol{x},\boldsymbol{q},t)\,\mathrm{d}\boldsymbol{q}\right)\mathbb{I}_d.$$

Here, as before, $\boldsymbol{q}\boldsymbol{q}^{\mathrm{T}}\in\mathbb{R}^{d\times d}_{\mathrm{sym}}$, with entries $(\boldsymbol{q}\boldsymbol{q}^{\mathrm{T}})_{i,j}=q_iq_j$ for $i,j=1,\ldots,d$, and $\mathbb{I}_d\in\mathbb{R}^{d\times d}_{\mathrm{sym}}$ is the identity matrix. However, in contrast with the case of FENE-type dumbbell models where D is a bounded ball in $\mathbb{R}^d$, in the classical Hookean model (and throughout this section) $D=\mathbb{R}^d$ and $U(s)=s$ for all $s\geq 0$, and it is this apparently trivial change that gives rise to nontrivial complications in the mathematical analysis of the classical Hookean dumbbell model. We note in passing that when $D = \mathbb{R}^d$ but $\lim_{s\to+\infty} U(s)/s = +\infty$, the existence of a large-data global-in-time weak solution can still be shown (cf. Barrett and Süli (2012)), however it is the classical Hookean model, with $U(s) = s$, that we shall be concerned with here, for two reasons: (a) it is precisely this model whose (formal) macroscopic closure yields the Oldroyd-B model; and (b) no other choice of U is known to result in exact macroscopic closure. We shall therefore focus on the classical Hookean dumbbell model, our aim being to make the process of its macroscopic closure, based on formal calculations, rigorous.

The Maxwellian M featuring in the Fokker–Planck equation is defined in this case by $M(\boldsymbol{q}) := Z^{-1}\exp(-\frac{1}{2}|\boldsymbol{q}|^2)$, where $Z := \int_D \exp(-\frac{1}{2}|\boldsymbol{q}|^2)\,\mathrm{d}\boldsymbol{q}$. Using the Maxwellian, we define the normalized probability density function $\widehat{\psi} := \psi/M$, $\widehat{\psi}_0 := \psi_0/M$, where ψ_0 is the initial value of ψ, i.e., $\psi(\boldsymbol{x}, \boldsymbol{q}, 0) = \psi_0(\boldsymbol{x}, \boldsymbol{q})$ for all $(\boldsymbol{x}, \boldsymbol{q}) \in \Omega \times D$. The Fokker–Planck equation satisfied by $\widehat{\psi}$ on $\Omega \times D \times (0, T]$ is then

$$M\partial_t\widehat{\psi} + M\boldsymbol{u}\cdot\nabla_x\widehat{\psi} + \mathrm{div}_q\left((\nabla_x\boldsymbol{u})\boldsymbol{q}M\widehat{\psi}\right) - \varepsilon M\Delta_x\widehat{\psi} = \frac{1}{2\lambda}\,\mathrm{div}_q\left(M\nabla_q\widehat{\psi}\right).$$

We supplement the resulting NSFP system with the following initial and boundary/decay conditions: $\boldsymbol{u} = \boldsymbol{0}$ on $\partial\Omega \times (0, T]$, $\boldsymbol{u}(\boldsymbol{x}, 0) = \boldsymbol{u}_0(\boldsymbol{x})$ for all $\boldsymbol{x} \in \Omega$, and

$$\begin{aligned}
\left|M\left[\frac{1}{2\lambda}\nabla_q\left(\frac{\psi}{M}\right) - (\nabla_x\boldsymbol{u})\boldsymbol{q}\frac{\psi}{M}\right]\right| \to 0 \ \text{ as } |\boldsymbol{q}| \to \infty \qquad & \text{on } \Omega \times (0, T],\\
\nabla_q\psi\cdot\boldsymbol{n} = 0 \qquad & \text{on } \partial\Omega \times D \times (0, T],\\
\psi(\boldsymbol{x}, \boldsymbol{q}, 0) = \psi_0(\boldsymbol{x}, \boldsymbol{q}) \qquad & \text{for all } (\boldsymbol{x}, \boldsymbol{q}) \in \Omega \times D,
\end{aligned}$$

where $\boldsymbol{n} = \boldsymbol{n}_{\partial\Omega}$ denotes the unit outward normal vector to $\partial\Omega$ and ψ_0 is such that

$$\psi_0 \geq 0, \quad \int_D \psi_0(\boldsymbol{x}, \boldsymbol{q})\,\mathrm{d}\boldsymbol{q} = 1 \ \text{ for a.e.} \boldsymbol{x} \in \Omega,$$

and $|\Omega| = 1$. The boundary/decay conditions ensure that these properties are propagated in time. As the specific value of the Deborah number $\lambda > 0$ plays no role in the discussion that follows, we shall set $\lambda = 1/2$ for ease of writing.

Let us also recall the following *Oldroyd-B system with stress-diffusion*:

$$\begin{aligned}
&\partial_t\boldsymbol{u} + \mathrm{div}_x(\boldsymbol{u}\otimes\boldsymbol{u}) - \nu\Delta_x\boldsymbol{u} + \nabla_x p = \mathrm{div}_x\,\mathbb{T}, \quad \mathrm{div}_x\,\boldsymbol{u} = 0,\\
&\partial_t\mathbb{T} + (\boldsymbol{u}\cdot\nabla_x)\,\mathbb{T} - \left[(\nabla_x\boldsymbol{u})\mathbb{T} + \mathbb{T}(\nabla_x^{\mathrm{T}}\boldsymbol{u})\right] - \varepsilon\Delta_x\mathbb{T} + 2\mathbb{T} = 2\mathbb{D}(\boldsymbol{u}),
\end{aligned} \tag{13}$$

posed on $\Omega \times (0, T]$; the penultimate term on the left-hand side of $(13)_2$ is usually referred to as the stress-diffusion term. At least formally, this model is the macroscopic closure of the Navier–Stokes–Fokker–Planck system upon multiplying the Fokker–Planck equation by $\boldsymbol{q}\boldsymbol{q}^{\mathrm{T}}$ and integrating over $\boldsymbol{q} \in D = \mathbb{R}^d$. Note, however, that this assertion has not been rigorously proved for weak solutions, except in two space dimensions (cf. Barrett and Süli (2018)). For further reference, we introduce the conformation tensor $\mathbb{S} := \mathbb{T} + \mathbb{I}_d$, which satisfies

$$\partial_t \mathbb{S} + (\boldsymbol{u} \cdot \nabla_x)\,\mathbb{S} - \big[(\nabla_x \boldsymbol{u})\mathbb{S} + \mathbb{S}(\nabla_x^{\mathrm{T}} \boldsymbol{u})\big] - \varepsilon \Delta_x \mathbb{S} + 2\,(\mathbb{S} - \mathbb{I}_d) = \mathbb{O}. \qquad (14)$$

To make a clear distinction between solutions to the (macroscopic) Oldroyd-B system $(13)_2$ and (14) on the one hand, and the tensors $\mathbb{S}$ and $\mathbb{T}$ defined through a solution $\widehat{\psi}$ to the (macro-micro) Navier–Stokes–Fokker–Planck system system via the formal macroscopic closure described above on the other, we denote the latter by

$$\mathbb{S}(\widehat{\psi}) := \int_D M\widehat{\psi}\, \boldsymbol{q}\boldsymbol{q}^{\mathrm{T}}\, \mathrm{d}\boldsymbol{q} \quad \text{and} \quad \mathbb{T}(\widehat{\psi}) := \mathbb{S}(\widehat{\psi}) - \left(\int_D M\widehat{\psi}\, \mathrm{d}\boldsymbol{q}\right) \mathbb{I}_d.$$

The main difficulty in proving the existence of a weak solution to the Navier–Stokes–Fokker–Planck system in the case of the classical Hookean dumbbell model is the lack of sufficiently strong a priori bounds on $\widehat{\psi}$ so as to ensure weak compactness of Kramers' polymeric extra-stress tensor $\mathbb{T}(\widehat{\psi})$ in L^1. The global existence of a weak solution is known only in two spatial dimensions: it was proved by Barrett and Süli (2018) through a rigorous derivation of the macroscopic closure of the NSFP model, i.e., by rigorously establishing the connection between the macro-micro NSFP model and the macroscopic Oldroyd-B model. This result was later extended to a larger class of data by La (2020), who introduced the concept of moment solutions to the Fokker–Planck equation. The global existence of a weak solution to the two-dimensional Oldroyd-B model (13) with stress-diffusion was shown by Barrett and Boyaval (2011). Global regularity was then shown by Constantin and Kliegl (2012). In the absence of stress-diffusion (i.e., when $\varepsilon = 0$), local well-posedness was shown in critical Besov spaces by Chemin and Masmoudi (2001) under a Beale–Kato–Majda type assumption, while Lions and Masmoudi (2000) proved global existence of a weak solution to the so called *corotational Oldroyd-B model* without stress-diffusion.

In view of the difficulties associated with proving the global existence of a weak solution to the classical Hookean NSFP system in three space dimensions, we relax the definition of weak solution and consider a class of generalized solutions, referred to as *generalized dissipative solutions*. The main idea is to allow for a defect to develop in the Navier–Stokes equation; this entails substituting the polymeric extra-stress tensor, $\mathbb{T}$, by $\mathbb{T} + \mathbb{M}_{NS}$, where $\mathbb{M}_{NS}$ is a positive semidefinite tensor-valued measure. This approach was suggested by Barrett and Süli (2018), who showed the global existence of a weak *subsolution* to the Hookean Navier–Stokes–Fokker–Planck model with centre-of-mass diffusion in three dimensions. The idea of Barrett and Süli (2018) was revisited in the recent paper Dçbiec and Süli (2023), giving a different argument: crucially, a *dissipation defect* measure, m_{OB}, was identified, which dominates the aforementioned measure $\mathbb{M}_{NS}$ and satisfies an appropriate energy inequality—namely, the energy inequality associated with the macroscopic Oldroyd-B model with stress-diffusion. Using this energy inequality it was then possible to prove a compatibility result: any generalized dissipative solution to the Hookean

NSFP system with centre-of-mass diffusion that has appropriate regularity is in fact a weak solution. We shall summarize the main results of Dębiec and Süli (2023) here.

Let $L^p_M(\Omega \times D)$, $p \in [1, \infty)$, denote the Maxwellian-weighted Lebesgue space with the norm

$$\|\phi\|_{L^p_M(\Omega\times D)} := \left(\int_{\Omega\times D} M|\phi|^p \,\mathrm{d}\boldsymbol{q}\,\mathrm{d}\boldsymbol{x}\right)^{1/p},$$

where $\Omega \subset \mathbb{R}^d$ is a bounded open Lipschitz domain and $D = \mathbb{R}^d$. Similarly, we introduce the Maxwellian-weighted Sobolev space $W^{1,2}_M(\Omega \times D)$ and its norm by

$$W^{1,2}_M(\Omega \times D) := \left\{\phi \in L^1_{\mathrm{loc}}(\Omega \times D) : \|\phi\|_{W^{1,2}_M(\Omega\times D)} < \infty\right\},$$

$$\|\phi\|_{W^{1,2}_M(\Omega\times D)} := \left\{\int_{\Omega\times D} M\left[|\phi|^2 + |\nabla_{\boldsymbol{x}}\phi|^2 + |\nabla_{\boldsymbol{q}}\phi|^2\right] \mathrm{d}\boldsymbol{q}\,\mathrm{d}\boldsymbol{x}\right\}^{1/2}.$$

Let $W^{1/2,4/3}(\Omega; \mathbb{R}^{d\times d})$ denote the space of those $\zeta \in L^{4/3}(\Omega; \mathbb{R}^{d\times d})$ for which

$$\|\zeta\|_{W^{1/2,4/3}(\Omega;\mathbb{R}^{d\times d})} := \|\zeta\|_{L^{4/3}(\Omega;\mathbb{R}^{d\times d})} + \left(\int_\Omega\int_\Omega \frac{|\zeta(\boldsymbol{x}) - \zeta(\boldsymbol{y})|^{4/3}}{|\boldsymbol{x} - \boldsymbol{y}|^{d+2/3}} \,\mathrm{d}\boldsymbol{y}\,\mathrm{d}\boldsymbol{x}\right)^{3/4}$$

is finite. Then, the fractional-order space $W^{1/2,4/3}_{\boldsymbol{n}}(\Omega; \mathbb{R}^{d\times d})$ is defined as the closure of $\{\zeta \in C^\infty(\overline{\Omega}; \mathbb{R}^{d\times d}) : \nabla_{\boldsymbol{x}}\zeta \cdot \boldsymbol{n} = 0 \text{ on } \partial\Omega\}$ in the norm of $W^{1/2,4/3}(\Omega; \mathbb{R}^{d\times d})$, where $\boldsymbol{n}$ is the unit outward normal vector to $\partial\Omega$. Similarly, we define the spaces

$$W^{1,2}_{\boldsymbol{n},\mathrm{div}}(\Omega; \mathbb{R}^d) := \overline{\{\boldsymbol{v} \in C^\infty(\overline{\Omega}; \mathbb{R}^d) \,:\, \boldsymbol{v}\cdot\boldsymbol{n} = 0 \text{ on } \partial\Omega,\ \operatorname{div}\boldsymbol{v} = 0\}}^{\|\cdot\|_{W^{1,2}(\Omega;\mathbb{R}^d)}},$$

$$W^{1,2}_{0,\mathrm{div}}(\Omega; \mathbb{R}^d) := \overline{\{\boldsymbol{v} \in C^\infty_0(\Omega; \mathbb{R}^d) \,:\, \operatorname{div}\boldsymbol{v} = 0\}}^{\|\cdot\|_{W^{1,2}(\Omega;\mathbb{R}^d)}},$$

$$L^2_{0,\mathrm{div}}(\Omega; \mathbb{R}^d) := \overline{W^{1,2}_{\boldsymbol{n},\mathrm{div}}(\Omega; \mathbb{R}^d)}^{\|\cdot\|_{L^2(\Omega;\mathbb{R}^d)}}.$$

For a normed linear space X, we denote by $M^{-1}X$ the space of all those ϕ for which $M\phi \in X$. By X' we denote the continuous dual of X. For $t \geq 0$, we define $\Omega_t := [0, t] \times \Omega$. By $\mathcal{M}^+(\overline{\Omega}; \mathbb{R}^{d\times d}_{\mathrm{sym}})$ we denote the set of all $\mathbb{R}^{d\times d}$-valued Borel measures on the closure of Ω, which take values in the cone of symmetric positive semidefinite $d \times d$ matrices. We shall identify this set with a subset of the dual of the Banach space of $\mathbb{R}^{d\times d}_{\mathrm{sym}}$-valued uniformly continuous functions defined on $\overline{\Omega}$.

Definition 1 (*Generalized dissipative solution to the Hookean NSFP system*) Let the initial data $(\boldsymbol{u}_0, \widehat{\psi}_0)$ satisfy $\boldsymbol{u}_0 \in L^2_{0,\mathrm{div}}(\Omega; \mathbb{R}^d)$, $d \in \{2, 3\}$, $\mathcal{F}(\widehat{\psi}_0) \in L^1_M(\Omega \times D)$,

$$\widehat{\psi}_0(\boldsymbol{x}, \boldsymbol{q}) \geq 0 \quad \text{for a.e. } (\boldsymbol{x}, \boldsymbol{q}) \in \Omega \times D, \qquad \int_D M(q)\widehat{\psi}_0(\boldsymbol{x}, \boldsymbol{q})\,\mathrm{d}\boldsymbol{q} = 1 \quad \text{for a.e.} \boldsymbol{x} \in \Omega.$$

A quadruple $(\boldsymbol{u}, \widehat{\psi}, \mathbb{M}_{NS}, m_{OB})$ is a generalized dissipative solution to the Navier–Stokes–Fokker–Planck system in the case of the classical Hookean dumbbell model provided that

$$\boldsymbol{u} \in C_{\text{weak}}([0,T]; L^2(\Omega;\mathbb{R}^d)) \cap L^2(0,T; W^{1,2}_{0,\text{div}}(\Omega;\mathbb{R}^d)),$$

$$\widehat{\psi} \in L^1(0,T; L^1_M(\Omega \times D)) \cap W^{1,2}(0,T; M^{-1}[W^{s,2}(\Omega \times D]'), \quad s > d+1,$$

$$\mathcal{F}(\widehat{\psi}) \in L^\infty(0,T; L^1_M(\Omega \times D)), \quad \sqrt{\widehat{\psi}} \in L^2(0,T; W^{1,2}_M(\Omega \times D)),$$

$$\widehat{\psi}(t,\boldsymbol{x},\boldsymbol{q}) \geq 0 \quad \text{for a.e. } (t,\boldsymbol{x},\boldsymbol{q}) \in (0,T)\times\Omega\times D,$$

$$\int_D M(q)\widehat{\psi}(t,\boldsymbol{x},\boldsymbol{q})\,\mathrm{d}\boldsymbol{q} = 1 \quad \text{for a.e. } (t,\boldsymbol{x}) \in (0,T)\times\Omega,$$

$$\mathbb{M}_{NS} \in L^\infty(0,T; \mathcal{M}^+(\overline{\Omega}; \mathbb{R}^{d\times d}_{\text{sym}})), \quad m_{OB} \in L^\infty(0,T; \mathcal{M}^+(\overline{\Omega})),$$

and the following identities are satisfied for all $t \in [0,T]$:

$$\begin{aligned}
&\int_{\Omega_t} \boldsymbol{u}\cdot\partial_t \boldsymbol{w}\,\mathrm{d}\boldsymbol{x}\,\mathrm{d}s + \int_{\Omega_t} (\boldsymbol{u}\otimes\boldsymbol{u}) : \nabla_x \boldsymbol{w}\,\mathrm{d}\boldsymbol{x}\,\mathrm{d}s - \nu\int_{\Omega_t} \nabla_x \boldsymbol{u} : \nabla_x \boldsymbol{w}\,\mathrm{d}\boldsymbol{x}\,\mathrm{d}s\\
&\quad = \int_{\Omega_t} \mathbb{S}(\widehat{\psi}) : \nabla_x \boldsymbol{w}\,\mathrm{d}\boldsymbol{x}\,\mathrm{d}s + \int_0^t \langle \mathbb{M}_{NS}(s), \nabla_x \boldsymbol{w}\rangle\,\mathrm{d}s\\
&\qquad + \int_\Omega \boldsymbol{u}(t)\cdot\boldsymbol{w}(t)\,\mathrm{d}\boldsymbol{x} - \int_\Omega \boldsymbol{u}_0\cdot\boldsymbol{w}(0)\,\mathrm{d}\boldsymbol{x}
\end{aligned}$$

for all $\boldsymbol{w} \in L^2(0,T; W^{1,2}_{0,\text{div}}(\Omega;\mathbb{R}^d)) \cap L^1(0,T; C^1(\overline{\Omega};\mathbb{R}^d)) \cap W^{1,1}(0,T; L^2(\Omega;\mathbb{R}^d))$;

$$\begin{aligned}
&\int_{\Omega_t}\int_D M\widehat{\psi}\partial_t\phi\,\mathrm{d}\boldsymbol{q}\,\mathrm{d}\boldsymbol{x}\,\mathrm{d}s + \int_{\Omega\times D} M\widehat{\psi}_0\phi(0)\,\mathrm{d}\boldsymbol{q}\,\mathrm{d}\boldsymbol{x} - \int_{\Omega\times D} M\widehat{\psi}(t)\phi(t)\,\mathrm{d}\boldsymbol{q}\,\mathrm{d}\boldsymbol{x}\\
&\quad = -\int_{\Omega_t}\int_D M\widehat{\psi}\boldsymbol{u}\cdot\nabla_x\phi\,\mathrm{d}\boldsymbol{q}\,\mathrm{d}\boldsymbol{x}\,\mathrm{d}s - \int_{\Omega_t}\int_D M\widehat{\psi}\,((\nabla_x\boldsymbol{u})\boldsymbol{q})\cdot\nabla_q\phi\,\mathrm{d}\boldsymbol{q}\,\mathrm{d}\boldsymbol{x}\,\mathrm{d}s\\
&\qquad + \varepsilon\int_{\Omega_t}\int_D M\nabla_x\widehat{\psi}\cdot\nabla_x\phi\,\mathrm{d}\boldsymbol{q}\,\mathrm{d}\boldsymbol{x}\,\mathrm{d}s + \int_{\Omega_t}\int_D M\nabla_q\widehat{\psi}\cdot\nabla_q\phi\,\mathrm{d}\boldsymbol{q}\,\mathrm{d}\boldsymbol{x}\,\mathrm{d}s
\end{aligned}$$

for all $\phi \in W^{1,1}(0,T; W^{s,2}(\Omega\times D))$ with $s > d+1$; and

$$\begin{aligned}
&\frac{1}{2}\int_\Omega |\boldsymbol{u}(t)|^2\,\mathrm{d}\boldsymbol{x} + \frac{1}{2}\int_\Omega \operatorname{tr}\mathbb{S}(\widehat{\psi}(t))\,\mathrm{d}\boldsymbol{x} + \langle m_{OB}(t), \mathbb{1}_{\overline{\Omega}}\rangle + \nu\int_{\Omega_t} |\nabla_x\boldsymbol{u}|^2\,\mathrm{d}\boldsymbol{x}\,\mathrm{d}s\\
&+\int_{\Omega_t} \operatorname{tr}\left(\mathbb{S}(\widehat{\psi}) - \mathbb{I}_d\right)\mathrm{d}\boldsymbol{x}\,\mathrm{d}s + \int_0^t \langle m_{OB}(s), \mathbb{1}_{\overline{\Omega}}\rangle\,\mathrm{d}s \leq \frac{1}{2}\int_\Omega |\boldsymbol{u}_0|^2\,\mathrm{d}\boldsymbol{x} + \frac{1}{2}\int_\Omega \operatorname{tr}\mathbb{S}(\widehat{\psi}_0)\,\mathrm{d}\boldsymbol{x}
\end{aligned}$$

for a.e. $t \in (0, T]$. Finally, we require that the following compatibility condition is satisfied: there exists a $\zeta \in L^1((0, T))$ such that, for a.e. $t \in (0, T)$,

$$|\langle \mathbb{M}_{NS}(t), \nabla_x \boldsymbol{w} \rangle| \leq \zeta(t) \|\boldsymbol{w}\|_{C^1(\overline{\Omega};\mathbb{R}^d)} \langle m_{OB}(t), \mathbb{1}_{\overline{\Omega}} \rangle \quad \forall\, \boldsymbol{w} \in C^1(\overline{\Omega}; \mathbb{R}^d).$$

The next three theorems, which are proved in Dębiec and Süli (2023), assert the global existence of a generalized dissipative solution and summarize particular instances when the defect measures appearing in the definition of generalized dissipative solution vanish.

Theorem 2 (Existence) *Let $d \in \{2, 3\}$. There exists at least one generalized dissipative solution to the Navier–Stokes–Fokker–Planck system in the case of the classical Hookean dumbbell model with $m_{OB} = \operatorname{tr} \mathbb{M}_{NS}$. This solution is entropy-dissipative in the sense that the following energy inequality is satisfied for a.e. $t \in [0, T]$:*

$$\begin{aligned}
&\frac{1}{2}\int_\Omega |\boldsymbol{u}(t)|^2 \,\mathrm{d}\boldsymbol{x} + \int_{\Omega\times D} M\mathcal{F}(\widehat{\psi}(t)) \,\mathrm{d}\boldsymbol{q}\,\mathrm{d}\boldsymbol{x} + \nu \int_{\Omega_t} |\nabla_x \boldsymbol{u}|^2 \,\mathrm{d}\boldsymbol{x}\,\mathrm{d}s \\
&\quad + 4\int_0^t \int_{\Omega\times D} M \left(\varepsilon \left| \nabla_x \sqrt{\widehat{\psi}} \right|^2 + \left| \nabla_q \sqrt{\widehat{\psi}} \right|^2 \right) \mathrm{d}\boldsymbol{q}\,\mathrm{d}\boldsymbol{x}\,\mathrm{d}s \\
&\leq \frac{1}{2}\int_\Omega |\boldsymbol{u}_0|^2 \,\mathrm{d}\boldsymbol{x} + \int_{\Omega\times D} M\mathcal{F}(\widehat{\psi}_0) \,\mathrm{d}\boldsymbol{q}\,\mathrm{d}\boldsymbol{x}.
\end{aligned}$$

Furthermore, when $d = 2$ and $\mathbb{S}(\widehat{\psi}_0) \in L^2(\Omega; \mathbb{R}^{2\times 2})$, $\mathbb{S}_0^{\mathrm{T}} = \mathbb{S}_0$, we also have $\mathbb{S}(\widehat{\psi}), \mathbb{M}_{NS} \in L^q(0, T; L^p(\Omega; \mathbb{R}^{2\times 2}))$ with $q = 2p/(p-2)$ and $2 \leq p < \infty$. Moreover, the pair $(\boldsymbol{u}, \bar{\mathbb{S}} := \mathbb{M}_{NS} + \mathbb{S}(\widehat{\psi}))$ is a unique weak solution of the Oldroyd-B system with stress diffusion.

Theorem 3 (Conditional regularity for $d = 2$) *Let $d = 2$ and $\partial\Omega \in C^{2,\alpha}$, $\alpha \in (0, 1)$. Suppose further that, in addition to the assumptions of Theorem 2, $\boldsymbol{u}_0 \in W^{1,2}_{0,\mathrm{div}}(\Omega; \mathbb{R}^d)$ and $\widehat{\psi}_0$ is such that $\mathbb{S}(\widehat{\psi}_0) \in W_{\boldsymbol{n}}^{1/2,4/3}(\Omega; \mathbb{R}^{2\times 2})$. Then, any generalized dissipative solution to the Hookean dumbbell model, as constructed in Theorem 2, satisfies $\mathbb{M}_{NS} \equiv \mathbb{O}$ and $m_{OB} \equiv 0$, $(\boldsymbol{u}, \widehat{\psi})$ is a weak solution to the Hookean dumbbell model and the pair $(\boldsymbol{u}, \mathbb{S}(\widehat{\psi}))$ is the unique weak solution of the Oldroyd-B system with stress diffusion.*

In three space dimensions, we are unable to provide explicit conditions in terms of the initial data, similar to those in Theorem 3, which guarantee that $\mathbb{M}_{NS} \equiv \mathbb{O}$ and $m_{OB} \equiv 0$; however, the following result still holds.

Theorem 4 (Conditional regularity for $d = 3$) *Let $d = 3$ and let $(\boldsymbol{u}, \widehat{\psi}, \mathbb{M}_{NS}, m_{OB})$ be a generalized dissipative solution to the classical Hookean dumbbell model. Suppose that $\partial_t \boldsymbol{u} \in L^1(0, T; L^2(\Omega; \mathbb{R}^3))$ and $\nabla_x \boldsymbol{u} \in L^1(0, T; C(\overline{\Omega}; \mathbb{R}^{3\times 3}))$; then, $\mathbb{M}_{NS} \equiv \mathbb{O}$, $m_{OB} \equiv 0$, and $(\boldsymbol{u}, \widehat{\psi})$ is a weak solution of the model.*

5 Concluding Remarks

While the discussion in the preceding sections was focused on mathematical models in the isothermal setting, and the solvent in which the polymer molecules are suspended was supposed to be a Newtonian (i.e., Navier–Stokes) fluid, many of the ideas that were presented extend to more general models. In this respect, we point the reader to the paper Bulíček et al. (2013) where, in the isothermal setting, NSFP-type systems were studied that involve a general implicit constitutive equation relating the viscous part of the Cauchy stress and the symmetric velocity gradient. There, we considered implicit relations that generate maximal monotone (possibly multi-valued) graphs, and the corresponding rate of dissipation was characterized by the sum of a Young function and its convex conjugate depending on the symmetric velocity gradient and the viscous part of the Cauchy stress. Such a framework is very general and includes, among others, classical power-law fluids, stress power-law fluids, fluids with activation criteria of Bingham or Herschel–Bulkley type, and shear-rate-dependent fluids with discontinuous viscosities as special cases. An important direction of future research concerns the mathematical analysis of kinetic models for nonisothermal flows of incompressible and compressible dilute polymeric fluids based on the developments reported in Dostalík et al. (2020), where a self-contained derivation of such models was presented, based on the identification of energy storage mechanisms and entropy production mechanisms in the fluid of interest.

Acknowledgements I wish to express my sincere gratitude to Miroslav Bulíček, Eduard Feireisl, Josef Málek and Vít Průša (Prague), Tomasz Dębiec and Aneta Wróblewska-Kamińska (Warsaw), Marvin Fritz (Linz), Chuhui He (Oxford & Shanghai), Yong Lu (Nanjing), Barbara Wohlmuth (Munich) and Ghozlane Yahiaoui (Oxford & London) for numerous stimulating discussions. This paper is dedicated to the memory of John W. Barrett (29 June 1955, Wimbledon – 30 June 2019, Wimbledon).

References

Barrett, J.W., Boyaval, S.: Existence and approximation of a (regularized) Oldroyd-B model. Math. Models Methods Appl. Sci. **21**(9), 1783–1837 (2011)

Barrett, J.W., Süli, E.: Existence of global weak solutions to some regularized kinetic models for dilute polymers. Multiscale Model. Simul. **6**(2), 506–546 (2007)

Barrett, J.W., Süli, E.: Existence and equilibration of global weak solutions to kinetic models for dilute polymers I: Finitely extensible nonlinear bead-spring chains. Math. Models Methods Appl. Sci. **21**(6), 1211–1289 (2011)

Barrett, J.W., Süli, E.: Existence and equilibration of global weak solutions to kinetic models for dilute polymers II: Hookean-type models. Math. Models Methods Appl. Sci. **22**(5), 1150024, 84pp (2012)

Barrett, J.W., Süli, E.: Existence of global weak solutions to finitely extensible nonlinear bead-spring chain models for dilute polymers with variable density and viscosity. J. Differ. Equ. **253**(12), 3610–3677 (2012)

Barrett, J.W., Süli, E.: Existence of global weak solutions to compressible isentropic finitely extensible bead-spring chain models for dilute polymers. Math. Models Methods Appl. Sci. **26**(3), 469–568 (2016a)

Barrett, J.W., Süli, E.: Existence of global weak solutions to compressible isentropic finitely extensible nonlinear bead-spring chain models for dilute polymers: the two-dimensional case. J. Differ. Equ. **261**(1), 592–626 (2016b)

Barrett, J.W., Süli, E.: Existence of global weak solutions to the kinetic Hookean dumbbell model for incompressible dilute polymeric fluids. Nonlinear Anal. Real World Appl. **39**, 362–395 (2018)

Bhave, A., Armstrong, R.C., Brown, R.A.: Kinetic theory and rheology of dilute, non-homogeneous polymer solutions. J. Chem. Phys. **95**(4), 2988–3000 (1991)

Bird, R.B., Curtiss, C.F., Armstrong, R.C., Hassager, O.: Dynamics of Polymeric Liquids. Kinetic Theory, vol. 2. Wiley (1987)

Bulíček, M., Málek, J., Süli, E.: Existence of global weak solutions to implicitly constituted kinetic models of incompressible homogeneous dilute polymers. Comm. Partial Differ. Equ. **38**(5), 882–924 (2013)

Chemin, J.-Y., Masmoudi, N.: About lifespan of regular solutions of equations related to viscoelastic fluids. SIAM J. Math. Anal. **33**(1), 84–112 (2001)

Constantin, P., Kliegl, M.: Note on global regularity for two-dimensional Oldroyd-B fluids with diffusive stress. Arch. Ration. Mech. Anal. **206**(3), 725–740 (2012)

Dçbiec, T., Süli, E.: On a class of generalised solutions to the kinetic Hookean dumbbell model for incompressible dilute polymeric fluids: existence and macroscopic closure. arXiv:2306.16901 (2023)

Doi, M., Edwards, S.: The Theory of Polymer Dynamics. Oxford University Press (1986)

Dostalík, M., Málek, J., PrÅša, V., Süli, E.: A simple construction of a thermodynamically consistent mathematical model for non-isothermal flows of dilute compressible polymeric fluids. Fluids **5**(3), 29pp (2020)

El-Kareh, A.W., Leal, L.G.: Existence of solutions for all Deborah numbers for a non-Newtonian model modified to include diffusion. J. Non-Newtonian Fluid Mech. **33**, 257–287 (1989)

Feireisl, E., Lu, Y., Süli, E.: Dissipative weak solutions to compressible Navier-Stokes-Fokker-Planck systems with variable viscosity coefficients. J. Math. Anal. Appl. **443**(1), 322–351 (2016)

Fritz, M., Süli, E., Wohlmuth, B.: Analysis of a dilute polymer model with a time-fractional derivative. SIAM J. Math. Anal. **56**(2), 2063–2089 (2024)

He, C.: Analysis of Navier–Stokes–Fokker–Planck systems for incompressible dilute polymeric fluids. D.Phil. Thesis. University of Oxford (2023)

He, C., Süli, E.: Existence of large-data global weak solutions to kinetic models of nonhomogeneous dilute polymeric fluids. Kinetic Relat. Model. **17**(1), 107–161 (2024)

Jourdain, B., Lelievre, T., Bris, C.: Existence of solution for a micro-macro model of polymeric fluid: the FENE model. J. Funct. Anal. **209**(1), 162–193 (2004)

La, J.: On diffusive 2D Fokker-Planck-Navier-Stokes systems. Arch. Ration. Mech. Anal. **235**(3), 1531–1588 (2020)

Lions, P.L., Masmoudi, N.: Global solutions for some Oldroyd models of non-Newtonian flows. Chinese Ann. Math. Ser. B **21**(2), 131–146 (2000)

Masmoudi, N.: Global existence of weak solutions to the FENE dumbbell model of polymeric flows. Invent. Math. **191**(2), 427–500 (2013)

Öttinger, H.C.: Stochastic Processes in Polymeric Fluids: Tools and Examples for Developing Simulation Algorithms. Springer (2012)

Pavliotis, G.A.: Stochastic Processes and Applications: Diffusion Processes, the Fokker–Planck and Langevin Equations. Springer (2014)

Renardy, M.: An existence theorem for model equations resulting from kinetic theories of polymer solutions. SIAM J. Math. Anal. **22**(2), 313–327 (1991)

Rouse, P.E.: A theory of the linear viscoelastic properties of dilute solutions of coiling polymers. J. Chem. Phys. **21**(1272) (1953)

Süli, E., Wróblewska-Kamińska, A.: The incompressible limit of compressible finitely extensible nonlinear bead-spring chain models for dilute polymeric fluids. J. Differ. Equ. **269**(6), 5110–5141 (2020)

Süli, E., Yahiaoui, G.: McKean–Vlasov diffusion and the well-posedness of the Hookean bead-spring-chain model for dilute polymeric fluids: small-mass limit and equilibration in momentum space. arXiv:1802.06268 (2018)

Warner, H.R.: Kinetic theory and rheology of dilute suspensions of finitely extendible dumbbells. Ind. Eng. Chem. Fundamen. **11**(3), 379–387 (1972)

Open Access This chapter is licensed under the terms of the Creative Commons Attribution 4.0 International License (http://creativecommons.org/licenses/by/4.0/), which permits use, sharing, adaptation, distribution and reproduction in any medium or format, as long as you give appropriate credit to the original author(s) and the source, provide a link to the Creative Commons license and indicate if changes were made.

The images or other third party material in this chapter are included in the chapter's Creative Commons license, unless indicated otherwise in a credit line to the material. If material is not included in the chapter's Creative Commons license and your intended use is not permitted by statutory regulation or exceeds the permitted use, you will need to obtain permission directly from the copyright holder.

Matrix Functions and Network Communicability Measures

Michele Benzi

Abstract In recent years, functions of matrices have played an increasingly important role in the analysis of graphs and networks, especially in the definition of powerful centrality and communicability measures and in the analysis of diffusion processes, both local and nonlocal. These techniques are having a significant impact in a variety of applications, ranging from social network analysis to chemical physics to the neurosciences. This contribution provides a succinct overview of how matrix functions can be used in the analysis of communicability in complex networks.

1 Introduction

Network science provides conceptual and quantitative tools for analyzing and understanding the structure, function and dynamics of large and complex interconnected systems. It has had a tremendous impact on virtually every area of knowledge, from the natural and life sciences to the social sciences, from engineering to the humanities, and it has become indispensable for understanding a wide range of phenomena arising in fields like sociology, biology, ecology, engineering, computer and information sciences, economics, finance, and many others (Barabasi 2002; Boccaletti et al. 2006; Brandes and Erlebach 2005; Estrada 2012d; Newman 2003, 2018).

Since the beginnings of social network analysis in the late 1940s, tools from graph theory, linear algebra, probability and statistics have been essential to this field (Rapoport 1963; Katz 1953; Seeley 1949). In particular, quantities defined in terms of matrices associated to graphs (such as the adjacency matrix and the graph Laplacian) have been used to define computable metrics for the structural analysis of both static and dynamic networks, such as centrality and communicability measures. The quantitative analysis of networks often translates into classical problems of numerical linear algebra such as matrix factorizations, solving sparse linear systems, computing eigenvalues and eigenvectors, and the evaluation of functions of matrices.

M. Benzi (✉)
Scuola Normale Superiore, Pisa, Italy
e-mail: michele.benzi@sns.it

© The Author(s) 2026

S. Oishi et al. (eds.), *Recent Developments in Industrial and Applied Mathematics*,
ICIAM2023 Springer Series 1, https://doi.org/10.1007/978-981-95-1446-5_11

These problems can be challenging in view of their possibly very large size, and require state-of-the-art techniques for their efficient solution.

This contribution surveys the use of matrix functions to define and investigate various notions of *communicability* in networks. Roughly speaking, communicability allows one to quantify how well (or poorly) two distinct nodes in a network can exchange information. This notion is closely related both to node centrality (which measures the importance of individual nodes) and to global measures of graph connectivity. Communicability can be defined in terms of graph walks (expressed by suitable functions of the adjacency matrix) but it can also be measured by means of diffusion processes and random walks on the graph, which are best expressed in terms of functions of the graph Laplacian.

For earlier (and more in-depth) surveys of matrix functions in network analysis the reader is referred to Estrada and Higham (2010) and to Benzi and Boito (2020).

2 Background on Graphs and Linear Algebra

This section provides the basic graph theory and linear algebra notions used in the sequel. Good reference works include Diestel (2000), Horn and Johnson (2013), Higham (2008); see also Estrada and Knight (2015).

2.1 Graphs and Matrices

A *directed graph* or *digraph*, G is defined by a set of n nodes (also called vertices) V and a set of edges $E = \{(i, j)|i, j \in V\}$. If $(i, j) \in E$ implies $(j, i) \in E$, then G is *undirected* and the edges are formed by unordered pairs of vertices.

In a directed graph, the *out-degree* of a vertex i, denoted by d_i^{out}, is the number of edges with i as the starting node, i.e., the number of edges in E of the form (i, k). The *in-degree* of node i is the number d_i^{in} of edges of the form (k, i). If G is undirected, then $d_i^{\text{out}} = d_i^{\text{in}} = d_i$, the *degree* of node i. An undirected graph G is d-*regular* if all the nodes have the same degree d.

A *walk* of length k in G is a list of nodes $i_1, i_2, \ldots, i_k, i_{k+1}$ such that for all $1 \le j \le k$ there is a (directed) edge from i_j to i_{j+1}. A walk is *closed* when $i_1 = i_{k+1}$. A walk with no repeated nodes is called a *path*, and a *cycle* is a closed walk with no repeated nodes, except for the first and last one. In a *weighted graph*, a weight $w_{ij} \in \mathbb{R}^+$ is assigned to each edge $(i, j) \in E$. If all the weights are taken equal to 1, the graph is *unweighted*. For ease of exposition, here G will be generally assumed to be unweighted, but most results extend to the weighted case with straightforward modifications.

A *loop* in a graph is an edge from a node i to itself. A graph is *simple* if it has no loops, no multiple edges, and its edges are unweighted. An undirected graph is *connected* if there exists a path between every pair of nodes. A directed graph is *strongly connected* if there exists a directed path between every pair of nodes.

In an undirected graph G, the *geodesic distance* between two nodes is the length of the shortest path joining them. If no such path exists, they distance between the nodes is set to infinity. The *diameter* of G is equal to the maximum distance between any two nodes in G; note that G has finite diameter if and only if it is connected.

A graph is called *bipartite* if its set of nodes V is the disjoint union of two non-empty sets V_1 and V_2 such that each edge connects a node from V_1 and a node from V_2. Given a directed graph, we can define an associated undirected bipartite graph as follows. Take two copies V and V' of the set of nodes of the original graph, and define edges according to the following rule: a node $i \in V$ and a node $j' \in V'$ are connected if and only if there is an edge from node i to node j in the directed graph.

The *adjacency matrix* $A = (a_{ij})$ associated with a simple graph G is an $n \times n$ matrix such that

$$a_{ij} = \begin{cases} 1 & \text{if } (i, j) \text{ is an edge in } G, \\ 0 & \text{otherwise.} \end{cases}$$

More generally, if the graph is weighted then $a_{ij} = w_{ij} > 0$ whenever $(i, j) \in E$. Recall that a square matrix A is *reducible* if there is a permutation matrix P such that

$$PAP^T = \begin{bmatrix} B & C \\ 0 & D \end{bmatrix}$$

with B and D square matrices. If no such P exists, A is said to be *irreducible*. It is well known that the adjacency matrix of a graph G is irreducible if and only if G is strongly connected.

Note that the in- and out-degrees of a node i can be computed as the i-th entries of the vectors $A^T\mathbf{1}$ and $A\mathbf{1}$, respectively, where $\mathbf{1}$ denotes the vector of all ones.

For an unweighted graph, the powers of the associated adjacency matrix A have an important property: each entry $[A^k]_{ij}$ counts the number of walks of length k from node i to node j. In particular, the diagonal entries $[A^k]_{ii}$ count the number of *closed* walks of length k passing through node i. This notion can easily be extended to weighted graphs.

If G is undirected, then its adjacency matrix A is symmetric and has nonnegative entries, with zeros on the main diagonal. The eigenvalues of A are real; we will label them in nondecreasing order as $\lambda_1 \geq \lambda_2 \geq \cdots \geq \lambda_n$. The matrix A can be diagonalized as $A = Q\Lambda Q^T$, where $\Lambda = \text{diag}(\lambda_1, \lambda_2, \ldots, \lambda_n)$, the matrix $Q = [\mathbf{q}_1, \mathbf{q}_2, \ldots, \mathbf{q}_n]$ is orthogonal and $\mathbf{q}_i$ is an eigenvector associated with λ_i. If G is connected, the Perron-Frobenius theorem (Horn and Johnson 2013) implies that $\lambda_1 > \lambda_2$. Moreover, the *dominant eigenvector* $\mathbf{q}_1$ can be chosen to have positive entries: we write $\mathbf{q}_1 > \mathbf{0}$.

If G is a strongly connected digraph, its adjacency matrix A is irreducible. Let $\rho(A)$ be the spectral radius of A. Then, again by the Perron-Frobenius theorem, $\lambda_1 = \rho(A)$ is a simple eigenvalue of A and the eigenvector associated with λ_1 can be chosen to be positive. If A is diagonalizable, then there exists an invertible matrix

X such that $A = X\Lambda X^{-1}$, where $\Lambda = \text{diag}(\lambda_1, \lambda_2, \ldots, \lambda_n)$ with $\lambda_1 \geq |\lambda_i|$ for $2 \leq i \leq n$, and the columns of X are corresponding eigenvectors. If, on the other hand, A is not diagonalizable, it can still be put into Jordan canonical form:

$$A = XJX^{-1} = X \begin{bmatrix} \lambda_1 & \mathbf{0}^T \\ \mathbf{0} & \hat{J} \end{bmatrix} X^{-1}.$$

Here J is the Jordan matrix of A, where for ease of notation the 1×1 block corresponding to the eigenvalue λ_1 appears first. The first column $\mathbf{x}_1$ of X is the dominant right eigenvector of A.

Besides the diagonal form and the Jordan form of A, another basic tool is the *singular value decomposition* (Golub and Van Loan 2013). Any matrix $A \in \mathbb{R}^{m\times n}$ of rank r can be decomposed as $A = U\Sigma V^T$, where $U \in \mathbb{R}^{m\times m}$ and $V \in \mathbb{R}^{n\times n}$ are orthogonal matrices, and $\Sigma \in \mathbb{R}^{m\times n}$ is zero except for its first r diagonal entries $\sigma_1, \ldots, \sigma_r$. Here $\sigma_1, \ldots, \sigma_r$ are positive real numbers and are called the *singular values* of A. The columns $\mathbf{u}_1, \mathbf{u}_2, \ldots, \mathbf{u}_m$ of U are the *left singular vectors* of A, whereas the columns $\mathbf{v}_1, \mathbf{v}_2, \ldots, \mathbf{v}_n$ of V are the *right singular vectors* of A. The singular values are uniquely determined by A, but the singular vectors, in general, are not. When A has rank $r < \min(m, n)$ it is often useful to consider the *compact singular value decomposition* of A. This is given by $A = U_r\Sigma_r V_r^T$, where $U_r = [\mathbf{u}_1, \mathbf{u}_2, \ldots, \mathbf{u}_r] \in \mathbb{R}^{m\times r}$, $V_r = [\mathbf{v}_1, \mathbf{v}_2, \ldots, \mathbf{v}_r] \in \mathbb{R}^{n\times r}$, and $\Sigma_r = \text{diag}(\sigma_1, \sigma_2, \ldots, \sigma_r) \in \mathbb{R}^{r\times r}$ is the leading $r \times r$ block of Σ.

If the undirected graph G is bipartite, with $V = V_1 \cup V_2$, there exists an ordering of the nodes such that the adjacency matrix of G takes the form

$$A = \begin{bmatrix} 0 & B \\ B^T & 0 \end{bmatrix},$$

where B is $|V_1| \times |V_2|$. It is well known (see, e.g., Golub and Van Loan 2013) that the nonzero eigenvalues of A are of the form $\lambda_i = \pm\sigma_i$, where the σ_i are the singular values of B.

If G is a digraph with n nodes, the associated undirected bipartite graph $\mathcal{G} = (\mathcal{V}, \mathcal{E})$ has $2n$ nodes and its adjacency matrix is of the form

$$\mathcal{A} = \begin{bmatrix} 0 & A \\ A^T & 0 \end{bmatrix},$$

where A is the $n \times n$ adjacency matrix of G.

The *Laplacian matrix* associated with an undirected, unweighted graph G is defined as $L = D - A$, where A is the adjacency matrix of G and $D = \text{diag}(d_1, \ldots, d_n)$ is the *degree matrix* of G.

For directed graphs, one may replace D by either the matrix of in-degrees $D^{\text{in}} = \text{diag}(d_1^{\text{in}}, \ldots, d_n^{\text{in}})$ or the matrix of out-degrees $D^{\text{out}} = \text{diag}(d_1^{\text{out}}, \ldots, d_n^{\text{out}})$, resulting in two distinct notions of Laplacian matrix, L_{in} and L_{out}, the second one being usually preferred.

For an undirected graph, the Laplacian L is a singular (since $L\mathbf{1} = \mathbf{0}$), symmetric, positive semidefinite M-matrix (Berman and Plemmons 1994). The algebraic multiplicity of the zero eigenvalue is equal to the number of connected components of the graph. In particular, the graph is connected if and only if the zero eigenvalue of the Laplacian is simple (i.e., if and only if L has rank $n - 1$). The smallest nonzero eigenvalue of the Laplacian of a connected graph is known as its *algebraic connectivity* (Fiedler 1973), and the corresponding eigenvector as its *Fiedler vector*. Loosely speaking, the algebraic connectivity of G measures how hard it is to disconnect G into two subgraphs of comparable size by removing selected edges: the larger the algebraic connectivity, the harder it is to disconnect G into two roughly equal subgraphs by removing a small number of edges. The sign of the entries of the Fiedler vector can be used to assign nodes of G to either subgraph. Hence, the graph Laplacian plays a role in the important problem of graph partitioning.

For a directed graph, both L_{in} and L_{out} are singular M-matrices, with $\mathbf{1}^T L_{\text{in}} = \mathbf{0}^T$ and $L_{\text{out}}\mathbf{1} = \mathbf{0}$. If G is strongly connected, then L_{in} and L_{out} have rank $n - 1$; the converse, in general, is not true.

The notion of *normalized Laplacian* is also widely used in the literature. The symmetric normalized Laplacian is defined as $\mathcal{L} = D^{-\frac{1}{2}} L D^{-\frac{1}{2}}$, where all nodes are assumed to have positive degree; if some nodes have zero degree, $D^{-1/2}$ is defined as the diagonal matrix such that $[D^{-1/2}]_{ii} = 1/\sqrt{d_i}$ if $d_i > 0$, and $[D^{-1/2}]_{ii} = 0$ if $d_i = 0$. As an effect of diagonal scaling, the eigenvalues of $\mathcal{L}$ belong to the interval $[0, 2]$. In the directed case, additional notions of normalized Laplacian $\mathcal{L}_{\text{in}} = I - D_{\text{in}}^{-1} A$ and $\mathcal{L}_{\text{out}} = I - A^T D_{\text{out}}^{-1}$ are also of interest.

2.2 *Functions of Matrices*

Several, essentially equivalent definitions of a matrix function exist (Higham 2008). A rather general definition is the one based on the diagonal or Jordan form. Let A be an $n \times n$ matrix (real or complex), assumed for now to be diagonalizable. Therefore $A = X \Lambda X^{-1}$, where $\Lambda = \text{diag}(\lambda_1, \lambda_2, \ldots, \lambda_n)$ is a diagonal matrix containing the eigenvalues of A, and X is invertible. Let $f(z)$ be a (real or complex) function that is well-defined on the eigenvalues of A. Then $f(A)$ is defined as

$$f(A) = X f(\Lambda) X^{-1},$$

where $f(\Lambda) = \text{diag}(f(\lambda_1), f(\lambda_2), \ldots, f(\lambda_n))$.

If A is not diagonalizable, let $A = X J X^{-1}$ be its Jordan decomposition. The Jordan matrix J has a block diagonal structure $J = \text{diag}(J_1, J_2, \ldots, J_m)$, with one or more blocks J_k associated with each distinct eigenvalue of A. Let J_k be one such block of size $\nu \times \nu$, with eigenvalue λ; then $f(J_k)$ is defined as

$$f(J_k) = \begin{bmatrix} f(\lambda) & f'(\lambda) & \frac{f''(\lambda)}{2!} & \cdots & \frac{f(\lambda)^{(\nu-1)}}{(\nu-1)!} \\ & f(\lambda) & f'(\lambda) & \cdots & \frac{f(\lambda)^{(\nu-2)}}{(\nu-2)!} \\ & & \ddots & \ddots & \vdots \\ & & & f(\lambda) & f'(\lambda) \\ & & & & f(\lambda) \end{bmatrix}.$$

Now set $f(J) = \operatorname{diag}(f(J_1), f(J_2), \ldots, f(J_m))$ and

$$f(A) = Xf(J)X^{-1}.$$

It can be shown that this definition is independent of the choice of X. Note that here $f(z)$ must be assumed to be differentiable a sufficient number of times at each eigenvalue of A. More precisely, denote with m_i the *index* of λ_i, i.e., size of the largest Jordan block in which the eigenvalue λ_i appears. If, for each distinct eigenvalue λ_i of A, the function f and its derivatives of order up to $m_i - 1$ are defined for $\lambda = \lambda_i$, f is said to be *defined on the spectrum of* A, and $f(A)$ is well-defined.

If $f(z)$ has a Taylor series expansion

$$f(z) = \sum_{k=0}^{\infty} a_k (z - z_0)^k$$

with radius of convergence R, this property can be used to give an alternative definition of $f(A)$. Suppose that each eigenvalue λ_i of A is either contained in the interior of the convergence disk $|z - z_0| < R$, or is on the boundary of the disk and the series for $f^{(m_i-1)}(\lambda)$ is convergent at $\lambda = \lambda_i$, with m_i the index of λ_i. Then $f(A)$ is well-defined and it is given by

$$f(A) = \sum_{k=0}^{\infty} a_k (A - z_0 I)^k.$$

Note that the converse is also true, i.e., if $f(A)$ can be expressed through the Taylor series of $f(z)$, then the eigenvalues of A satisfy the conditions stated above.

The notion of matrix function can be extended to rectangular matrices via the singular value decomposition (Hawkins and Ben-Israel 1973). Take $A \in \mathbb{R}^{m \times n}$ and let $A = U_r \Sigma_r V_r^T$ be its compact singular value decomposition. If f is a scalar function such that $f(\sigma_i)$ is defined for $i = 1, \ldots, r$, then the *generalized matrix function* $f^\diamond$ induced by f is defined as

$$f^\diamond(A) = U_r f(\Sigma_r) V_r^T,$$

where $f(\Sigma_r) = \operatorname{diag}(f(\sigma_1), \ldots, f(\sigma_r))$. If A is square, $f^\diamond(A)$ does not necessarily coincide with the standard matrix function $f(A)$; however, if A is symmetric positive definite (or semidefinite and $f(0) = 0$), then $f^\diamond(A)$ reduces to $f(A)$.

3 Matrix Functions for Communicability Measures

A fundamental problem in the structural analysis of networks is that of quantifying how well different parts of a network, such as individual nodes or groups of nodes, are connected to one another. This problem is of great importance, for instance, in the design and analysis of robust infrastructural networks, in the study of urban traffic, in anatomy and medicine (especially brain networks), in the study of vulnerability of financial networks, in bibliometry, in community detection, in the study of the chemico-physical properties of molecular networks, and in many other fields. Matrix functions provide an elegant mathematical framework for tackling this problem.

3.1 Network Communicability

The notion of *communicability* between two nodes, first introduced by Estrada and Hatano (2008), provides an effective measure of the connectivity between pairs of nodes. The underlying intuition is that two nodes i and j are well-connected if there are many walks in G connecting i and j, with shorter walks being given more weight than longer ones. Having in mind the interpretation of the entries of A^k in terms of graph walks, Estrada and Hatano proposed the following definition of communicability between two nodes $i, j \in V$:

$$[\mathrm{e}^{\beta A}]_{ij} = \delta_{ij} + [A]_{ij} + \frac{1}{2!}[A^2]_{ij} + \cdots = \sum_{k=0}^{\infty} \frac{\beta^k}{k!}[A^k]_{ij} \,. \tag{1}$$

Here $\beta > 0$ is the *inverse temperature*, which can be used to tune the relative contributions of short and long walks (Estrada and Hatano 2007; Estrada et al. 2012). When $i = j$ we obtain a measure of the centrality of node i, known as *subgraph centrality*, see Estrada and Rodríguez-Velázquez (2005). In other words, $[\mathrm{e}^{\beta A}]_{ii}$ is a measure of the importance, or influence, of node i in the graph. Of course, other functions besides the matrix exponential can be used. For example, the off-diagonal entries of the resolvent $f(A) = (I - \alpha A)^{-1}$, with $0 < \alpha < \frac{1}{\lambda_1}$, can be used to define the *Katz communicability* between nodes (Katz 1953). An *eigenvector-based communicability* can be obtained by taking the product $x_1(i)x_1(j)$ of the ith and jth entries of the (positive) dominant eigenvector $\mathbf{x}_1$ of a (strongly) connected network, normalized so that $\|\mathbf{x}_1\|_2 = 1$. Again, for $i = j$ we obtain a well-known centrality measure, the *eigenvector centrality* of node i (Bonacich 1987). Eigenvector communicability can be interpreted as a measure of communicability that takes into account longer walks: for a connected undirected graph, it can be obtained as the limit of the expression (1), normalized by $\mathrm{e}^{\beta\lambda_1}$, as $\beta \to \infty$. The same applies to the Katz communicability (multiplied by $(1 - \alpha\lambda_1)$) in the limit $\alpha \to \frac{1}{\lambda_1}^{-}$. At the other extreme we have the simplest and most local notion of communicability between nodes, taking the values 1 or 0 depending on whether or not there is an edge between nodes i and j. This can be seen as the limit, as $\beta \to 0+$ or $\alpha \to 0+$, of the two parameter-dependent communicability measures based on $e^{\beta A}$ and $(I - \alpha A)^{-1}$, respectively; see the analysis in Benzi and Klymko (2015). For intermediate values of α and β, these communicability measures are able to take into account the effect of short, medium, and long range interaction between nodes.

The exponential-based communicability (1) has been used very effectively in a number of fields. For example, it has been used to develop algorithms for the community detection problem (Estrada 2011; Estrada and Hatano 2009; Xu 2020) and, more generally, for the identification of sparse and dense subgraphs of large graphs (Fioriti and Chinnici 2014). Other applications of exponential communicability include: the analysis of urban traffic flow and other transportation networks (Akbarzadeh and Estrada 2018; Mussone and Notari 2019); the study of anatomical networks (Ostachuk 2024) and especially brain networks, including the identification of anomalies in the brains of patients affected by stroke, Type 2 diabetes and other conditions (Betzel et al. 2016; Crofts and Higham 2009; Crofts et al. 2011; Gravert et al. 2017; de Reus and van den Heuvel 2014; Pião Azevedo et al. 2024; Veergossen et al. 2019); the modeling of cancer spread (Gladilin 2017); models of concept learning by weighted networks (Koponen and Nousiainen 2014); the characterization of water molecules in the high-density and low-density phases (Benzi et al. 2023; Faccio et al. 2022, 2023); the modeling of fracture networks in rocks (Santiago et al. 2016); the study of the spread of contagion in financial networks (Bartesaghi et al. 2020; Silva et al. 2017); the analysis of the influence of papers in citation networks (Zhao et al. 2019); and many others.

3.2 *Total Node Communicability as a Centrality Measure*

Exponential-based communicability can also be used to introduce a node centrality measure, the *total communicability* of a node, defined as the sum of the communicabilities of a node with all the nodes in the network, including itself (Benzi and Klymko 2013):

$$\mathrm{TC}(i) = \sum_{j=1}^{n} [e^{\beta A}]_{ij} = [e^{\beta A}\mathbf{1}]_i \tag{2}$$

(recall that the row sums of a matrix function $f(A)$ are the components of the vector $f(A)\mathbf{1}$). A significant advantage of this centrality measure over those based on the diagonal entries of $f(A)$ (like subgraph centrality) is the fact that using Krylov subspace methods, the action of a matrix function on a vector can be computed much faster than the n diagonal entries of $[f(A)]_{ii}$ for large n; see for example (Benzi et al. 2023).

It must be mentioned, however, that centrality measures based on the diagonal entries $[f(A)]_{ii}$ can have a higher discriminating power than those based on the row sums $[f(A)\mathbf{1}]_i$. Indeed, if a graph G is d-regular, then $A\mathbf{1} = d\mathbf{1}$ and therefore also

$f(A)\mathbf{1} = f(d)\mathbf{1}$. Hence, all nodes in G get the same centrality score and no measure of this kind is able to distinguish one node from the others (the same holds, of course, for degree and eigenvector centrality). On the other hand, there are examples of regular graphs for which the subgraph centrality $C(i) = [e^A]_{ii}$ is able to distinguish the nodes (Estrada and Rodríguez-Velázquez 2005). Generally speaking, however, real-world networks are not regular and, very often, subgraph centrality and total node communicability will identify the same nodes as the most central. Thus, in the case of large networks the use of the more easily computable total communicability can be advantageous.

A global measure of the ability of a network to exchange information (broadly understood) can be obtained by summing up all the total communicabilities of the nodes in G; normalizing by the number of nodes results in the *total network communicability* (see Benzi and Klymko 2013):

$$\mathrm{TC}(G) = \frac{1}{n}\sum_{i=1}^{n}\sum_{j=1}^{n}[e^{\beta A}]_{ij} = \frac{1}{n}\mathbf{1}^T e^{\beta A}\mathbf{1}. \tag{3}$$

Not surprisingly this quantity, which can be interpreted as the *average communicability* of a node, is maximized by K_n, the complete graph with n nodes. The total network communicability can be interpreted as a measure of network connectivity, and in this regard it plays a role similar to that of the *Estrada index*:

$$EE(G) = \sum_{i=1}^{n}[e^{\beta A}]_{ii} = \mathrm{Tr}\,(e^{\beta A}),$$

which is, however, more expensive to compute. Both measures can be used to compare different network structures (with the same number of nodes and edges), for example to design highly efficient networks, see the next subsection.

3.3 Network Modification Techniques

Network efficiency and robustness are of great importance in the design and realization of actual networks, both in nature and in different fields of engineering. A network with high communicability is likely to be efficient in the routing of messages or in the transportation of goods, and also to be robust against localized loss of connectivity (e.g., edge removal) due either to random accidents or to deliberate, malicious attacks. In some situations high connectivity can have undesirable effects, e.g., in the spreading of diseases, financial crises or computer viruses. In public health campaigns and law enforcement or anti-terrorism operations, it is desirable to disrupt or weaken an existing highly connected network. Highly sparse networks that display excellent connectivity properties are known in the computer science literature as *expander graphs*; we refer to Hoory et al. (2006) for an outstanding survey.

A natural application of total network communicability and similar global connectivity measures is in the comparison of different network designs, or in the modification of an existing network in order to achieve a desired goal. The following problems are often considered:

1. *Network rewiring*: given a sparse network $G = (V, E)$ with $|E| = m$, choose which k existing edges ($k \ll m$) to rewire so as to increase, or decrease, the total network communicability as much as possible.
2. *Network updating*: given $G = (V, E)$ as above and a budget of $k \ll m$ edges, choose which k pairs of unconnected nodes in G to connect so as to increase the total network communicability as much as possible.
3. *Network downdating*: given G as above, choose which k existing edges ($k \ll m$) to delete so as to decrease the total communicability as little or as much as possible; the first problem corresponds to the goal of having similar connectivity at a lower cost (fewer edges), the second to the goal of disrupting or weakening an existing network in order to drastically reduce its effectiveness with minimal effort.

In all these problems, the total network communicability can be used as an objective function $f : G \longrightarrow \mathbb{R}$ to evaluate the relative effectiveness of different rewiring, updating, or downdating strategies. This is not the only measure of connectivity available; for example, the spectral radius of A, the algebraic connectivity or the Estrada index can also be used, and these measures are all correlated. A nice feature of the total communicability is that it is easy to approximate, often at a cost of $O(n)$ (i.e., linear in the number of nodes). This fact is exploited in Arrigo and Benzi (2016a), where several heuristics are implemented and compared in the solution of the updating/downdating problems described above (the rewiring problem is similar and can be dealt with using similar techniques). These heuristics only require the identification of a small fraction of nodes in G having the highest centrality (for example, eigenvector or total communicability centrality) and are faster, sometimes by orders of magnitude, than previously proposed methods. As shown in Arrigo and Benzi (2016a), a large increase in total communicability (as well as in other measures) is obtained by connecting highly central nodes that are not already connected by an edge; conversely, deleting edges that connect highly central nodes tends to induce large drops in total communicability. If the goal is to delete as many edges as possible without disconnecting the network and with minimal decrease in total communicability, the best strategy is the removal of edges connecting peripheral nodes (i.e. nodes not in the subset of highly central nodes). The connectivity of the network after each edge removal can be checked in time that is typically linear in n, for a sparse network; see Arrigo and Benzi (2016a) and the references therein for details.

Edge rewiring, addition and deletion translate into low-rank modifications of the adjacency matrix. The problem of updating/downdating a matrix function after a low-rank change has been investigated in Beckermann et al. (2018) and Pozza and Tudisco (2018), with an eye to applications in network analysis, such as assessing the stability of node rankings under network modifications.

3.4 Communicability Distance

Communicability between nodes can also be used to define a metric (on undirected graphs) which is often more useful than the standard (geodesic) distance. Estrada (2012a, 2012c) has shown that letting

$$d_c(i,j) = \sqrt{C(i,i) + C(j,j) - 2C(i,j)}, \quad \text{where} \quad C(i,j) = [\mathrm{e}^A]_{ij}, \tag{4}$$

defines a distance on the set of nodes of an undirected graph G. To see this, let $A = QDQ^T$ be the spectral decomposition of the adjacency matrix A, and let $\mathbf{u}_j$ denote the jth column of Q^T. Letting $\mathbf{x}_j = \mathrm{e}^{D/2}\mathbf{u}_j$, we have

$$d_c(i,j)^2 = (\mathbf{x}_i - \mathbf{x}_j)^T(\mathbf{x}_i - \mathbf{x}_j) = \|\mathbf{x}_i - \mathbf{x}_j\|_2^2,$$

showing that $d_c(i,j)$ is just the Euclidean distance of the vectors $\mathbf{x}_i, \mathbf{x}_j \in \mathbb{R}^n$. A similar distance based on the resolvent $(I - \alpha A)^{-1}$ instead of the matrix exponential has been introduced in Chebotarev (2012).

The exponential-based communicability distance has several interesting applications, for example in the analysis of urban traffic modeling (Akbarzadeh and Estrada 2018; Silver et al. 2018). The related notion of *communicability angles* has been shown to be able to identify the critical (i.e., most diffusive) links in consensus dynamics (Estrada et al. 2015). A notion of *spatial efficiency* based on the communicability geometry is introduced and applied to a variety of real-world networks in Estrada and Hatano (2016). Extensions of communicability geometry to multiplex networks are described in Estrada (2019), with applications to the study of the neuronal network of the worm *C. elegans*.

3.5 Communicability in Digraphs

In the case of directed networks it is necessary to take into account the directionality of communication between nodes. A simple approach is to define the communicability of two nodes as was done in the undirected case, via (1). The lack of symmetry in A, and therefore in e^A (or, more generally, in $f(A)$) is just a reflection of the fact that in a directed graph, sending information from node i to node j may be easier (or harder) than from node j to node i. Taking row sums of $f(A)$ corresponds to a total communicability measure that corresponds to the centrality of nodes as *hubs* (i.e., broadcasters of information), while taking column sums of $f(A)$ results in a centrality measure of nodes as *authorities* (i.e., receivers of information). Note that using the diagonal entries (which are more expensive to compute) does not allow one to distinguish these two roles played by the nodes.

In some situations it is desirable to distinguish further between four types of communicability: broadcaster-to-receiver, broadcaster-to-broadcaster, receiver-to-receiver, and receiver-to-broadcaster. Such measures have been first introduced in Benzi et al. (2013), using the bipartite model of a directed graph described in Sect. 2. Let

$$\mathcal{A} = \begin{bmatrix} 0 & A \\ A^T & 0 \end{bmatrix} \tag{5}$$

be the adjacency matrix of the bipartite graph $\mathcal{G} = (\mathcal{V}, \mathcal{E})$ obtained from the original digraph with adjacency matrix A. In this model, the first n nodes contained in $\mathcal{V}$ are regarded as the original nodes of the digraph when they play the role of broadcasters of information, while the last n copies contained in $\mathcal{V}$ correspond to the same nodes regarded as receivers.

As shown in Benzi et al. (2013, Proposition 1), the exponential of the symmetric matrix $\mathcal{A}$ is given by

$$\mathrm{e}^{\mathcal{A}} = \begin{bmatrix} \cosh(\sqrt{AA^T}) & U\sinh(\Sigma)V^T \\ V\sinh(\Sigma)U^T & \cosh(\sqrt{A^TA}) \end{bmatrix}, \tag{6}$$

where $A = U\Sigma V^T$ is an SVD of A. The entries of this matrix are nonnegative and may be used to quantify how effectively the nodes communicate when they are acting as broadcasters or receivers of information in G. Specifically, it is shown in Benzi et al. (2013) that the main diagonal of the first diagonal block $\cosh(\sqrt{AA^T})$ contains the centralities of the nodes in the original network in their role of hubs, whereas the diagonal entries of the second block $\cosh(\sqrt{A^TA})$ provide the centralities for the nodes in their role of authorities. When the largest singular value of A is well-separated from the remaining ones, these rankings are close to those given by Kleinberg's HITS algorithm (Kleinberg 1999); see Benzi et al. (2013). Similarly to the off-diagonal entries of the matrix exponential of an undirected graphs, the off-diagonal entries of these diagonal blocks are measures of the communicability of pairs of nodes, both acting as broadcasters (respectively, receivers) of information. Finally, the entries in the off diagonal block $U\sinh(\Sigma)V^T$ (a generalized matrix function!) and its transpose $V\sinh(\Sigma)U^T$ provide information on the communicability of pairs of nodes considered as hub-authority pairs.

As before, one can also define global communicability measures for digraphs that take into account the dual role of each node in G. Indeed, the *total hub communicability* and *total authority communicability* of G, defined as $\mathrm{T_hC}(A) = \mathbf{1}^T\cosh(\sqrt{AA^T})\mathbf{1}$ and $\mathrm{T_aC}(A) = \mathbf{1}^T\cosh(\sqrt{A^TA})\mathbf{1}$, respectively, account for the overall ability of the network of exchanging information when all its nodes are playing the same role of broadcasters ($\mathrm{T_hC}(A)$) or receivers ($\mathrm{T_aC}(A)$). Using these measures, network updating/downdating strategies based on edge modifications have been introduced in Arrigo and Benzi (2016b).

4 Temporal Communicability

Many real-life complex systems evolve in time: taking this additional feature into account leads to models based on *temporal networks*. Given a (fixed) set of n nodes, a temporal network is an ordered sequence $\{G^{[k]}\}_{k=1,\ldots,M}$ of graphs sharing the given set of nodes. For simplicity, the graphs here are assumed to be simple, undirected, and unweighted. Each graph $G^{[k]}$ records the state of the network at time t_k, where $t_1 < t_2 < \cdots < t_M$ are time points, and is represented by an associated symmetric adjacency matrix $A^{[k]}$.

Communicability measures based on counting walks can be extended to temporal networks through the use of *dynamic walks*. A dynamic walk is a sequence of edges connecting nodes in the network, with the added constraint that the sequence of edges must respect the time ordering. Nodes participating in many dynamic walks are capable of communicating well with other nodes in the network, and can therefore spread or receive information effectively not only across a given graph but also over time.

Denote with $\rho(A^{[k]})$ the spectral radius of the kth adjacency matrix, and by $\rho^* = \max_k\{\rho(A^{[k]})\}$ the maximum spectral radius over all the adjacency matrices. The *dynamic communicability matrix* is defined in Grindrod et al. (2011) as

$$Q = (I - \alpha A^{[1]})^{-1}(I - \alpha A^{[2]})^{-1} \cdots (I - \alpha A^{[M]})^{-1}, \tag{7}$$

where the parameter α satisfies $0 < \alpha < 1/\rho^*$, so that each resolvent in (7) can be expanded in a convergent power series in the corresponding adjacency matrix. Writing each resolvent as a geometric series and expanding the product in (7), we see that Q contains all terms of the form

$$\alpha^{\ell}(A^{[t_{k_1}]}A^{[t_{k_2}]} \cdots A^{[t_{k_\ell}]}), \tag{8}$$

where $t_{k_1} \leq t_{k_2} \leq \cdots \leq t_{k_\ell}$. Each entry (i, j) in the product $A^{[t_{k_1}]}A^{[t_{k_2}]} \cdots A^{[t_{k_\ell}]}$ counts the number of dynamic walks of length ℓ from node i to node j, where the k-th edge of the walk comes from time-step t_k. Hence, Q_{ij} is a weighted sum of dynamic walks of all possible lengths from node i to node j, where walks of length ℓ are weighted by a factor α^{ℓ}. Hence, Q_{ij} is a measure of communicability between nodes i and j, with node i in the role of broadcaster and node j as the receiver. The i-th row sum

$$\mathrm{BC}(i) = \sum_{j=1}^{n} Q_{ij} = [Q \cdot \mathbf{1}]_i$$

is a measure of how well node i broadcasts information to the rest of the network, and is a measure of *broadcast centrality*. On the other hand, the j-th column sum

$$\mathrm{RC}(j) = \sum_{i=1}^{n} Q_{ij} = [Q^T \cdot \mathbf{1}]_j$$

measures how well node j receives information from all other nodes in the network and provides a measure of *receive centrality*. It is worth noting that, in general, the broadcast and receive centrality of each node are different, since the matrix Q is not symmetric. Observing that each adjacency matrix is symmetric, we have

$$Q^T = \left((I - \alpha A^{[1]})^{-1}(I - \alpha A^{[2]})^{-1} \cdots (I - \alpha A^{[M]})^{-1}\right)^T = \\ = (I - \alpha A^{[M]})^{-1}(I - \alpha A^{[M-1]})^{-1} \cdots (I - \alpha A^{[1]})^{-1},$$

showing that broadcast and receive centralities are related by a reversal of the time ordering. An application of dynamic communicability to a real-world temporal network arising in the study of a contagion model is described in Chen et al. (2017).

It is interesting to note that using the matrix exponential instead of the resolvent produces a temporal communicability matrix that fails to satisfy certain combinatorial and consistency requirements: as a consequence, not all walks of the same length are given equal weight, as shown in Arrigo et al. (2022).

5 Modeling Diffusion on Graphs

Many types of phenomena in network analysis can be modeled as diffusion processes on graphs. Diffusion, both standard and non-local, yields another perspective on network connectivity and communicability.

5.1 Functions of the Graph Laplacian: The Heat Equation

So far, the main focus of this survey has been on functions of the adjacency matrix, which have a direct interpretation in terms of graph walks. In many situations, however, it is natural to consider functions of $L = D - A$, the graph Laplacian of G. More specifically, the Laplacian (or one of its normalized versions) is used in the study of diffusion processes and random walks on graphs, both time-discrete and time-continuous. Such random walks represent an essential tool for exploring network structure. Similarly, the behavior of the solution of evolution equations involving graph Laplacians, such as the heat equation, provide important information on the network. Among the simplest such equations is the initial value problem for the heat equation on G:

$$\frac{\mathrm{d}\mathbf{x}}{\mathrm{d}t} = -L\,\mathbf{x}, \quad t > 0; \quad \mathbf{x}(0) = \mathbf{x}_0. \tag{9}$$

This is a system of linear ordinary differential equations, where the unknown vector-valued function $\mathbf{x}(t)$ can be interpreted as the concentration of some substance on the nodes of G, or maybe as the "temperature" of each node, at time t. The initial distribution $\mathbf{x}_0$ is usually a nonnegative vector, often normalized so that $\|\mathbf{x}_0\|_1 = 1$. In this case, it can be interpreted as a probability distribution on the nodes of G. It is well known that the solution of (9) is unique and is given for all $t \geq 0$ by

$$\mathbf{x}(t) = \mathrm{e}^{-tL}\mathbf{x}_0 = S(t)\,\mathbf{x}_0\,,$$

where we have set $S(t) = \mathrm{e}^{-tL}$.

The family $\{S(t)\}_{t\geq 0}$ forms a continuous one-parameter commutative semigroup with $-L$ as its infinitesimal generator, i.e.,

$$S'(0) = \lim_{t\to 0+} \frac{1}{t}\,(S(t) - I) = -L\,.$$

It is easily seen that for all $t \geq 0$, $S(t)$ is a *doubly stochastic* matrix; i.e., it is a nonnegative matrix with row and columns sums equal to 1 (recall that $S(t)$ is symmetric, since G is undirected). This implies that for all $t > 0$ the solution vector $\mathbf{x}(t) = S(t)\mathbf{x}_0$ is a probability distribution on G; in particular, the ℓ^1-norm of $\mathbf{x}(t)$ remains constant and equal to 1. Hence, the evolution of $\mathbf{x}(t)$ describes a continuous time random walk on G. For a connected graph, L is irreducible and there is a unique steady-state distribution $\mathbf{x}_* = \lim_{t\to\infty}\mathbf{x}(t)$. This asymptotic distribution is easily found to be the unique positive solution to the homogeneous linear system $L\mathbf{x} = \mathbf{0}$ satisfying $\|\mathbf{x}_*\|_1 = 1$, i.e., the uniform distribution $\mathbf{x}_* = \frac{1}{n}\mathbf{1}$. In other words, at steady state the system is in "thermal equilibrium". The rate of convergence of $\mathbf{x}(t)$ to $\mathbf{x}_*$ depends on the topology of the graph and is largely determined by the algebraic connectivity of G. This can be seen by considering the eigendecomposition of the graph Laplacian:

$$L = Q\Lambda Q^T = \sum_{k=1}^{n} \lambda_k\, \mathbf{q}_k\mathbf{q}_k^T\,. \tag{10}$$

For a connected graph the eigenvalues of L satisfy $0 = \lambda_1 < \lambda_2 \leq \cdots \leq \lambda_n$ and the eigenvector associated with the simple zero eigenvalue, normalized in the ℓ^2-norm, is $\mathbf{q}_1 = \frac{1}{\sqrt{n}}\mathbf{1}$. Therefore

$$\mathrm{e}^{-tL}\mathbf{x}_0 = \frac{1}{n}(\mathbf{1}^T\mathbf{x}_0)\mathbf{1} + \sum_{k=2}^{n} \mathrm{e}^{-t\lambda_k}(\mathbf{q}_k^T\mathbf{x}_0)\mathbf{q}_k \longrightarrow \frac{1}{n}\mathbf{1} \quad \text{for} \quad t\to\infty \tag{11}$$

if $\mathbf{x}_0$ is nonnegative and has unit 1-norm.

It is clear from (11) that the rate of convergence to equilibrium depends primarily on the smallest nonzero eigenvalue λ_2 of L, i.e., on the algebraic connectivity of G. The larger is λ_2, the faster the convergence to equilibrium. Among al possible graphs with n nodes the fastest convergence to steady-state is obtained, of course, for the complete graph K_n nodes, in which every node is connected to any other node. Indeed, the Laplacian of this graph has only two distinct eigenvalues, 0 and n (the maximum possible value), and equilibrium is reached extremely fast. The already mentioned expander graphs provide more interesting examples of highly connected graphs; these are graphs that are at the same time highly sparse and have a relatively large value of λ_2.

The use of diffusion models based on the *fractional graph Laplacian* has also been proposed, due to its ability to generate random walks that can efficiently explore large networks (Riascos and Mateos 2014, 2021). For $\alpha \in (0, 1)$, the fractional graph Laplacian is defined by

$$L^\alpha = Q\Lambda^\alpha Q^T\,, \tag{12}$$

where Λ^α is the diagonal matrix with entries given by $\lambda_1^\alpha = 0, \lambda_2^\alpha, \ldots, \lambda_n^\alpha$ (cf. (10)). In this definition, the principal (positive) branch of the αth root λ_i is chosen. If G is connected, L^α is a full matrix: in other words, L^α is a *non-local operator*. It turns out that the matrix L^α, like L itself, is a singular M-matrix. If Δ denotes the diagonal matrix containing the diagonal entries of L^α, the matrix $S = I - L^\alpha\Delta^{-1}$ is column stochastic and is the transition matrix of a discrete time random walk on G. Since S is dense, the random walk on G is not constrained to move from node to neighboring node, but has a nonzero probability of "jumping" from one node to any other node, including very distant ones. Furthermore, the probability of making long-range jumps may be non-negligible, making the random walk associated with $S = I - L^\alpha\Delta^{-1}$ more efficient in the exploration of a large and sparse network compared to the "local" random walk associated with $I - LD^{-1}$, especially when the diameter of G is not small. Similar considerations apply to the continuous-time non-local diffusion process

$$\frac{\mathrm{d}\mathbf{x}}{\mathrm{d}t} = -L^\alpha\,\mathbf{x}, \quad t > 0, \quad \mathbf{x}(0) = \mathbf{x}_0\,, \tag{13}$$

or the analogous one based on the normalized fractional Laplacian. These results can be extended, at least in part, to directed graphs using the nonsymmetric Laplacian $L_{\text{out}} = D - A$ (where D is the diagonal matrix of outdegrees); see Benzi et al. (2020). (It should be mentioned that a different and sometimes advantageous approach to speeding up network navigation has been proposed in Estrada et al. (2017, 2018) by means of the *k-path Laplacian*, a natural generalization of the graph Laplacian first introduced in Estrada (2012b).)

It is natural to use the exponentials e^{-tL} (or e^{-tL^α}, for $0 < \alpha < 1$) to define diffusion-based communicability measures. In the case of a d-regular graph, for which $L = dI - A$, the Laplacian-based communicability between nodes i and j is just

$$[\mathrm{e}^{-tL}]_{ij} = \mathrm{e}^{-td}[\mathrm{e}^{tA}]_{ij}\,,$$

and therefore it reduces, up to a node-independent factor, to the (standard) exponential-based communicability (1), with $\beta = t$. On the other hand for non-regular graphs, or in the case of the fractional Laplacian, this commuicability measure is different from (1). It has a simple physical interpretation: it is the amount of substance found on node j at time t given that all the substance was initially concentrated on node i. In terms of (continuous time) random walks, it is the probability of being at node j at time t starting from node i at time $t = 0$.

For a connected, undirected graph, the diagonal entries of $S(t) = \mathrm{e}^{-tL}$ decrease monotonically from the value 1 for $t = 0$ to the asymptotic value $\frac{1}{n}$ for $t \to \infty$, while the off-diagonal entries increase, also monotonically, from the value 0 for $t = 0$ to the value $\frac{1}{n}$ as $t \to \infty$ (see (11). To prove monotonicity, note that

$$\frac{\mathrm{d}}{\mathrm{d}t}\mathrm{e}^{-tL} = -L\mathrm{e}^{-tL} = -\sum_{k=2}^{n} \lambda_k \mathrm{e}^{-t\lambda_k}\, \mathbf{q}_k \mathbf{q}_k^T\,,$$

therefore

$$\frac{\mathrm{d}}{\mathrm{d}t}[\mathrm{e}^{-tL}]_{ii} = -\sum_{k=2}^{n} \lambda_k \mathrm{e}^{-t\lambda_k}\, q_k(i)^2 < 0\,,$$

whereas for $i \neq j$ there holds for all $t > 0$

$$\frac{\mathrm{d}}{\mathrm{d}t}[\mathrm{e}^{-tL}]_{ij} = -\sum_{k=2}^{n} \lambda_k \mathrm{e}^{-t\lambda_k}\, q_k(i)q_k(j) > -\sum_{k=2}^{n} \lambda_k\, q_k(i)q_k(j) = -[L]_{ij} \geq 0\,.$$

Clearly, relatively large values of $[\mathrm{e}^{-tL}]_{ij}$ for small t indicate that nodes i and j exhibit good communicability. In other words, a fast convergence of $[\mathrm{e}^{-tL}]_{ij}$ to the limit $\frac{1}{n}$ indicates a good communicability between the two nodes. Similar considerations apply if L is replaced by L^α with $0 < \alpha < 1$.

6 Computational Aspects

For small to moderate size networks (say, up to a few thousand nodes), it is possible to compute each individual entry of the matrix function $f(A)$ or $f(L)$ using a variety of methods. For undirected graphs, using the eigendecomposition of the matrix is often the easiest way to carry out the desired calculation. For directed graphs the computation of matrix functions is somewhat more delicate, but a number of techniques are available for handling small to moderate size problems, see Higham (2008). Similarly, for generalized matrix functions one can simply compute the (compact) SVD of A to obtain $f^\diamond(A)$.

In the case of large-scale networks, on the other hand, it is generally impossible to compute every entry of a matrix function. Even if A is sparse, the matrix $f(A)$ is generally a dense matrix and therefore impossible to store for large n. Fortunately, in the case of large-scale graphs it is seldom required to compute all the entries of a matrix function; as we have seen, typical calculations require computing the *action* of a matrix function on a vector, or a subset of selected entries (for example the ones on the main diagonal). For these tasks there are efficient algorithms which use the matrix A (or L) only in the form of matrix-vector products. Moreover, such computations may not need to be very accurate: for instance, centrality measures are generally used to identify the top-ranked nodes, so there is no need to compute the centrality indices of all the nodes with high accuracy. Bounds on selected entries of $f(A)$ are often sufficient. Methods based on quadrature rules and the Lanczos process (Golub and Meurant 2010) can be used to obtain bounds or estimates on the individual entries of $f(A)$ (see, e.g., Benzi and Boito 2010), while computing the action of a matrix function on a vector can be done efficiently by means of Krylov subspace methods. In the case of generalized matrix functions, efficient techniques have been developed in Arrigo et al. (2016). We refer to Benzi and Boito (2020) for a more detailed discussion of these aspects.

7 Conclusions

As shown in this concise survey, communicability is by now an established tool in the analysis and design of large, complex networks from a variety of domains, with new applications being found all the time. Efficient algorithms exist for the computation of communicability measures, based on state-of-the art developments in numerical linear algebra. Progress in this field will bring about the ability to perform computations on networks of increasing size and complexity.

Future work is likely to focus on the development of matrix and tensor functions arising in the study of temporal networks, and on the extension of existing techniques to more sophisticated network models, such as multiplex and other multilayer networks. The case of temporal networks is also in need of further development.

Acknowledgements I would like to thank Paola Boito (University of Pisa) for very useful input.

References

Akbarzadeh, M., Estrada, E.: Communicability geometry captures traffic flow in cities. Nat. Hum. Behav. **2**, 645–652 (2018)

Arrigo, F., Benzi, M.: Updating and downdating techniques for optimizing network communicability. SIAM J. Sci. Comput. **38**, B25–B49 (2016a)

Arrigo, F., Benzi, M.: Edge modification criteria for enhancing the communicability of digraphs. SIAM J. Matrix Anal. Appl. **37**, 443–468 (2016b)

Arrigo, F., Benzi, M., Fenu, C.: Computation of generalized matrix functions. SIAM J. Matrix Anal. Appl. **37**, 836–860 (2016)

Arrigo, F., Higham, D.J., Noferini, V., Wood, R.: Dynamic Katz and related network measures. Linear Algebra Appl. **655**, 159–185 (2022)

Barabasi, A.L.: Linked: The New Science of Networks. Perseus, Cambridge, UK (2002)

Bartesaghi, P., Benzi, M., Clemente, G.P., Grassi, R., Estrada, E.: Risk-dependent centrality in economic and financial networks. SIAM J. Financ. Math. **11**, 526–565 (2020)
Beckermann, B., Kressner, D., Schweitzer, M.: Low-rank updates of matrix functions. SIAM J. Matrix Anal. Appl. **39**, 539–565 (2018)
Benzi, M., Bertaccini, D., Durastante, F., Simunec, I.: Non-local network dynamics via fractional graph Laplacians. J. Complex Netw. **8**, cnaa017 (2020)
Benzi, M., Boito, P.: Quadrature rule-based bounds for functions of adjacency matrices. Linear Algebra Appl. **433**, 637–652 (2010)
Benzi, M., Boito, P.: Matrix functions in network analysis. GAM Mitt. **43**, e202000012 (2020)
Benzi, M., Estrada, E., Klymko, C.: Ranking hubs and authorities using matrix functions. Linear Algebra Appl. **438**, 2447–2474 (2013)
Benzi, M., Faccio, C., Daidone, I., Zanetti-Polzi, L.: Structural analysis of water networks. J. Complex Netw. **11**, cnad001 (2023)
Benzi, M., Klymko, C.: Total communicability as a centrality measure. J. Complex Netw. **1**, 124–149 (2013)
Benzi, M., Klymko, C.: On the limiting behavior of parameter-dependent network centrality measures. SIAM J. Matrix Anal. Appl. **36**, 686–706 (2015)
Berman, A., Plemmons, R.J.: Nonnegative Matrices in the Mathematical Sciences. Classics in Applied Mathematics, vol. 9. Society for Industrial and Applied Mathematics, Philadelphia, PA (1994)
Betzel, R.F., Gu, S., Medaglia, J.D., Pasqualetti, F., Bassett, D.S.: Optimally controlling the human connectome: the role of network topology. Sci. Rep. **6** (2016) Article 30770
Boccaletti, S., Latora, V., Moreno, Y., Chavez, M., Hwang, D.-U.: Complex networks: structure and dynamics. Phys. Rep. **424**, 175–308 (2006)
Bonacich, P.: Power and centrality: a family of measures. Am. J. Sociol. **92**, 1170–1182 (1987)
Brandes, U., Erlebach, T. (eds.): Network Analysis: Methodological Foundations. Lecture Notes in Computer Science, vol. 3418. Springer, Berlin (2005)
Chebotarev, P.: The walk distances in graphs. Discr. Appl. Math. **160**, 1484–1500 (2012)
Chen, I., Benzi, M., Chang, H.H., Hertzberg, V.S.: Dynamic communicability and epidemic spread: a case study on an empirical dynamic contact network. J. Complex Netw. **5**, 274–302 (2017)
Crofts, J.J., Higham, D.J.: A weighted communicability measure applied to complex brain networks. J. Royal Soc. Interface **6**, 411–414 (2009)
Crofts, J.J., Higham, D.J., Bosnell, R., Jbabdi, S., Matthews, P.M., Behrens, T.E.J., Johansen-Berg, H.: Network analysis detects changes in the contralesional hemisphere following stroke. Neuroimage **54**, 161–169 (2011)
de Reus, M.A., van den Heuvel, M.P.: Simulated rich club lesioning in brain networks: a scaffold for communication and integration? Front. Hum. Neurosc. (2014). https://doi.org/10.3389/fnhum.2014.00647
Diestel, R.: Graph Theory. Springer-Verlag, Berlin (2000)
Estrada, E.: Community detection based on network communicability. Chaos **21** (2011). Article 016103
Estrada, E.: Complex networks in the Euclidean space of communicability distances. Phys. Rev. E **85** (2012a). Article 066122
Estrada, E.: Path Laplacian matrices: introduction and application to the analysis of consensus in networks. Linear Algebra Appl. **436**, 3373–3391 (2012b)
Estrada, E.: The communicability distance in graphs. Linear Algebra Appl. **436**, 4317–4328 (2012c)
Estrada, E.: The Structure of Complex Networks: Theory and Applications. Oxford University Press (2012d)
Estrada, E.: Communicability geometry of multiplexes. New J. Phys. **21** (2019). Article 015004
Estrada, E., Hameed, E., Hatano, N., Langer, M.: Path Laplacian operators and superdiffusive processes on graphs. I. One-dimensional case. Linear Algebra Appl. **523**, 307–334 (2017)
Estrada, E., Hameed, E., Langer, M., Puchalska, A.: Path Laplacian operators and superdiffusive processes on graphs. II. Two-dimensional lattice. Linear Algebra Appl. **555**, 373–397 (2018)

Estrada, E., Hatano, N.: Statistical-mechanical approach to subgraph centrality in complex networks. Chem. Phys. Lett. **439**, 247–251 (2007)

Estrada, E., Hatano, N.: Communicability in complex networks. Phys. Rev. E **77** (2008). Article 036111

Estrada, E., Hatano, N.: Communicability graph and community structures in complex networks. Appl. Math. Comput. **214**, 500–511 (2009)

Estrada, E., Hatano, N.: Communicability angle and the spatial efficiency of networks. SIAM Rev. **58**, 692–715 (2016)

Estrada, E., Hatano, N., Benzi, M.: The physics of communicability in complex networks. Phys. Rep. **514**, 89–119 (2012)

Estrada, E., Higham, D.J.: Network properties revealed through matrix functions. SIAM Rev. **52**, 696–714 (2010)

Estrada, E., Knight, P.A.: A First Course in Network Science. Oxford University Press (2015)

Estrada, E., Rodríguez-Velázquez, J.A.: Subgraph centrality in complex networks. Phys. Rev. E **71** (2005). Article 056103

Estrada, E., Vargas-Estrada, E., Ando, H.: Communicability angles reveal the critical links in network consensus dynamics. Phys. Rev. E **92** (2015). Article 052809

Faccio, C., Benzi, M., Daidone, I., Zanetti-Polzi, L.: Low- and high-density forms of liquid water revealed by a new medium-range order descriptor. J. Mol. Liquids **355**, 118922 (2022)

Faccio, C., Fonte, N., Daidone, I., Zanetti-Polzi, L.: Enhanced connectivity and mobility in liquid water: implications for the high density liquid structure and its onset. J. Mol. Liquids **392**, 123425 (2023)

Farahat, A., LoFaro, T., Miller, J., Rae, G., Ward, L.A.: Authority rankings from HITS, PageRank, and SALSA: existence, uniqueness, and effect of initialization. SIAM J. Sci. Comput. **27**, 1181–1201 (2006)

Fiedler, M.: Algebraic connectivity of graphs. Czech. Math. J. **23**, 298–305 (1973)

Fioriti, V., Chinnici, M.: Identifying sparse and dense sub-graphs in large graphs with a fast algorithm. Europhys. Lett. **108** (2014). Article 50006

Gladilin, E.: Graph-theoretical model of global human interactome reveals enhanced long-range communicability in cancer networks. PLoS ONE **12** (2017). Article e0170953. https://doi.org/10.1371/journal.pone.0170953

Golub, G.H., Meurant, G.: Matrices, Moments and Quadrature with Applications. Princeton University Press, Princeton, NJ (2010)

Golub, G.H., Van Loan, C.F.: Matrix Computations, 4th edn. Johns Hopkins University Press, Baltimore and London (2013)

Gravert, M.M., Dolan, R.J., Behrens, T.E.J.: A map of abstract relational knowledge in the human hippocampal-entorhinal cortex, eLife (2017). Article **6**, e17086 (2017). https://doi.org/10.7554/eLife.17086

Grindrod, P., Parsons, M.C., Higham, D.J., Estrada, E.: Communicability across evolving networks. Phys. Rev. E **83** (2011). Article 046120

Hawkins, J.B., Ben-Israel, A.: On generalized matrix functions. Linear Multilinear Algebra **1**, 163–171 (1973)

Higham, N.J.: Functions of Matrices: Theory and Computation. Society for Industrial and Applied Mathematics, Philadelphia, PA (2008)

Hoory, S., Linial, N., Wigderson, A.: Expander graphs and their applications. Bull. Am. Math. Soc. (New Ser.) **43**, 439–561 (2006)

Horn, R.A., Johnson, C.R.: Matrix Analysis, 2nd edn. Cambridge University Press (2013)

Katz, L.: A new status index derived from sociometric analysis. Psychometrica **18**, 39–43 (1953)

Kleinberg, J.: Authoritative sources in a hyperlinked environment. J. ACM **46**, 604–632 (1999)

Koponen, I.T., Nousiainen, M.: Concept networks in learning: finding key concepts in learners' representations of the interlinked structure of scientific knowledge. J. Complex Netw. **2**, 187–202 (2014)

Mussone, L., Notari, R.: A comparative analysis of underground and bus transit networks through graph theory. Environ. Plan. B: Urban Anal. Sci. (2019). Research Article https://doi.org/10.1177/2399808319879460

Newman, M.: The structure and function of complex networks. SIAM Rev. **45**, 167–256 (2003)

Newman, M.: Networks, 2nd edn. Oxford University Press (2018)

Ostachuk, A.: A network analysis of arthropod evolution and the potential of the primitive. Sci. Rep. **14** (2024). Article 503

Pião Azevedo, C., Salgado, P.A., Azevedo-Perdicoúlis, T.-P., Lopes dos Santos, P.: Determination of effective connectivity of brain activity in the resting brain. In: Cunha, A., Paiva, A., Pereira, S. (eds.) Wireless Mobile Communication and Health Care, pp. 17–32. Springer (2024)

Pozza, S., Tudisco, F.: On the stability of network indices defined by means of matrix functions. SIAM J. Matrix Anal. Appl. **39**, 1521–1546 (2018)

Rapoport, A.: Mathematical models of social interaction. In: Handbook of Mathematical Psychology, vol. 2, pp. 493–579. Wiley, New York (1963)

Riascos, A.P., Mateos, J.L.: Fractional dynamics on networks. Emergence of anomalous diffusion and Lévy flights. Phys. Rev. E **90** (2014). Article 032809

Riascos, A.P., Mateos, J.L.: Random walks on weighted networks: a survey of local and non-local dynamics. J. Complex Netw. **9** (2021), cnab032

Santiago, E., Velasco-Hernandez, J.X., Romero-Salcedo, M.: A descriptive study of fracture networks in rocks using complex network metrics. Comput. Geosci. **88**, 97–114 (2016)

Seeley, J.R.: The net of reciprocal influence: a problem in treating sociometric data. Can. J. Psychol. **3**, 234–240 (1949)

Silva, T.C., Stancato Souza, S.R., Miranda Tabak, B.: Monitoring vulnerability and impact diffusion in financial networks. J. Econ. Dyn. Contr. **76**, 109–135 (2017)

Silver, G., Akbarzadeh, M., Estrada, E.: Tuned communicability metrics in networks. The case of alternative routes for urban traffic. Chaos, Solitons Fractals **116**, 402–413 (2018)

Veergossen, L.W., Schram, M.T., de Jong, J.J., Stehouwer, C.D., Schaper, N.C., Henry, R.M., van der Kallen, C.J., Dagnelie, P.C., van Boxtel, M.P., Eussen, S.J., Backes, W.H., Jansen, J.F.: White matter connectivity abnormalities in prediabetes and Type 2 diabetes: the Maastricht study. Diabetes Care (2019). Article dc190762

Xu, Y.: A spectral method to detect community structure based on the communicability modularity. Phys. A: Stat. Mech. Appl. **537** (2020). Article 122751

Zhao, W., Yin, Z., Fan, T., Luo, J.: Research on influence spread of scientific research team based on scientific factor quantification of big data. Int. J. Distr. Sens. Netw. (2019). Research Article https://doi.org/10.1177/1550147719842158

Open Access This chapter is licensed under the terms of the Creative Commons Attribution 4.0 International License (http://creativecommons.org/licenses/by/4.0/), which permits use, sharing, adaptation, distribution and reproduction in any medium or format, as long as you give appropriate credit to the original author(s) and the source, provide a link to the Creative Commons license and indicate if changes were made.

The images or other third party material in this chapter are included in the chapter's Creative Commons license, unless indicated otherwise in a credit line to the material. If material is not included in the chapter's Creative Commons license and your intended use is not permitted by statutory regulation or exceeds the permitted use, you will need to obtain permission directly from the copyright holder.

Persistence Module and Schubert Calculus

Yasuaki Hiraoka, Kohei Yahiro, and Chenguang Xu

Abstract A multiplication on persistence diagrams is introduced by means of Schubert calculus. The key observation behind this multiplication comes from the fact that the representation space of persistence modules has the structure of the Schubert decomposition of a flag. In particular, isomorphism classes of persistence modules correspond to Schubert cells, thereby the Schubert calculus naturally defines a multiplication on persistence diagrams. The meaning of the multiplication on persistence diagrams is carried over from that on Schubert calculus, i.e., algebro-geometric intersections of varieties of persistence modules.

1 Introduction

Persistence module (Edelsbrunner et al. 2002) is one of the most established methods in topological data analysis (Carlsson 2009), and its mathematical structures have been studied in detail. For example, the interval decomposition of persistence modules (Zomorodian and Carlsson 2005) and its stability property (Cohen-Steiner et al. 2007) provide us with reliable data descriptors called persistence diagrams, which have been applied in various scientific research (Chan et al. 2013; Giusti et al. 2015; Hiraoka et al. 2016; Saadatfar et al. 2017). Furthermore, vectorizations of persistence diagrams (Bubenik 2015; Kusano et al. 2016, 2018; Reininghaus et al.

Y. Hiraoka (✉) · K. Yahiro
Kyoto University Institute for Advanced Study, Kyoto University, Kyoto, Japan
e-mail: hiraoka.yasuaki.6z@kyoto-u.ac.jp

K. Yahiro
e-mail: yahiro.kohei.5r@kyoto-u.ac.jp

Y. Hiraoka
Inamori Research Institute for Science, Kyoto, Japan

RIKEN Center for Advanced Intelligence Project, Kyoto, Japan

C. Xu
Department of Mathematics, Kyoto University, Kyoto, Japan
e-mail: xu.chenguang.k34@kyoto-u.jp

© The Author(s) 2026

S. Oishi et al. (eds.), *Recent Developments in Industrial and Applied Mathematics*, ICIAM2023 Springer Series 1, https://doi.org/10.1007/978-981-95-1446-5_12

2015) provide us with additive operations and metrics on persistence diagrams. This theoretical extension clarifies several important geometric and probabilistic aspects of persistence diagrams (Hiraoka et al. 2018), and accordingly, they are utilized in machine learning applications.

From this historical background, it is natural to ask next whether we can introduce a proper multiplication operation on persistence diagrams. In this paper, we show that Schubert calculus (Fulton 1997; Ikeda 2018) can be well-defined on persistence diagrams and provides a multiplication on them. Here, we remark that Schubert calculus is studied as intersection theory of flag varieties in algebraic geometry, and there are many combinatorial research including Young diagrams, Schubert polynomials, and so on.

The key to introducing Schubert calculus on persistence diagrams comes from a flag structure equipped in persistence modules. In the setting of graded modules (Knudson 2008; Zomorodian and Carlsson 2005), we know that allowable basis change operation should respect the birth times of generators, and this rule restricts basis change matrices to be (upper) triangular. It follows from this fact that a natural structure of flags in a vector space will be induced in persistence modules. From this correspondence, we see that isomorphism classes of persistence modules can be equivalently regarded as Schubert cells, and accordingly, we have a multiplication on persistence diagrams induced by that on Schubert varieties. Furthermore, the meaning of the multiplication on persistence diagrams is carried over from that on Schubert calculus, i.e., algebro-geometric intersections of varieties of persistence modules.

This paper is organized as follows. Section 2 summarizes some preliminaries about flag varieties and the Schubert decomposition, and then shows the correspondence between isomorphism classes of persistence modules and Schubert cells. In Sect. 3, we show a multiplication on persistence diagrams based on Schubert calculus, as well as a connection to Schubert polynomials. Several examples of multiplications on persistence diagrams are also shown there.

2 Persistence Module and Schubert Decomposition

2.1 Flag Variety

Let $E = \mathbb{C}^n$ be an n-dimensional ambient $\mathbb{C}$-vector space. A (complete) flag is a sequence of vector subspaces

$$V_\bullet : \{\mathbf{0}\} = V_0 \subset V_1 \subset \cdots \subset V_n = E, \quad \dim V_i = i.$$

As a dual notation, a flag can also be expressed as

$$V^\bullet : E = V^0 \supset V^1 \supset \cdots \supset V^n = \{\mathbf{0}\}, \quad \operatorname{codim} V^i = i.$$

The set of all flags in E is denoted by $\mathcal{F}_n$, called the flag variety. We can construct a surjective map

$$\begin{aligned}\pi : \mathrm{GL}_n(\mathbb{C}) &\to \mathcal{F}_n\\ (\boldsymbol{v}_1 \dots \boldsymbol{v}_n) &\mapsto (V_i := \langle \boldsymbol{v}_1, \dots, \boldsymbol{v}_i \rangle)_{i=1}^n,\end{aligned}$$

where $\mathrm{GL}_n(\mathbb{C})$ is the general linear group, $(\boldsymbol{v}_1 \dots \boldsymbol{v}_n)$ is an $n \times n$ non-singular matrix expressed by its column vectors, and $\langle \boldsymbol{v}_1, \dots, \boldsymbol{v}_i \rangle$ is the subspace spanned by $\{\boldsymbol{v}_1, \dots, \boldsymbol{v}_i\}$. Also, it is easy to see that $\mathrm{GL}_n(\mathbb{C})$ naturally acts on $\mathcal{F}_n$, and this action makes $\mathcal{F}_n$ a homogeneous space. Hence, we have $\mathrm{GL}_n(\mathbb{C})/B \simeq \mathcal{F}_n$, where B is the subset of all upper triangular matrices. This bijection provides $\mathcal{F}_n$ with a structure of a projective variety (Humphreys 1975).

2.2 *Schubert Decomposition*

Definition 2.1 Let S_n denote the symmetric group on n elements. For $w \in S_n$, we define

(i) $V_\bullet^w \in \mathcal{F}_n$ to be a flag associated with w given by

$$V_i^w := \langle \boldsymbol{e}_{w(1)}, \dots, \boldsymbol{e}_{w(i)} \rangle, \quad 1 \le i \le n,$$

(ii) $F_w^\bullet$ to be the flag complement to $V_\bullet^w$ indexed by codimension:

$$F_w^i := \langle \boldsymbol{e}_{w(i+1)}, \dots, \boldsymbol{e}_{w(n)} \rangle, \quad 0 \le i \le n-1,$$

(iii) $\mathcal{U}_w$ to be the set of flags in $\mathcal{F}_n$ in general position with $F_w^\bullet$:

$$\mathcal{U}_w := \{V_\bullet \in \mathcal{F}_n \ : \ F_w^i \cap V_i = \{0\}, 1 \le i \le n\},$$

(iv) $U_w \subseteq \mathrm{GL}_n(\mathbb{C})$ to be the set of matrices satisfying the following conditions:

$$\begin{aligned}U_w = \{\xi \in \mathrm{GL}_n(\mathbb{C}) \ : \ &\xi_{w(j),j} = 1 \text{ for } 1 \le j \le n \text{ and}\\ &\xi_{i,k} = 0 \text{ for } 1 \le i \le n,\ k > w^{-1}(i)\},\end{aligned}$$

(v)

$$U_w^+ := \{\xi \in U_w \ : \ \xi_{jk} = 0,\ j > w(k)\},$$

$$U_w^- := \{\xi \in U_w \ : \ \xi_{jk} = 0,\ j < w(k)\}.$$

Definition 2.2 For $w \in S_n$, the images $X_w^\circ := \pi(U_w^+)$ and $\Omega_w^\circ := \pi(U_w^-)$ are called the *Schubert cell* and *opposite Schubert cell*, respectively.

Then, we have the following decomposition of the flag variety $\mathcal{F}_n$.

Theorem 2.3 (Fulton 1997; Ikeda 2018) *The flag variety can be decomposed in the following two ways:*

$$\mathcal{F}_n = \bigsqcup_{w \in S_n} X_w^\circ = \bigsqcup_{w \in S_n} \Omega_w^\circ. \tag{1}$$

Both are called the Schubert cell decomposition of $\mathcal{F}_n$.

For $w \in S_n$, let us define $d_w(i, j) = \#\{s \leq j : w(s) \geq i\}$. Then, the Bruhat order on S_n is given by

$$v \leq w \overset{\text{def}}{\Longleftrightarrow} d_v(i, j) \leq d_w(i, j) \quad \text{for } \forall i, j$$

for $v, w \in S_n$.

Definition 2.4 For $w \in S_n$, $X_w = \bigsqcup_{u \leq w} X_u^\circ$ and $\Omega_w = \bigsqcup_{u \geq w} \Omega_u^\circ$ are called the Schubert variety and the opposite Schubert variety, respectively.

It is known that those varieties are realized as the Zariski closure $X_w = \overline{X_w^\circ}$, and $\Omega_w = \overline{\Omega_w^\circ}$.

2.3 Correspondence Between Schubert Cells and Persistence Modules

Let $\mathbb{C}[z]$ be the polynomial ring with a single variable z, and $\mathrm{mod}(\mathbb{C}[z])$ be the category of non-negatively graded $\mathbb{C}[z]$-modules. In this paper, we take the following definition of persistence modules.

Definition 2.5 A (one-parameter) persistence module M is defined as a finitely generated object in $\mathrm{mod}(\mathbb{C}[z])$.

Remark 2.6 (Zomorodian and Carlsson 2005) In this setting, the interval decomposition

$$M \simeq \left(\bigoplus_{i=1}^{p} (z^{b_i})/(z^{d_i}) \right) \oplus \left(\bigoplus_{i=p+1}^{p+q} (z^{b_i}) \right)$$

of the persistence module M is derived by the structure theorem for finitely generated graded modules over $\mathbb{C}[z]$ (PID). Here, (z^b) denotes the ideal generated by a monomial z^b. The multiset $D = \{[b_i, d_i) : i = 1, \ldots, p\} \sqcup \{[b_i, \infty) : i = p+1, \ldots, p+q\}$ of intervals in $\mathbb{R}$ is called the persistence diagram or barcode. For each interval, b_i and d_i are called the birth time and death time, respectively.

For a persistence module M, we can construct a minimal free resolution

$$0 \longrightarrow F_1 \xrightarrow{\partial_1} F_0 \xrightarrow{\epsilon} M \longrightarrow 0, \tag{2}$$

with some free modules $F_0, F_1 \in \mathrm{mod}(\mathbb{C}[z])$. Therefore, we have

$$M \simeq F_0/\ker\epsilon = F_0/\mathrm{im}\partial_1,$$

implying that the parametrization on isomorphism classes is given by the embedding $F_1 \xrightarrow{\partial_1} F_0$. Hereafter, we regard F_1 as a submodule in F_0. Recall that the free modules F_0 and F_1 take the forms

$$F_0 \cong \bigoplus_{i=1}^{m} z^{b_i}\mathbb{C}[z], \quad F_1 \cong \bigoplus_{i=1}^{n} z^{d_i}\mathbb{C}[z], \tag{3}$$

where $\boldsymbol{b} : b_1 \leq \cdots \leq b_m$ and $\boldsymbol{d} : d_1 \leq \cdots \leq d_n$ are birth and death sequences, respectively. They satisfy the following conditions

(i) $m \geq n$: any generator of M needs to be born before the death, but can also persist to infinity;
(ii) $b_i \leq d_i$ for $i = 1, \ldots, n$: the number of existing death generators shall not surpass that of birth generators.

If $m > n$, by adding death generators distant enough, the number of two types of generators can be equalized without losing the essential structure. So we only consider the case $m = n$ (i.e., $\mathrm{rank} M = 0$). Furthermore, for the sake of simplicity, we assume the following setting.

Assumption 2.7 The birth and death sequence is strict ($b_1 < \cdots < b_n, d_1 < \cdots < d_n$), and separated ($b_n < d_1$).

In fact, we can obtain a similar result without this assumption by replacing B with a general parabolic subgroup in the following discussion. We abbreviate the condition in Assumption 2.7 as $\boldsymbol{b} < \boldsymbol{d}$.

For reformulating one-parameter persistence modules as flags, we first set an ambient free module as

$$F = \bigoplus_{i=1}^{n} \mathbb{C}[z] = \left\{ \sum_{i=1}^{n} a_i e_i : a_i \in \mathbb{C}[z] \right\},$$

where $\{e_i : i = 1, \ldots, n\}$ is a basis of F, and $F_1 \subseteq F_0$ are regarded as submodules in F.

Now let us set a basis of F_0 as $F_0 = \langle \xi_1, \ldots, \xi_n \rangle$, where ξ_i's are homogeneous with $\deg \xi_i = b_i$. The matrix formulation of F_0 in F is given by

$$\begin{aligned}(\xi_1 \cdots \xi_n) &= (e_1 \cdots e_n) \begin{pmatrix} \lambda_{11} z^{b_1} & \cdots & \lambda_{1n} z^{b_n} \\ \vdots & & \vdots \\ \lambda_{n1} z^{b_1} & \cdots & \lambda_{nn} z^{b_n} \end{pmatrix} \\ &= (e_1 \cdots e_n) \cdot \Lambda \cdot \mathrm{diag}(z^{b_1}, \ldots, z^{b_n}),\end{aligned} \tag{4}$$

where $\mathrm{diag}(a_1, \ldots, a_n)$ expresses the diagonal matrix with a_i at the ith entry. Since $\langle \xi_1, \ldots, \xi_n \rangle$ is a basis, the matrix $\Lambda = (\lambda_{ij})$ is an element of $\mathrm{GL}_n(\mathbb{C})$.

Let us define a collection of submodules of F_0 as

$$\mathcal{K}(\boldsymbol{b}, \boldsymbol{d}) = \mathcal{K} := \left\{ K \subseteq F_0 : K \cong \bigoplus_{i=1}^{n} z^{d_i} \mathbb{C}[z] \right\}.$$

First, let us remark on the correspondence between base changes on submodules in $\mathcal{K}(\boldsymbol{b}, \boldsymbol{d})$ and flags in $\mathbb{C}^n$. Let $K = \langle \eta_1, \ldots, \eta_n \rangle = \langle \eta'_1, \ldots, \eta'_n \rangle$ be two expressions using different generators with $\deg(\eta_i) = \deg(\eta'_i) = d_i$ for $K \in \mathcal{K}(\boldsymbol{b}, \boldsymbol{d})$. The basis change matrix $C(z)$ for $(\eta'_1 \ldots \eta'_n) = (\eta_1 \ldots \eta_n) \cdot C(z)$ takes the following upper triangular form:

$$\begin{aligned}C(z) &= \mathrm{diag}(z^{d_1}, \ldots, z^{d_n})^{-1} \cdot C \cdot \mathrm{diag}(z^{d_1}, \ldots, z^{d_n}) \\ &= \begin{pmatrix} z^{-d_1} & & 0 \\ & \ddots & \\ 0 & & z^{-d_n} \end{pmatrix} \begin{pmatrix} c_{11} & c_{12} & \cdots & c_{1n} \\ 0 & c_{22} & \cdots & c_{2n} \\ \vdots & \vdots & \ddots & \vdots \\ 0 & 0 & \cdots & c_{nn} \end{pmatrix} \begin{pmatrix} z^{d_1} & & 0 \\ & \ddots & \\ 0 & & z^{d_n} \end{pmatrix}.\end{aligned} \tag{5}$$

When we set an ordering of the basis of F_0 as $(\xi_1 \cdots \xi_n)$, the submodule $K = \langle \eta_1, \ldots, \eta_n \rangle \in \mathcal{K}(\boldsymbol{b}, \boldsymbol{d})$ can be represented as a matrix form

$$\begin{aligned}(\eta_1 \cdots \eta_n) &= (\xi_1 \cdots \xi_n) \begin{pmatrix} a_{11} z^{d_1 - b_1} & \ldots & a_{1n} z^{d_n - b_1} \\ \vdots & & \vdots \\ a_{n1} z^{d_1 - b_n} & \ldots & a_{nn} z^{d_n - b_n} \end{pmatrix} \\ &= (\xi_1 \cdots \xi_n) \mathrm{diag}(z^{-b_1}, \cdots, z^{-b_n}) \cdot A \cdot \mathrm{diag}(z^{d_1}, \ldots, z^{d_n}),\end{aligned} \tag{6}$$

where $d_j \neq b_i$ for $a_{ij} \neq 0$ due to the minimality of the free resolution. We note $A = (a_{ij}) \in \mathrm{GL}_n(\mathbb{C})$. Using this formula and (7), the basis change $(\eta'_1 \ldots \eta'_n) = (\eta_1 \ldots \eta_n) C(z)$ becomes

$$\begin{aligned}(\eta'_1 \ldots \eta'_n) &= (\eta_1 \ldots \eta_n) C(z) \\ &= (\xi_1 \cdots \xi_n) \mathrm{diag}(z^{-b_1}, \cdots, z^{-b_n}) \cdot A \cdot C \cdot \mathrm{diag}(z^{d_1}, \ldots, z^{d_n}).\end{aligned} \tag{7}$$

Namely, the basis change of K can be performed by acting an upper triangular matrix C to the matrix representation A.

Proposition 2.8 *Assume $\boldsymbol{b} < \boldsymbol{d}$. Then, there exists a bijective correspondence:*

$$\mathcal{K}(\boldsymbol{b}, \boldsymbol{d}) \cong \mathrm{GL}_n(\mathbb{C})/B \cong \mathcal{F}_n.$$

Proof The second bijection is already shown in Sect. 2.1, so we will show the first bijection. Given a matrix $A = (a_{ij}) \in \mathrm{GL}_n(\mathbb{C})$, we want to associate it with an element in $\mathcal{K}(\boldsymbol{b}, \boldsymbol{d})$. Death generators $\eta_1, \dots, \eta_n$ can be obtained from the birth generators via

$$(\eta_1 \cdots \eta_n) = (\xi_1 \cdots \xi_n) \begin{pmatrix} a_{11} z^{d_1 - b_1} & \dots & a_{1n} z^{d_n - b_1} \\ \vdots & & \vdots \\ a_{n1} z^{d_1 - b_n} & \dots & a_{nn} z^{d_n - b_n} \end{pmatrix}. \tag{8}$$

Since $\deg \xi_i = b_i$, it is easy to verify $\deg \eta_i = d_i$ for $i = 1, \dots, n$. Because $\{\xi_1, \dots, \xi_n\}$ is a basis of F_0 and $A \in \mathrm{GL}_n(\mathbb{C})$, $\{\eta_1, \dots, \eta_n\}$ is $\mathbb{C}[z]$-linearly independent. Define a submodule $K_A := \langle \eta_1, \dots, \eta_n \rangle \subset F_0$. This operation gives us a map

$$\begin{aligned} \varphi \colon \mathrm{GL}_n(\mathbb{C}) &\to \mathcal{K}(\boldsymbol{b}, \boldsymbol{d}) \\ A &\mapsto K_A \end{aligned}.$$

Given any submodule $K \in \mathcal{K}(\boldsymbol{b}, \boldsymbol{d})$, as K is free, we can choose a basis $K = \langle \eta_1, \dots, \eta_n \rangle$ and construct a matrix A as in Eq. (6), implying that φ is surjective. Then, the map φ induces an equivalence relation $\sim$ on $\mathrm{GL}_n(\mathbb{C})$:

$$\text{for any } A,\ A' \in \mathrm{GL}_n(\mathbb{C}),\ A \sim A' \text{ if and only if } \varphi(A) = \varphi(A').$$

Thus $\mathrm{GL}_n(\mathbb{C})/\sim$ is bijective to $\mathcal{K}(\boldsymbol{b}, \boldsymbol{d})$. Let A and A' be such two equivalent matrices and $\{\eta_1, \dots, \eta_n\}, \{\eta'_1, \dots, \eta'_n\}$ be the corresponding two generators. By (7), they satisfy the relation

$$(\eta'_1 \dots \eta'_n) = (\xi_1 \cdots \xi_n)\mathrm{diag}(z^{-b_1}, \cdots, z^{-b_n}) \cdot A \cdot C \cdot \mathrm{diag}(z^{d_1}, \dots, z^{d_n}),$$

where the shape of C is specified in (7) by setting $z = 1$. Comparing this equation with the defining equation of $\{\eta'_1, \dots, \eta'_n\}$:

$$(\eta'_1 \cdots \eta'_n) = (\xi_1 \cdots \xi_n)\mathrm{diag}(z^{-b_1}, \cdots, z^{-b_n}) \cdot A' \cdot \mathrm{diag}(z^{d_1}, \dots, z^{d_n})$$

implies $A \cdot C = A'$ for some C. It is easy to verify that the collection of all valid shapes of C is exactly B, from which the bijection follows. □

We recall that π denotes the surjection $\pi : \mathrm{GL}_n(\mathbb{C}) \to \mathcal{F}_n$. Combining with the Schubert cell decomposition of the flag variety in (1), it follows that $\mathcal{K}(\boldsymbol{b}, \boldsymbol{d})$ has a cell decomposition:

$$\mathcal{K}(\boldsymbol{b}, \boldsymbol{d}) = \bigsqcup_{w \in S_n} \mathcal{K}_w^\circ(\boldsymbol{b}, \boldsymbol{d}), \quad \mathcal{K}_w^\circ(\boldsymbol{b}, \boldsymbol{d}) = \varphi \circ \pi^{-1}(X_w^\circ). \tag{9}$$

Theorem 2.9 *The Schubert cell decomposition above gives a parametrization of isomorphism classes of persistence modules with the birth sequence* $\boldsymbol{b}$ *and death sequence* $\boldsymbol{d}$.

Proof The isomorphism class of a persistence module M is determined by its minimal free resolution

$$0 \longrightarrow F_1 \xrightarrow{\partial_1} F_0 \xrightarrow{\epsilon} M \longrightarrow 0,$$

so $M \simeq F_0/K_A$ for some $K_A \in \mathcal{K}(\boldsymbol{b}, \boldsymbol{d})$. Since $K_A = K_{A'}$ if and only if $A \sim A'$, the isomorphism class $[M]$ uniquely assigns a Schubert cell X_w°, where w is determined by $A \in U_w^+$. Conversely, it is easy to see that, for a given Schubert cell X_w°, we can obtain an isomorphism class of a persistence module corresponding to that Schubert cell. □

From this theorem, for a persistence module $M \simeq F_0/K_A$ with $A \in U_w^+$, we obtain the interval decomposition:

$$M \simeq \bigoplus_{i=1}^{n} (z^{b_{w(i)}})/(z^{d_i}).$$

which subsequently determines a persistence diagram

$$D_w^\circ(\boldsymbol{b}, \boldsymbol{d}) = \{(b_{w(i)}, d_i) : i = 1, \ldots, n\}.$$

Theorem 2.9 and (9) show the bijective correspondence among the Schubert cells X_w°, $\mathcal{K}_w^\circ(\boldsymbol{b}, \boldsymbol{d})$, and persistence diagrams $D_w^\circ(\boldsymbol{b}, \boldsymbol{d})$ for given birth and death sequences $\boldsymbol{b}$ and $\boldsymbol{d}$. We also note that the set of persistence diagrams can be regarded as a poset by the Bruhat order.

3 Schubert Calculus on Persistence Modules

In this section, we first briefly review the formulation of the Chow ring and Schubert polynomials by following Fulton (1997), Ikeda (2018). Based on this preparation, we introduce a multiplication on persistence diagrams in Sect. 3.3.

3.1 Chow Ring

The length of a permutation $w \in S_n$ is given by $\ell(w) = \#\{(i, j) \in [n] \times [n] \ : \ i < j, w(i) > w(j)\}$. An element $w_0 \in S_n$ given by

$$w_0(i) = n - i + 1, \quad i = 1, \dots, n$$

is the unique permutation having the longest length $\ell(w_0) = \frac{n(n-1)}{2}$. For $w \in S_n$, we also define $w^\vee := w_0 w$ (i.e., $i \mapsto n - w(i) + 1$).

We assign a formal symbol σ_w, called a Schubert class, to each $w \in S_n$ and consider the following additive groups

$$A^k(\mathcal{F}_n) := \bigoplus_{w \in S_n, \ell(w)=k} \mathbb{Z}\sigma_w, \quad A^*(\mathcal{F}_n) := \bigoplus_{k=0}^{\frac{n(n-1)}{2}} A^k(\mathcal{F}_n). \tag{10}$$

We first recall the following theorem

Theorem 3.1 (Kleiman 1974) *Let $Y \subseteq \mathcal{F}_n$ be an irreducible variety of codimension k, and $w \in S_n$ be a permutation with length $\ell(w) = k$. Then for a generic $g \in \mathrm{GL}_n(\mathbb{C})$,*

$$\#(Y \cap gX_w)$$

is a finite number irrelevant to the g chosen.

Following the same notation in Theorem 3.1, we define the symbol of an irreducible variety Y by

$$[Y] := \sum_{w \in S_n, \ell(w)=k} \#(Y \cap gX_w)\sigma_w \in A^k(\mathcal{F}_n). \tag{11}$$

Then, we can show $[\Omega_w] = \sigma_w$ and $[X_w] = \sigma_{w^\vee}$. Theorem 3.1 also guarantees that, for $u, v, w \in S_n$ with $\ell(u) = \ell(v) + \ell(w)$, the number

$$c^u_{wv} := \#(\Omega_w \cap g\Omega_v \cap X_u) \tag{12}$$

is a finite number irrelevant to the generic g chosen.

The above argument leads to the definition of the product of Schubert classes by

$$\sigma_w \cdot \sigma_v = \sum_{\substack{u \in S_n \\ \ell(u)=\ell(w)+\ell(v)}} c^u_{wv}\sigma_u.$$

The additive group $A^*(\mathcal{F}_n)$ equipped with this product is called the Chow ring. The coefficients c^u_{wv} are called *Littlewood-Richardson coefficients*. There exists no known rule to compute them in general cases, and limited rules are discovered only in several special cases. For example, when one of the multipliers is of form σ_{r_i} for a transposition r_i (only exchanging $i \leftrightarrow i+1$), the coefficients are determined by Monk's formula.

Theorem 3.2 (Monk 1959) *Let $t_{ab} \in S_n$ denote the transposition between a and b. For any $w \in S_n$ and $1 \leq i < n$,*

$$\sigma_w \sigma_{r_i} = \sum_{\substack{a \leq i < b \\ \ell(wt_{ab}) = \ell(w)+1}} \sigma_{wt_{ab}} \tag{13}$$

holds.

Since the symmetric group S_n is generated by $r_1, \ldots, r_{n-1}$, we can compute any products of Schubert classes by the repeated applications of Monk's formula in principle.

Let n be a positive integer. Recall that a polynomial $f \in \mathbb{Z}[z_1, \ldots, z_n]$ is called a *symmetric polynomial* if for any $w \in S_n$, $f(z_{w(1)}, \ldots, z_{w(n)}) = f(z_1, \ldots, z_n)$. In particular, we define the *elementary symmetric polynomial* of degree k to be

$$e_k = \sum_{1 \leq i_1 < \cdots < i_k \leq n} z_{i_1} \cdots z_{i_k}.$$

Let us consider an ideal generated by elementary symmetric polynomials up to degree n:

$$I_n := \langle e_k \mid 1 \leq k \leq n \rangle.$$

Then, we can identify the Chow ring $A^*(\mathcal{F}_n)$ with the following quotient ring.

Theorem 3.3 (Fulton 1997) *The Chow ring $A^*(\mathcal{F}_n)$ is isomorphic to the following graded ring*

$$\mathcal{R}_n := \mathbb{Z}[z_1, \ldots, z_n]/I_n.$$

3.2 Schubert Polynomial

For $m > n \in \mathbb{N}$, we define an embedding $\phi_{nm} : S_n \to S_m$ by $\phi_{nm}(w) = w'$, where

$$w'(i) = \begin{cases} w(i), & i = 1, \ldots, n \\ i, & i = n+1, \ldots, m. \end{cases}$$

Then (S_n, ϕ_{nm}) is a direct system over $\mathbb{N}$, leading to the infinite symmetric group

$$S_\infty := \varinjlim_n S_n = \bigsqcup_n S_n / \sim$$

as the direct limit. For a permutation $w \in S_\infty$, we write $w \in S_n$ if w fixes all numbers after n.

We also note that a natural embedding

$$\begin{aligned}\iota_n : \mathbb{C}^n &\hookrightarrow \mathbb{C}^{n+1} \\ (x_1, \ldots, x_n) &\mapsto (x_1, \ldots, x_n, 0)\end{aligned}$$

induces an embedding $\iota_n : \mathcal{F}_n \to \mathcal{F}_{n+1}$, so any flag can be regarded as a flag in a space with a higher dimension. Similarly, we define a composition

$$\iota_{nm} := \iota_{m-1} \circ \cdots \circ \iota_n : \mathcal{F}_n \to \mathcal{F}_m.$$

The pullback of this map induces a morphism between Chow rings:

$$\iota_{nm}^* : A^*(\mathcal{F}_m) \to A^*(\mathcal{F}_n).$$

Then, this naturally defines the inverse limit

$$\begin{aligned}A^*(\mathcal{F}_\infty) &:= \varprojlim_n A^*(\mathcal{F}_n) \\ &= \bigoplus_{r=0}^{\infty} \varprojlim_n A^r(\mathcal{F}_n).\end{aligned}$$

Let us denote Schubert classes in $A^*(\mathcal{F}_n)$ as $\sigma_w^{(n)}$ for $w \in S_n$ by specifying n. Then, it follows from the fact (e.g., Proposition 10.3 in Ikeda (2018))

$$\iota_n^*(\sigma_w^{(n+1)}) = \begin{cases} \sigma_w^{(n)}, & w \in S_n \\ 0, & w \notin S_n \end{cases}$$

that the inverse limit can be written as

$$\begin{aligned}A^*(\mathcal{F}_\infty) &= \bigoplus_{r=0}^{\infty} \bigoplus_{\substack{w \in S_\infty \\ \ell(w)=r}} \mathbb{Z}\hat{\sigma}_w \\ &= \bigoplus_{w \in S_\infty} \mathbb{Z}\hat{\sigma}_w,\end{aligned}$$

where

$$\hat{\sigma}_w = (\sigma_w^{(n)}) \in \varprojlim_n A^r(\mathcal{F}_n)$$

denotes the limit of the Schubert class for $r = \ell(w)$.

Now, let us construct a polynomial ring with infinitely many variables. We note that any polynomial in $\mathbb{Z}[z_1, \ldots, z_n]$ can be naturally embedded in $\mathbb{Z}[z_1, \ldots, z_m]$ for

$m \geq n$. Hence, we obtain a directed system whose direct limit

$$\mathbb{Z}[z] := \varinjlim_{n} \mathbb{Z}[z_1, \ldots, z_n]$$

is the polynomial ring in which we introduce Schubert polynomials below.

Theorem 3.4 (Fulton 1997; Ikeda 2018) *The map*

$$\Psi : \mathbb{Z}[z] \to A^*(\mathcal{F}_\infty) = \bigoplus_{r=0}^{\infty} \varprojlim_{n} A^r(\mathcal{F}_n)$$
$$f_r(z) \mapsto (f_r^{(n)} \bmod I_n),$$

is bijective, where $f_r(z) \in \mathbb{Z}[z]$ is a homogeneous polynomial with degree r and $f_r^{(n)} = f_r(z_1, \ldots, z_n, 0, \ldots)$. Here, the ring structure is induced by the isomorphism in Theorem 3.3.

Definition 3.5 The Schubert polynomial $\mathfrak{S}_w$ for $w \in S_\infty$ is defined by

$$\Psi(\mathfrak{S}_w) = \hat{\sigma}_w.$$

The Schubert polynomials can explicitly be described by using the difference operator:

$$\partial_i f := \frac{f - r_i f}{z_i - z_{i+1}}$$

defined on $f \in \mathbb{Z}[z]$. Let us define $\delta(n) = (n-1, \ldots, 1, 0)$.

Proposition 3.6 (Fulton 1997; Ikeda 2018) *Suppose $w \in S_\infty$ is expressed in $w \in S_n$ by $w = w_0 r_{i_1} \ldots r_{i_k}$ with $k = \ell(w_0) - \ell(w)$, where w_0 is the longest element in S_n. Then the polynomial*

$$\partial_{i_k} \ldots \partial_{i_1} z^{\delta(n)}$$

is defined independently of the choice of n and $(i_1, \ldots, i_k)$, and coincides with the Schubert polynomial $\mathfrak{S}_w$.

For later use, we show the Hasse diagrams and Schubert polynomials for S_2 and S_3 (Fig. 1).

Similar to Theorem 3.2, the Schubert polynomials also satisfy the Monk formula:

Theorem 3.7 *For $w \in S_\infty$ and $i \geq 1$,*

$$\mathfrak{S}_w \mathfrak{S}_{r_i} = \sum_{\substack{a \leq i < b \\ \ell(w t_{ab}) = \ell(w)+1}} \mathfrak{S}_{w t_{ab}}.$$

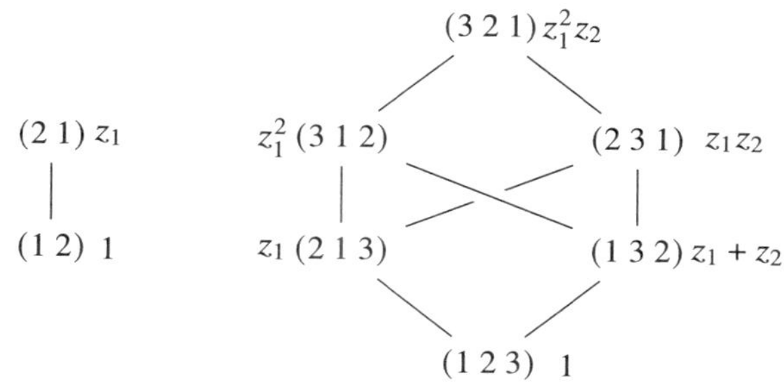

Fig. 1 Hasse diagram and Schubert polynomials for S_2 (left) and S_3(right)

3.3 *Multiplication on Persistence Diagrams*

In Sects. 3.1 and 3.2, we defined the product structure on Schubert classes (and Schubert polynomials) which bijectively correspond to persistence diagrams through flag varieties. This means that we can naturally introduce a product structure on persistence diagrams, which has a natural geometric meaning induced by that of Schubert varieties, i.e., the intersection of two isomorphism classes of persistence modules in $\mathcal{K}(\boldsymbol{b}, \boldsymbol{d})$. In this section, we show several examples of the products of persistence diagrams.

From Theorem 2.9, we have the correspondence between the Schubert variety X_w° and the persistence diagram

$$D_w^\circ(\boldsymbol{b}, \boldsymbol{d}) = \{(b_{w(i)}, d_i) : i = 1, \ldots, n\}$$

for each $w \in S_n$. In parallel to the definition $X_w = \sqcup_{u \leq w} X_u^\circ$, we introduce a symbol

$$D_w(\boldsymbol{b}, \boldsymbol{d}) = \bigsqcup_{u \leq w} D_u^\circ(\boldsymbol{b}, \boldsymbol{d}).$$

This definition is reasonable since $D_w(\boldsymbol{b}, \boldsymbol{d})$ contains the information of degenerate cases. Also, we recall $[X_w] = \sigma_{w^\vee}$. From these correspondences, we naturally identify the persistence diagrams $D_w(\boldsymbol{b}, \boldsymbol{d})$ with $\sigma_{w^\vee}$ from now on. Similarly, we also associate $\hat{D}_w(\boldsymbol{b}, \boldsymbol{d})$ with $\hat{\sigma}_w$ as a limiting object $n \to \infty$. In the following, we simply abbreviate $D_w = D_w(\boldsymbol{b}, \boldsymbol{d})$. We remark that D_{w_0} becomes the identity element of the product.

Example 3.8 For $n = 2$, let us consider the persistence diagram $D_w(\boldsymbol{b}, \boldsymbol{d})$ with $w = (12) \in S_2$ (see Fig. 2) and the product $D_w \cdot D_w$. Expressing the product in the Schubert classes lead to

$$\sigma_{(12)^\vee} \cdot \sigma_{(12)^\vee} = \sigma_{(21)} \cdot \sigma_{(21)}.$$

We apply Monk's formula (13) with $w = r_1 = (21)$. In this case, the only choice for $(a, b) = (1, 2)$ gives $\ell(wt_{12}) \neq \ell(w) + 1$. Therefore,

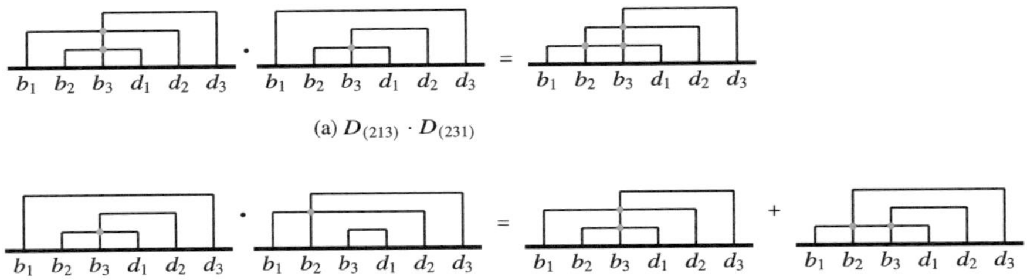

(a) $D_{(213)} \cdot D_{(231)}$

(b) $D_{(231)} \cdot D_{(312)}$

Fig. 2 Illustration of $D_{(213)} \cdot D_{(231)}$ and $D_{(231)} \cdot D_{(312)}$

$$\sigma_{(12)^\vee} \cdot \sigma_{(12)^\vee} = 0,$$

meaning $D_w \cdot D_w = 0$.

On the other hand, the same product using the Schubert polynomials leads to

$$\mathfrak{S}_{(12)^\vee} \cdot \mathfrak{S}_{(12)^\vee} = \mathfrak{S}_{(21)} \cdot \mathfrak{S}_{(21)} = z_1^2 = \mathfrak{S}_{(312)} = \mathfrak{S}_{(132)^\vee}.$$

The reason for this discrepancy is caused by the fact that Monk's formula in Theorem 3.2 fixes the total dimension n, while that in Theorem 3.7 can consider n to be large enough. More explicitly, the Schubert class $\sigma_{(12)^\vee}$ expresses a flag with codimension one in $\mathbb{C}^2$, and then $\sigma_{(12)^\vee} \cdot \sigma_{(12)^\vee}$ corresponds to the intersection with codimension two, leading to zero in the product. On the other hand, in the setting of Schubert polynomials, we embed w in S_3 (or S_n for $n > 2$), whereby w is understood as $w(213)$. Then, the total space becomes $\mathbb{C}^3$, and hence the codimension two intersection can be nontrivial. As a result, we have $\hat{D}_w \cdot \hat{D}_w = \hat{D}_{(132)}$.

Example 3.9 For $n = 3$, let us consider the product $D_{(213)} \cdot D_{(231)}$. From $\sigma_{(213)^\vee} \cdot \sigma_{(231)^\vee} = \sigma_{(231)} \cdot \sigma_{(213)}$, let us set $w = (231)$ and $r_1 = (213)$. Then, for $(a, b) = (1, 2), (1, 3)$ in the Monk's formula (13), we have $\ell(wt_{12}) = \ell(w) + 1$ and $\ell(wt_{13}) \neq \ell(w) + 1$. Thus

$$\sigma_{(213)^\vee} \cdot \sigma_{(231)^\vee} = \sigma_{(321)} = \sigma_{(123)^\vee}.$$

We also have the same result using the Schubert polynomials:

$$\mathfrak{S}_{(213)^\vee} \cdot \mathfrak{S}_{(231)^\vee} = \mathfrak{S}_{(231)} \cdot \mathfrak{S}_{(213)} = z_1 z_2 \cdot z_1 = \mathfrak{S}_{(321)} = \mathfrak{S}_{(123)^\vee}.$$

Therefore, we have $D_{(213)} \cdot D_{(231)} = \hat{D}_{(213)} \cdot \hat{D}_{(231)} = D_{(123)}$.

Similarly, we can show that

$$D_{(231)} \cdot D_{(312)} = \hat{D}_{(231)} \cdot \hat{D}_{(312)} = D_{(213)} + D_{(132)}.$$

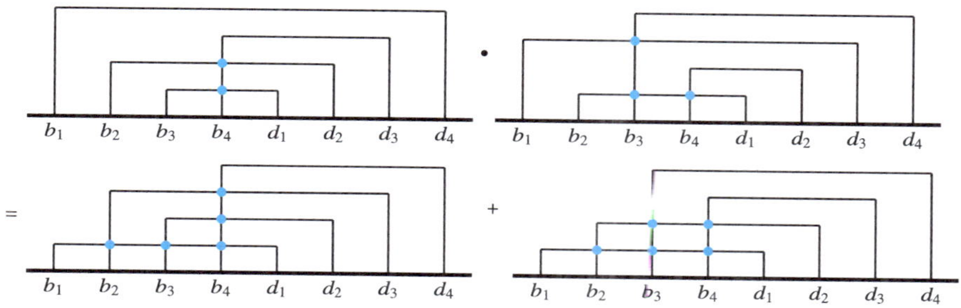

Fig. 3 Illustration of $D_{(3241)} \cdot D_{(2413)}$

Example 3.10 For $n = 4$, the following product can be calculated in a similar way as above:

$$D_{(3241)} \cdot D_{(2413)} = D_{(1324)} + D_{(1243)}.$$

Using the same settings as above, this relation can be graphically expressed as Fig. 3.

4 Discussion

In this paper, we defined a multiplication on persistence diagrams by means of Schubert calculus. The meaning of this multiplication stems from algebro-geometric intersections of varieties of persistence modules. More precisely, by interpreting (11) and (12) from persistence modules viewpoints, this multiplication characterizes the intersection of *generic persistence modules* in a representation space. This paper only focuses on the algebraic side of the multiplication, but its meaning in topological data analysis should also be addressed in future. For example, the multiplication structure studied in this paper may be applicable to define higher moments of persistence diagrams, which will be important to uncover statistical topological properties of data. Also, deepening geometric understandings of the multiplication over homological generators in persistence diagrams will be useful in data science community. Here, we remark that the paper Curry et al. (2024) studies a genericity of merge trees in the context of inverse problems of persistence diagrams, so it may be useful to shed new light on the above problem.

Acknowledgements The authors would like to express their sincere gratitude to Hiroyuki Ochiai for stimulating discussions. YH and KY are supported by the Japan Society for the Promotion of Science, Grant-in-Aid for Transformative Research Areas (A) (22A201).

References

Bubenik, P.: Statistical topological data analysis using persistence landscapes. J. Mach. Learn. Res. **16**, 77–102 (2015)

Carlsson, G.: Topology and data. Bull. Am. Math. Soc. **46**, 255–308 (2009)

Chan, J.M., Carlsson, G., Rabadan, R.: Topology of viral evolution. Proc. Natl. Acad. Sci. **110**, 18566–18571 (2013)

Cohen-Steiner, D., Edelsbrunner, H., Harer, J.: Stability of persistence diagrams. Discret. Comput. Geom. **37**, 103–120 (2007)

Curry, J., DeSha, J., Garin, A., Hess, K., Kanari, L., Mallery, B.: From trees to barcodes and back again II: Combinatorial and probabilistic aspects of a topological inverse problem. Comput. Geom. **116** (2024)

Edelsbrunner, H., Letscher, D., Zomorodian, A.: Topological persistence and simplification. In: Proceedings 41st Annual Symposium on Foundations of Computer Science, vol. 28, pp. 511–533 (2002)

Fulton, W.: Young Tableaux: With Applications to Representation Theory and Geometry. Cambridge University Press (1997)

Giusti, C., Pastalkova, E., Curto, C., Itskov, V.: Clique topology reveals intrinsic geometric structure in neural correlations. Proc. Natl. Acad. Sci. **112**, 13455–13460 (2015)

Hiraoka, Y., Nakamura, T., Hirata, A., Escolar, E., Matsue, K., Nishiura, Y.: Hierarchical structures of amorphous solids characterized by persistent homology. Proc. Natl. Acad. Sci. **113**, 7035–7040 (2016)

Hiraoka, Y., Shirai, T., Trinh, K.D.: Limit theorems for persistence diagrams. Ann. Appl. Probab. **28**, 2740–2780 (2018)

Humphreys, J.: Linear Algebraic Groups. Springer (1975)

Ikeda, T.: Lectures on Enumerative Geometry. University of Tokyo Press (2018). In Japanese

Kleiman, S.: The transversality of a general translate. Compos. Math. **28**, 287–297 (1974)

Knudson, K.: A refinement of multi-dimensional persistence. Homol. Homotopy Appl. **10**, 259–281 (2008)

Kusano, G., Fukumizu, K., Hiraoka, Y.: Persistence weighted Gaussian kernel for topological data analysis. In: Proceedings of the 33rd International Conference on Machine Learning. PMLR 48, pp. 2004–2013 (2016)

Kusano, G., Fukumizu, K., Hiraoka, Y.: Kernel method for persistence diagrams via kernel embedding and weight factor. J. Mach. Learn. Res. **18**, 1–41 (2018)

Monk, D.: The geometry of flag manifolds. Proc. Lond. Math. Soc. **59**, 253–286 (1959)

Reininghaus, J., Huber, S., Bauer, U., Kwitt, R.: A stable multi-scale kernel for topological machine learning. In: Proceedings of the IEEE Conference on Computer Vision and Pattern Recognition, pp. 4741–4748 (2015)

Saadatfar, M., Takeuchi, H., Robins, V., Nicolas, F., Hiraoka, Y.: Pore configuration landscape of granular crystallization. Nat. Commun. **8**, 15082 (2017)

Zomorodian, A., Carlsson, G.: Computing persistent homology. Discret. Comput. Geom. **33**, 249–274 (2005)

Open Access This chapter is licensed under the terms of the Creative Commons Attribution 4.0 International License (http://creativecommons.org/licenses/by/4.0/)), which permits use, sharing, adaptation, distribution and reproduction in any medium or format, as long as you give appropriate credit to the original author(s) and the source, provide a link to the Creative Commons license and indicate if changes were made.

The images or other third party material in this chapter are included in the chapter's Creative Commons license, unless indicated otherwise in a credit line to the material. If material is not included in the chapter's Creative Commons license and your intended use is not permitted by statutory regulation or exceeds the permitted use, you will need to obtain permission directly from the copyright holder.

Feasibility Problems and Optimization for Natural Channels

J. M. Martínez

Abstract The Saint-Venant equations are widely used to model real river flows. Solving these equations requires the use of sophisticated partial differential equation (PDE) solvers. In this paper, it is proposed that, under different conditions on data availability, the solution of the Saint-Venant equations can be represented by feasibility models that depend on a moderate number of parameters. In particular, a new non-rectangular model for the shape of cross sections is proposed. This method can be used under different assumptions on the knowledge of the variables influencing the evolution of the system.

Keywords Models for natural channels · Saint-Venant equations · Feasibility problems · Optimization

1 Introduction

Mathematical models are formal constructions designed to mimic physical phenomena. Often, mathematical models take the form of partial differential equations, in which quantities evolve over time according to spatial positions and obey well established physical laws. Solving partial differential equations requires knowledge of initial conditions and appropriate boundary conditions. In addition, in most practical situations, partial differential equations involve parameters that must be carefully estimated in order to obtain reliable results.

This work was supported by FAPESP (grants 2013/07375-0, 2016/01860-1, and 2018/24293-0) and CNPq (grants 302538/2019-4 and 302682/2019-8).

J. M. Martínez (✉)
Department of Applied Mathematics, Institute of Mathematics, Statistics, and Scientific Computing (IMECC), State University of Campinas, Campinas, SP, Brazil
e-mail: martinez@ime.unicamp.br

© The Author(s) 2026

S. Oishi et al. (eds.), *Recent Developments in Industrial and Applied Mathematics*, ICIAM2023 Springer Series 1, https://doi.org/10.1007/978-981-95-1446-5_13

The Saint-Venant Equations (Chaudhry 2022; Saint-Venant 1871) are often used to describe the evolution of fluid flows through real rivers. These equations are given by:

$$\frac{\partial A}{\partial t} + \frac{\partial Q}{\partial x} = q \tag{1}$$

and

$$\frac{\partial Q}{\partial t} + \frac{\partial}{\partial x}\left(\frac{Q^2}{A}\right) + gA\frac{\partial z}{\partial x} + \frac{n_g^2 Q|Q|P^{4/3}}{A^{7/3}} = 0. \tag{2}$$

Throughout this paper we assume that lengths are expressed in meters and time is expressed in seconds. Assume that $x \in [x_{min}, x_{max}]$ and $t \in [t_{min}, t_{max}]$. The variables x and t represent position and time, respectively. In addition, $z_b(x, t)$ denotes bed elevation, $z(x, t)$ is surface elevation, $h(x, t) = z(x, t) - z_b(x, t)$ is the depth of the river at (x, t), $A(x, t)$ is the area of the wetted cross section, $Q(x, t)$ is the flow-rate, $P(x, t)$ is the wetted perimeter, $q(x, t)$ is the lateral contribution per unit of channel length, and g is the acceleration of gravity, assumed to be 9.81 m/s^2. Equation (1) describes the conservation of mass and the Eq. (2) represents the conservation of linear momentum.

In addition to (1) and (2), admissibility conditions are usually required to be satisfied in real situations. Most frequently, admissibility conditions correspond to available observations or estimations and can be represented by equalities or inequalities involving the state variables. For simplicity, we will represent admissibility conditions in the form

$$(Q, A, z, h, q) \in \mathcal{Z}, \tag{3}$$

where $\mathcal{Z}$ is a set in an adequate space of functions.

The coefficient n_g represents the roughness. Many papers describe the Saint-Venant equations using the Manning roughness coefficient $\mathtt{n} = n_g/\sqrt{g}$. Following Yen (1992), we prefer the formulation with n_g because the units of this coefficient are $m^{1/6}$, which seems more natural than the units of of $\mathtt{n}$, which are $m^{-1/3}s$. The roughness coefficient depends on x due to the morphological differences of the river along its course and sediment deposition can also affect the roughness coefficients over time (Birgin et al. 2024). In addition, the variability of n_g with respect to position and time appears in real flows due to the uncertainty inherent in the variation of the wetted perimeter, which is fractal in nature (Gouyet 1996). Accordingly, if we assume geometrically simple cross-sections, the associated error must be absorbed by the variability of n_g. Sometimes, n_g is known as Yen-Manning coefficient (Birgin et al. 2024).

There are many numerical methods for solving the Saint-Venant equations, using appropriate initial and boundary conditions. See, for example, Lai and Khan (2012a, b, c), LeVeque (1992), Porto (2000), Simões et al. (2017), Ying et al. (2004). The efficiency of different methods is generally confirmed by their behavior in paradigmatic simplified situations or by comparison with real measured depth data (Emmett et al. 1979; Ferreira et al. 2017; Meade et al. 1979). In the latter case, the numerical solution of the Saint-Venant equations must be be complemented by optimization procedures for the estimation of Manning coefficients (Agresta et al. 2021; Askar and Al-jumaily 2008; Ayvaz 2013; Birgin and Martínez 2022; Birgin et al. 2024; Ding et al. 2004; Ding and Wang 2005; Guta and Prasad 2018; Marcus et al. 1992; Pappenberger et al. 2005).

Several difficulties are associated with solving the Saint-Venant equations for real rivers using PDE algorithms: The boundary and/or initial conditions may be unknown or known with high uncertainty; the geometry of cross sections is usually unknown; real rivers are not rectilinear, which affects the common assumption of constant Manning coefficients; uncertainty of the lateral contributions motivates to neglect this factor during the numerical resolution of the PDE's; and numerical solvers may become unstable or require unreliable small time steps to achieve stability. On the other hand, partial knowledge of the solution, due to available observations or estimates, is not employed in the process of solving numerically the equations. Roughly speaking, numerical integration of the PDE's may use information that is not available and ignore information that comes from reliable measurements. These inconveniences are generally overcome by the techniques of PDE-Constrained Optimization. In this case, unknown quantities are estimated by solving optimization problems in which the computation of the objective function or the constraints requires solving a partial differential equation (Antil et al. 2019).

A different approach for solving (1) and (2) under different conditions consists of assuming that the state variables $A(x, t)$ and $Q(x, t)$ are given by analytical expressions depending on a vector of parameters $c \in \mathbb{R}^n$, which must be determined by imposing the fulfillment of the equations in a sample of points (x, t) and the satisfaction of known conditions or constraints. Known conditions and constraints may or may not include initial and boundary conditions. This approach is similar to the classical idea of collocation methods (Ascher and Petzold 1998; Iserles 1996) for which many useful variations have been proposed, even recently (Wang and Chen 2009). When the postulated models for the state variables are neural networks, we are in presence of PINN (Physical Informed Neural Networks), an area that is currently the subject of intense research (see Raissi et al. 2019 and references therein). If one has qualitative knowledge of the real phenomenon, it is possible to propose models with a moderate number of parameters and specific analytic forms, so that the "overparameterization" inherent in neural network approaches (Wright 2023) does not occur.

In this paper we will postulate that cross sections of natural channels can be approximated in the form shown in Fig. 1. This shape can be described with only 4 parameters, which are assumed to vary linearly between x_{min} and x_{max}. Thus, the desired geometry for all cross sections of the river can be described with only 8 parameters to be estimated. It is unclear whether the intuition reflected in Fig. 1 could be expressed by a neural work with a moderate number of weights. Consequently, we will postulate that the shape of the cross section of the channel depends on a set of parameters $c \in \mathbb{R}^n$ in the manner described in Sects. 2 and 3. The bed elevation $z_b(x)$, the inlet discharge $Q(x_{min}, t)$, the lateral contribution $q(x, t)$, and the surface elevation $z(x, t)$ are also postulated to be functions of the parameters

$c_i, i = 1, \ldots, n$. Given the bed and surface elevations, and the shape of the cross section, we can compute the area $A(x, t)$ as a function of the parameter c. Then, we use the conservation of mass equation to calculate $Q(x, t)$ (see (7)). With this information at hand, we can can decide whether there are positive Manning coefficients $n_g(x, t)$ for which the conservation of linear momentum holds. We consider a solution of the problem to have been found if there are coefficients c such that the values $n_g(x, t)$ lie between reasonable given bounds for all points (x, t) chosen in a large sample.

Analogous approaches were used in Birgin and Martínez (2023) and Martínez (2024). In both papers, the cross sections were assumed to be rectangular. This assumption greatly simplifies the computations, but is unreliable in real situations, and it is unclear in what way this affects the accuracy of the predictions. In addition, Birgin and Martínez (2023) and Martínez (2024) assume that the cross sections are given by analytical polynomial expressions, while in the present paper the expression for $A(x, t)$ is derived from a simple analytical expression for $z(x, t)$. This postulation is better adapted to reality and is motivated by physical "well-balanced" considerations: Typically, in stationary situations, $z(x, t)$ tends to be constant, but due to irregularities in $z_b(x, t)$, this is not the case for $A(x, t)$. Then it is more natural to write simple analytic expressions for $z(x, t)$ than it is for for $A(x, t)$.

This paper is organized as follows. In Sect. 2 we introduce the basic definitions that allow us to express our problem as a set of infinitely many inequalities. In Sect. 3 we define the analytical expressions for modeling the cross sections of the channel. In Sect. 4 we introduce the set of inequalities that define the feasibility problem. In Sect. 5 an algorithm is defined for checking feasibility at an arbitrary point (x, t). Optimization procedures are described in Sect. 6 and numerical experiments are presented in Sect. 7. Finally, in Sect. 8 we state some conclusions and lines for future research.

Notation

The symbol $\|\cdot\|$ denotes Euclidean norm.

2 Basic Definitions

We are interested in finding model-based solutions of the Saint-Venant equations subject to admissibility conditions. So, we want to find model functions that represent $A(x, t)$,$Q(x, t)$, $q(x, t)$, $z(x, t)$, $z_b(x, t)$, $h(x, t)$, $n_g(x, t)$, and $P(x, t)$ and satisfy (1) and (2) in a region given by $x \in [x_{min}, x_{max}]$ and $t \in [t_{min}, t_{max}]$. These models will depend on a vector of parameters $c \in \mathbb{R}^n$ and, besides fulfilling the Saint Venant equations, will satisfy conditions imposed by different assumptions.

We propose that the surface elevation of the river $z(x, t)$, the bed elevation $z_b(x)$, the upstream flow-rate $Q(x_{min}, t)$ and the lateral contribution $q(x, t)$ are represented by analytical functions that, with some abuse of notation, we denote $z(c, x, t)$, $z_b(c, x)$, $Q(c, x_{min}, t)$, and $q(c, x, t)$, respectively, where $c \in \mathbb{R}^n$ is a vector of unknown parameters. Moreover, the shape of cross sections will come from analytical expressions that also depend on c. As a consequence, other quantities that describe the channel evolution will also depend explicitly on the unknowns c. For example, the depth (difference between surface elevation and bed elevation) will depend of the parameters c according to

$$h(c, x, t) = z(c, x, t) - z_b(c, x). \tag{4}$$

The cross area computed using the bed elevation, the surface elevation, and the shape of the cross section at position x and time t will be denoted $A(c, x, t)$ and the generated flow-rate will be denoted $Q(c, x, t)$.

By (1)–(2), if $A(c, x, t)$ and $Q(c, x, t)$ conform a solution of the Saint-Venant equations, we have that

$$\frac{\partial A}{\partial t}(c, x, t) + \frac{\partial Q}{\partial x}(c, x, t) = q(c, x, t) \tag{5}$$

and

$$\begin{aligned} &\frac{\partial Q}{\partial t}(c, x, t) + \frac{\partial}{\partial x}\left(\frac{Q(c, x, t)^2}{A(c, x, t)}\right) + gA(c, x, t)\frac{\partial z}{\partial x}(c, x, t) \\ &+\frac{n_g(c, x, t)^2 Q(c, x, t)|Q(c, x, t)|P(c, x, t)^{4/3}}{A(c, x, t)^{7/3}} = 0. \end{aligned} \tag{6}$$

By (5), for all $x_{min}, x \in \mathbb{R}$,

$$Q(c, x, t) = Q(x_{min}, t) - \int_{x_{min}}^{x} \left[\frac{\partial A}{\partial t}(c, \xi, t) - q(c, \xi, t)\right] d\xi. \tag{7}$$

In our approach, the operators $\frac{\partial}{\partial x}$ and $\frac{\partial}{\partial t}$ that appear in (5) and (6) will be replaced with difference operators as described below.

Assume that $\Delta x > 0$ and $\Delta t > 0$ are given, and F is a function of the variables x and t. We will denote:

$$F_x(x, t, \Delta x) = \frac{F(x + \Delta x, t) - F(x - \Delta x, t)}{2\Delta x}, \; F_t(x, t, \Delta t) = \frac{F(x, t + \Delta t) - F(x, t - \Delta t)}{2\Delta t}, \text{ and}$$

$$F_{tt}(x, t, \Delta t) = \frac{F(x, t + \Delta t) - 2F(x, t) + F(x, t - \Delta t)}{(\Delta t)^2}.$$

An exception to this notation is the case of Q_x, which, as a consequence of (5), will be defined by

$$Q_x(c, x, t) = q(c, x, t) - A_t(c, x, t). \tag{8}$$

In order to simplify the notation, we will denote $F_x(x, t)$, $F_t(x, t)$, and $F_{tt}(x, t)$ instead of $F_x(x, t, \Delta x)$, $F_t(x, t, \Delta t)$, and $F_{tt}(x, t, \Delta t)$ whenever this simplification does not lead to confusion.

In principle, motivated by (7), we would like to postulate:

$$Q(c, x, t) = Q(x_{min}, t) - \int_{x_{min}}^{x} [A_t(c, \xi, t) - q(c, \xi, t)]d\xi. \tag{9}$$

However, as the integral in (9) may be hard to compute, we will define, for every function of the variables x, and t,

$$\mathcal{S}_{x_{min}}^{x} F(x, t) = \int_{x_{min}}^{x} F_{spline}(\xi, t)d\xi$$

where $F_{spline}(x, t)$ is a spline function defined by the points $(x_{\min} + k\Delta x, F(x_{\min} + k\Delta x, t)$, for $k = 0, 1, \ldots, n_x$, where n_x is the first index j such that $x_{\min} + j\Delta x \geq x$. Roughly speaking, $\mathcal{S}_{x_{min}}^{x}$ is a numerical approximation of $\int_{x_{min}}^{x}$.

So, motivated by (9), we postulate:

$$Q(c, x, t) = Q(x_{min}, t) - \mathcal{S}_{x_{min}}^{x} [A_t(c, x, t) - q(c, x, t)]. \tag{10}$$

Consequently,

$$Q_t(c, x, t) = Q_t(x_{min}, t) - \mathcal{S}_{x_{min}}^{x} [A_{tt}(c, x, t) - q_t(c, x, t)]. \tag{11}$$

Since $\frac{\partial}{\partial x}\left(\frac{Q^2}{A}\right) = \frac{2Q\frac{\partial Q}{\partial x}A - Q^2\frac{\partial A}{\partial x}}{A^2}$, the difference form of (6) will be expressed as:

$$Q_t(c, x, t) + \frac{2QQ_x(c, x, t)A(c, x, t) - Q(c, x, t)^2 A_x(c, x, t)}{A(c, x, t)^2} + gA(c, x, t)z_x(c, x, t)$$

$$+ \frac{n_g(c, x, t)^2 Q(c, x, t)|Q(c, x, t)|P(c, x, t)^{4/3}}{A(c, x, t)^{7/3}} = 0. \tag{12}$$

Let us define

$$U(c, x, t) = Q_t(c, x, t) + \frac{2Q(c, x, t)Q_x(c, x, t)A - Q(c, x, t)^2 A_x(c, x, t)}{A(c, x, t)^2} + gA(c, x, t)z_x(c, x, t) \tag{13}$$

and

$$W(c, x, t) = n_g(c, x, t)^2 P(c, x, t)^{4/3}. \tag{14}$$

Therefore, (12) can be expressed as

$$U(c, x, t) + W(c, x, t)\frac{Q(c, x, t)|Q(c, x, t)|}{A(c, x, t)^{7/3}} = 0. \tag{15}$$

3 Analytical form of the Cross Sections

Recall that we start suggesting an analytical expression $z(c, x, t)$ for the surface elevation. Using (4) and the analytical expression $z_b(c, x)$, we deduce an analytical expression $h(c, x, t)$ for the depth. Then, using the analytical expressions that describe the cross section, we compute the cross area $A(c, x, t)$. This process is described in the present section.

3.1 The Function $\tilde{h}(c, x, s)$

Unlike many papers dealing with Saint-Venant equations, we will not assume that cross areas are rectangular. Instead, we will assume that, for each spatial position x, analytical functions $M(c, x)$, $B(c, x)$, $D(c, x)$, and $E(c, x)$ define the way in which the shape of the cross section varies with respect to the position x. Firstly, we will describe the way in which the depth is expressed as a function of the wetted half-width of the channel fluid surface. It will be assumed that $M(c, x)$, $B(c, x)$, $D(c, x)$, $E(c, x)$ are non-negative and $D(c, x) \leq E(c, x)$. The theoretical depth will be zero if the half-width s of the channel section is smaller than $D(c, x)$. Then, the depth increases monotonically from $D(c, x)$ to $E(c, x)$ according to an $\int$-like function with asymptote $M(c, x)$ and inflection point in $s = D(c, x) + \sqrt{\frac{1}{3B(c,x)}}$. Finally, if $s > E(c, x)$ the theoretical depth becomes linear, preserving continuity and differentiability. (That is, the derivative of the linear function coincides with the derivative of the $\int$-like function when $s = E(c, x)$.) The cross section will be assumed to be symmetric with respect to $s = 0$. See Fig. 1.

We assume that $M(c, x) > 0$, $B(c, x) > 0$, and $E(c, x) > D(c, x) > 0$ for all $x \in \mathbb{R}$. Given a position $x \in [x_{min}, x_{max}]$, we will assume that the depth at this position depends on the half-width s of the channel by means of the function $\tilde{h}(c, x, s)$ defined below.

$$\tilde{h}(c, x, s) = 0 \text{ if } 0 \leq s \leq D(c, x), \tag{16}$$

$$\tilde{h}(c, x, s) = \frac{M(c, x)B(c, x)[s - D(c, x)]^2}{B(c, x)[s - D(c, x)]^2 + 1} \text{ if } s \in [D(c, x), E(c, x)], \tag{17}$$

and

$$\tilde{h}(c, x, s) = \tilde{h}(c, x, E(c, x)) + \frac{\partial \tilde{h}}{\partial s}(c, x, E(c, x))(s - E(c, x)) \text{ if } s > E(c, x). \tag{18}$$

See Fig. 1.

Consequently, if $s \in [0, D(c, x)]$,

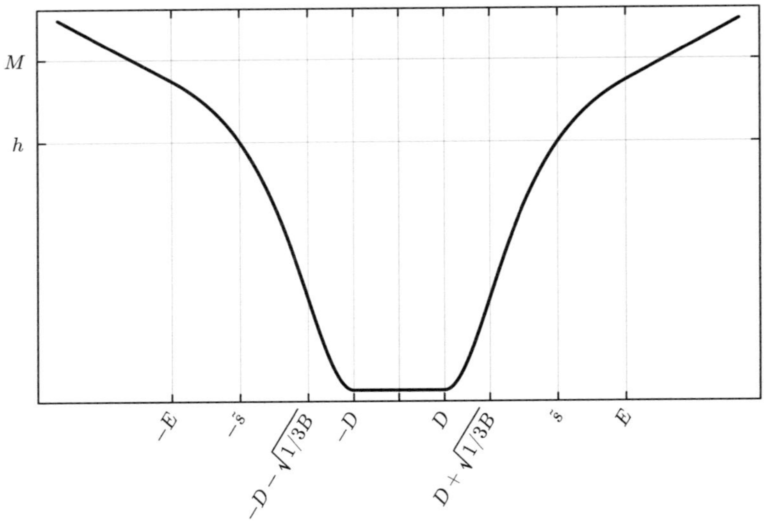

Fig. 1 Cross section with $D = 2$, $B = 1/12$, $M = 10$, $E = 10$, $h = 7.5$

$$\frac{\partial \tilde{h}}{\partial s}(c, x, s) = 0. \tag{19}$$

If $s \in (D(c, x), E(c, x))$,

$$\frac{\partial \tilde{h}}{\partial s}(c, x, s) = \frac{2M(c, x)B(c, x)(s - D(c, x))}{[B(c, x)(s - D(c, x))^2 + 1]^2}, \tag{20}$$

and

$$\frac{\partial^2 \tilde{h}}{\partial s^2}(c, x, s) = \frac{2M(c, x)B(c, x)[1 - 3B(c, x)(s - D(c, x))^2]}{[B(c, x)(s - D(c, x))^2 + 1]^3}. \tag{21}$$

Therefore, $\tilde{h}(c, x, s)$ is strictly increasing for $s \geq D(c, x)$, $\tilde{h}(c, x, s)$ is convex if $s \leq D(c, x) + \sqrt{\frac{1}{3B(c,x)}}$, and is concave if $s \geq D(c, x) + \sqrt{\frac{1}{3B(c,x)}}$. Finally, $\tilde{h}(c, x, s)$ is linear if $s \geq E(c, x)$ and $\lim_{s\to\infty} h(s) = \infty$.

3.2 The Half-Width Function $\tilde{s}(c, x, t)$

Given a position x, the width of the basis of the channel will be $2D(c, x)$, whereas the width at the flow surface, which depends of the time t, will be $2\tilde{s}(c, x, t)$

Assume that we have analytical expressions for the surface elevation $z(c, x, t)$ and the bed elevation $z_b(c, x)$. From (4) we obtain an analytical formula for $h(c, x, t)$.

However, we primarily want an analytical expression for $A(c, x, t)$. Then, we need to establish an analytical description of the wetted half-width $\tilde{s}(c, x, t)$ in terms of $h(c, x, t)$. In turn, $A(c, x, t)$ will be obtained from the wetted half-width of the channel by means of integration in Sect. 3.3.

We define $\tilde{s}(c, x, t) > 0$ by the unique solution of the equation

$$\tilde{h}(c, x, \tilde{s}(c, x, t)) = h(c, x, t). \tag{22}$$

Recall that $h(c, x, t) > 0$ and the function $\tilde{h}(c, x, s)$ is strictly increasing and goes from 0 to infinity as s grows, so that $\tilde{s}(c, x, t)$ is well defined.

Let us show how to compute $\tilde{s}(c, x, t)$. Given $h(c, x, t) > 0$ we have two candidates for being $\tilde{s}(c, x, t)$. One of them comes from isolating s in the equation $\tilde{h}(c, x, s) = 0$ using (17) and the other comes from isolation using (18). In the first case, if the solution s is in $[D(c, x), E(c, x)]$ and we are done. Note that, according to (17), $\tilde{h}(c, x, s) < M(c, x)$ for all $s \in [D(c, x), E(c, x)]$. So, if $h(c, x, t) \geq M(c, x)$, we have that $\tilde{s}(c, x, t) \geq E(c, x)$. When there is no solution using (17), the isolation of s using (18) gives the desired value of $\tilde{s}(c, x, t)$.

Using (17) we have:

$$\frac{M(c, x)B(c, x)[s - D(c, x)]^2}{B(c, x)[s - D(c, x)]^2 + 1} = h(c, x, t).$$

Thus,

$$M(c, x)B(c, x)[s - D(c, x)]^2] = h(c, x, t)B(c, x)[s - D(c, x)]^2 + h(c, x, t).$$

Then,

$$[M(c, x)B(c, x) - h(c, x, t)B(c, x)][s - D(c, x)]^2] = h(c, x, t).$$

So,

$$[s - D(c, x)]^2 = \frac{h(c, x, t)}{[M(c, x)B(c, x) - h(c, x, t)B(c, x)]}.$$

Therefore,

$$s = D(c, x) + \sqrt{\frac{h(c, x, t)}{[M(c, x) - h(c, x, t)]B(c, x)}}. \tag{23}$$

If $M(c, x) \leq h(c, x, t)$ or if the right-hand side of (23) is bigger than $E(c, x)$ the solution given by (23) must be discarded, and $\tilde{s}(c, x, t)$ must be given using (18). Namely, in this case:

$$h(c, x, t) = \tilde{h}(c, x, E(c, x)) + \frac{d}{ds}\tilde{h}(c, x, E(c, x))(s - E(c, x)),$$

thus,

$$\tilde{s}(c,x,t) = E(c,x) + \frac{h(c,x,t) - \tilde{h}(c,x,E(c,x))}{\frac{d}{ds}\tilde{h}(c,x,E(c,x))}. \tag{24}$$

Summing up, if $h(c,x,t) \leq M(c,x)$ and $D(c,x) + \sqrt{\frac{h(c,x,t)}{[M(c,x)-h(c,x,t)]B(c,x)}}$ $\leq E(c,x)$ we have that

$$\tilde{s}(c,x,t) = D(c,x) + \sqrt{\frac{h(c,x,t)}{[M(c,x) - h(c,x,t)]B(c,x)}}. \tag{25}$$

Otherwise,

$$\tilde{s}(c,x,t) = E(c,x) + \frac{[h(c,x,t) - \tilde{h}(c,x,E(c,x))](B(c,x)(E(c,x) - D(c,x))^2 + 1)^2}{2M(c,x)B(c,x)(E(c,x) - D(c,x))}. \tag{26}$$

3.3 The Function $A(c, x, t)$

If the wetted cross section is defined by $\tilde{h}(c,x,t,s)$ and $\tilde{s}(c,x,t)$, the wetted area is given by

$$A(c,x,t) = 2\tilde{s}(c,x,t)h(c,x,t) - 2\int_0^{\tilde{s}(c,x,t)} \tilde{h}(c,x,t,\xi)d\xi. \tag{27}$$

So,

$$A(c,x,t) = 2\tilde{s}(c,x,t)h(c,x,t) - 2\int_{D(x)}^{\tilde{s}(c,x,t)} \tilde{h}(c,x,t,\xi)d\xi, \tag{28}$$

Therefore, if $h(c,x,t) \leq M(c,x)$ and $\tilde{s}(c,x,t) = D(c,x) + \sqrt{\frac{h(c,x,t)}{[M(c,x)-h(c,x,t)]B(c,x)}}) \leq E(c,x)$ we have that

$$A(c,x,t) = 2\tilde{s}(c,x,t)h(c,x,t) - 2\int_{D(c,x)}^{D(c,x)+\sqrt{\frac{h(c,x,t)}{[M(c,x)-h(c,x,t)]B(c,x)}}} \frac{M(c,x)B(c,x)[\xi - D(c,x)]^2}{B(c,x)[\xi - D(c,x)]^2 + 1} d\xi, \tag{29}$$

So, if $h(c,x,t) \leq M(c,x)$ and $\tilde{s}(c,x,t) \leq E(c,x)$,

$$A(c,x,t) = 2\tilde{s}(c,x,t)h(c,x,t)$$

$$-2\frac{M(c,x)}{\sqrt{B(c,x)}}\left(\sqrt{B(c,x)}(\tilde{s}(c,x,t)-D(c,x))-\arctan[\sqrt{B(c,x)}(\tilde{s}(c,x,t)-D(c,x))]\right). \tag{30}$$

Finally, if $h(c,x,t) > M(c,x)$ or $\tilde{s}(c,x,t) > E(c,x)$,

$$A(c,x,t) = 2\tilde{s}(c,x,t)h(c,x,t)$$

$$-2\frac{M(c,x)}{\sqrt{B(c,x)}}\left(\sqrt{B(c,x)}(E(c,x)-D(c,x))-\arctan[\sqrt{B(c,x)}(E(c,x)-D(c,x)\right)$$

$$-[\tilde{h}(c,x,E(c,x))+h(c,x,t)][\tilde{s}(c,x,t)-E(c,x)]. \tag{31}$$

4 Basic Inequalities of the Feasibility Problem

Given the analytical expressions $z(c,x,t)$, $z_b(c,x)$, $Q(c,x,t)$, $B(c,x)$, $M(c,x)$, $D(c,x)$, and $E(c,x)$, the function $A(c,x,t)$ is defined by (30).

The coefficients $c \in \mathbb{R}^n$ will be obtained aiming the fulfillment of the inequations below whenever $x \in [x_{min}, x_{max}]$ and $t \in [t_{min}, t_{max}]$:

We will generate solutions of the Saint-Venant equations that satisfy

$$A(c,x,t) > 0, z(c,x,t) > z_b(c,x,t), \tag{32}$$

and

$$Q(c,x,t)U(c,x,t) \le -\eta, \tag{33}$$

where $U(c,x,t)$ is defined in (13) and $\eta > 0$ is a positive tolerance.

In addition we will assume that the admissibility conditions (3) can be expressed in the form

$$Z([Q, A, z, h, Q_x, Q_t, A_x, A_t, z_x, z_t, h_x, h_t, q, q_x, q_t, z_b, [z_b]_x](x,t)) \le 0, \tag{34}$$

where Z is assume values in $\mathbb{R}^d$. Note that this expression allows one to impose mixed boundary conditions and initial conditions when they are available, as well as estimations of the state variables at different points of the domain. Formally, if there is no admissibility constraint at (x,t) we can define $Z(x,t) = 0$.

The requirement (32) will be satisfied if we also impose, for some tolerance $\eta_h > 0$,

$$z(c,x,t) - z_b(c,x) \ge \eta_h. \tag{35}$$

In order to maintain lateral inflows between reasonable values we require that

$$2\tilde{s}(c, x, t)\eta_1 \leq q(c, x, t) \leq 2\tilde{s}(c, x, t)\eta_2, \tag{36}$$

where $\eta_1, \eta_2 > 0$ and $2\tilde{s}(c, x, t)$ is the width of the wetted surface defined by (25) and (26).

Sensible values for the tolerances η_1 and η_2 come from considerations on standard precipitation measurements. Roughly speaking, the cross area is the volume of fluid in the channel considering a length of 1 m. Therefore, reasonable bounds for $q(c, x, t)$ correspond to the lateral contribution that may be expected for positions between x meters and $x + 1$ meters. This contribution is the product between the lateral contribution that may be expected on a square of 1 m^2 times the width of the channel. Usually, lateral contributions due to rain is measured in terms of the height (in millimeters) of a column of rain water whose basis is a square with side equal to one meter. Therefore, if measured rain is (say) L mm per hour, this corresponds to $2/3600$ mm per second, So, in the slice of the channel with 1 m length, the additional elevation of the fluid during one second is the width of the channel (in meters) times $L/3600$.

The tolerance $\eta_h > 0$ is employed to discard solutions in which cross area vanishes. Note that null areas are not admissible for well-definiteness of the moment Eq. (6). The value $\eta_h = 0.01$ (1 cm) seems to be reasonable.

The inequality (33) will be replaced by the fulfillment of

$$Q(c, x, t) \geq \eta_Q \text{ and } U(c, x, t) \leq -\eta_U \tag{37}$$

or

$$Q(c, x, t) \leq -\eta_Q \text{ and } U(c, x, t) \geq \eta_U. \tag{38}$$

The fulfillment of (37) or (38) will be called *Moment feasibility*.

A reasonable value of η_Q is 0.01 (m^3/s) and a reasonable value for η_U is 10^{-4} (m^3/s^2).

We still need to impose constraints on $W(c, x, t)$, defined in (14).

By (20), when $\tilde{s}(c, x, t) \leq E(c, x)$,

$$P(c, x, t) = 2D(c, x) + 2\int_{D(c,x)}^{\tilde{s}(x,t)} \sqrt{1 + \left(\frac{\partial \tilde{h}}{\partial s}(c, x, s)\right)^2}\, ds.$$

So,

$$P(c, x, t) = 2D(c, x) + 2\int_{D(c,x)}^{\tilde{s}(x,t)} \sqrt{1 + \left(\frac{2M(c, x)B(c, x)(s - D(c, x))}{[B(c, x)(s - D(c, x))^2 + 1]^2}\right)^2}\, d\sigma$$

$$= 2D(c, x) + 2\int_0^{\tilde{s}(x,t)-D(c,x)} \sqrt{1 + \left(\frac{2M(c, x)B(c, x)\sigma}{[B(c, x)\sigma^2 + 1]^2}\right)^2} d\sigma. \tag{39}$$

When $\tilde{s}(c, x, t) > E(c, x)$, we have that

$$P(c, x, t) = 2D(c, x) + 2\int_0^{E(c,x)-D(c,x)} \sqrt{1 + \left(\frac{2M(c, x)B(c, x)\sigma}{[B(c, x)\sigma^2 + 1]^2}\right)^2} d\sigma$$
$$+ \sqrt{(\tilde{s}(c, x, t) - E(c, x))^2 + (\tilde{h}(c, x, E(c, x)) - h(c, x, t))^2}. \tag{40}$$

However, due to the fractal nature of the perimeter (Gouyet 1996), and the fact that the integral involved in (39) does not admit an analytical expression, we decided to rely on reasonable bounds for $P(c, x, t)$. More precisely, we find it sensible to impose:

$$2[D(c, x) + \sqrt{(\tilde{s}(c, x, t) - D(c, x))^2 + h(c, x, t)^2}] \leq P(c, x, t) \leq 2[\tilde{s}(c, x, t) + h(c, x, t)]. \tag{41}$$

The requirement (41) is equivalent to

$$(2[D(c, x) + \sqrt{(\tilde{s}(c, x, t) - D(c, x))^2 + h(c, x, t)^2}])^{\frac{4}{3}} \leq P(c, x, t)^{\frac{4}{3}} \leq (2[\tilde{s}(c, x, t) + h(c, x, t)])^{\frac{4}{3}}. \tag{42}$$

If, in addition, we require that

$$\eta_{ming} \leq n_g(c, x, t) \leq \eta_{maxng}, \tag{43}$$

we obtain the condition

$$\eta_{ming}^2(2[D(c, x) + \sqrt{(\tilde{s}(c, x, t) - D(c, x))^2 + h(c, x, t)^2}])^{\frac{4}{3}}$$
$$\leq n_g(c, x, t)^2 P(c, x, t)^{\frac{4}{3}} \leq \eta_{maxng}^2(2[\tilde{s}(c, x, t) + h(c, x, t)])^{\frac{4}{3}}. \tag{44}$$

Thus, by (14),

$$\eta_{ming}^2(2[D(c, x) + \sqrt{(\tilde{s}(c, x, t) - D(c, x))^2 + h(c, x, t)^2}])^{\frac{4}{3}}$$
$$\leq W(c, x, t) \leq \eta_{maxng}^2(2[\tilde{s}(c, x, t) + h(c, x, t)])^{\frac{4}{3}}. \tag{45}$$

But, by (15),

$$W(c, x, t) = -\frac{U(c, x, t)A(c, x, t)^{7/3}}{Q(c, x, t)|Q(c, x, t)|}. \tag{46}$$

Therefore, (45) induces the following constraints:

$$-\frac{U(c,x,t)A(c,x,t)^{7/3}}{Q(c,x,t)|Q(c,x,t)|} \geq \eta_{ming}^2 (2[D(c,x) + \sqrt{(\tilde{s}(c,x,t) - D(c,x))^2 + h(c,x,t)^2}])^{\frac{4}{3}} \tag{47}$$

and

$$-\frac{U(c,x,t)A(c,x,t)^{7/3}}{Q(c,x,t)|Q(c,x,t)|} \leq \eta_{maxng}^2 (2[\tilde{s}(c,x,t) + h(c,x,t)])^{\frac{4}{3}}. \tag{48}$$

Equivalently,

$$U(c,x,t) \leq -\frac{\eta_{ming}^2 (2[D(c,x) + \sqrt{(\tilde{s}(c,x,t) - D(c,x))^2 + h(c,x,t)^2}])^{\frac{4}{3}} Q(c,x,t)|Q(c,x,t)|}{A(c,x,t)^{7/3}} \tag{49}$$

and

$$U(c,x,t) \geq -\frac{\eta_{maxng}^2 (2[\tilde{s}(c,x,t) + h(c,x,t)])^{\frac{4}{3}} Q(c,x,t)|Q(c,x,t)|}{A(c,x,t)^{7/3}} \tag{50}$$

Note that (49) and (50) imply that $Q(c,x,t)$ and $U(c,x,t)$ must have opposite signs. Thus, in practice, the constraints (37) and (38) can be ignored.

Note that, under the fulfillment of these constraints, $W(c,x,t)$ is fully determined by (46) but $n_g(c,x,t)$ and $P(c,x,t)$ are not. Instead, these quantities are merely subject to the constraint (14):

$$n_g(c,x,t)^2 P(c,x,t)^{4/3} = W(c,x,t).$$

This constraint is compatible with the conditions (41) and (43). If $W(c,x,t)$ is computed by (46) and satisfies (45), there exist $n_g(c,x,t)$ and $P(c,x,t)$ satisfying (14), (41) and (43). This fact is proved in the following lemma.

Lemma 4.1 *Assume that (45) holds. Then, there exist* $n_g(c,x,t)$ *and* $P(c,x,t)$ *satisfying (42) and (43).*

Proof Let us write

$$P_{min} = \log((2[D(c,x) + \sqrt{(\tilde{s}(c,x,t) - D(c,x))^2 + h(c,x,t)^2}])^{\frac{4}{3}}),$$

$$P_{max} = \log((2[\tilde{s}(c,x,t) + h(c,x,t)])^{\frac{4}{3}}),$$

$$P = \log(P(c,x,t)^{\frac{4}{3}}),$$

$$N = \log(n_g(c,x,t)^2),$$

$$N_{min} = \log(n_{ming}),$$

$$N_{max} = \log(n_{maxg}),$$

and

$$L = \log(W(c, x, t)).$$

Then, (14) is equivalent to

$$N + P = L,$$

(42) is equivalent to

$$P_{min} \le P \le P_{max},$$

and (43) is equivalent to

$$N_{min} \le N \le N_{max}.$$

By (45) we have that

$$N_{min} + P_{min} \le L \le N_{max} + P_{max}. \tag{51}$$

Consider the equation

$$(N_{min} + P_{min}) + t(N_{max} - N_{min} + P_{max} - P_{min}) = L.$$

The solution of this equation is

$$t = \frac{L - (N_{min} + P_{min})}{(N_{max} + P_{max}) - (N_{min} + P_{min})}$$

By (51) we have that $0 \le t \le 1$. Therefore, defining

$$N = N_{min} + t(N_{max} - N_{min}) \text{ and } P = P_{min} + t(P_{max} - P_{min})$$

we have that

$$N_{min} \le N \le N_{max}, P_{min} \le P \le P_{max}, \text{ and } N + P = L. \tag{52}$$

This implies the thesis of the Lemma. □

5 Checking Feasibility

A tentative choice of the parameters c may produce models that could represent solutions of the channel equations. Therefore, for each tentative c, given a point (x, t) we must check whether the basic inequalities required for define solutions are fulfilled.

Given parameters $c \in \mathbb{R}^n$, a position $x \in [x_{min}, x_{max}]$ and a time instant $t \in [t_{min}, t_{max}]$ the process for checking if basic inequalities are satisfied is given by the following algorithm.

Algorithm CHECK
Assume that the tolerances $\eta_h, \eta_1, \eta_2, \eta_Q, \eta_U, \eta_{maxng} > 0$, and $\eta_{ming} \geq 0$ are given.

Step 1 *Establish the shape of the cross area* (63), (64)
Compute $D(c, x)$, $B(c, x)$, $M(c, x)$ and $E(c, x)$ by means of (63) and (64) or by means of alternative models.

Step 2 *Compute depth and check*
Compute $z(c, x, t)$ and $z_b(c, x)$ using (65) and (67) or by means of alternative models.
Compute $h(c, x, t) = z(c, x, t) - z_b(c, x)$.
If (35) (which depends on η_h) does not hold, discard c and stop the execution of CHECK.

Step 3 *Compute wetted area* (25), (26), (30), (31)
If $h(c, x, t) \leq M(c, x)$ and $D(c, x) + \sqrt{\frac{h(c,x,t)}{[M(c,x)-h(c,x,t)]B(c,x)}} \leq E(c, x)$ compute

$$\tilde{s}(c, x, t) = D(c, x) + \sqrt{\frac{h(c, x, t)}{[M(c, x) - h(c, x, t)]B(c, x)}}$$

and

$$A(c, x, t) = 2\tilde{s}(c, x, t)h(c, x, t)$$

$$- 2\frac{M(c, x)}{\sqrt{B(c, x)}}\left(\sqrt{B(c, x)}(\tilde{s}(c, x, t) - D(c, x)) - \arctan[\sqrt{B(c, x)}(\tilde{s}(c, x, t) - D(c, x))]\right).$$

Otherwise, compute

$$\tilde{s}(c, x, t) = E(c, x) + \frac{[h(c, x, t) - \tilde{h}(c, x, E(c, x))](B(c, x)(E(c, x) - D(c, x))^2 + 1)^2}{2M(c, x)B(c, x)(E(c, x) - D(c, x))}.$$

and

$$A(c, x, t) = 2\tilde{s}(c, x, t)h(c, x, t)$$

$$- 2\frac{M(c,x)}{\sqrt{B(c,x)}}\left(\sqrt{B(c,x)}(E(c,x) - D(c,x)) - \arctan[\sqrt{B(c,x)}(E(c,x) - D(c,x)\right)$$

$$- [\tilde{h}(c,x,E(c,x)) + h(c,x,t)][\tilde{s}(c,x,t) - E(c,x)].$$

Step 4 *Compute lateral contribution and check*

Compute $q(c, x, t)$ by means of (68) or by means of alternative models.

If (36) (which depends on η_1 and η_2) does not hold, discard c and stop the execution of CHECK.

Step 5 *Compute flow-rate* (10), (66)

Step 5.1

Compute inlet discharge $Q(c, x_{min}, t)$ using (66) or using alternative models.

Step 5.2

Compute $Q(c, x, t)$ using (10).

Step 6 *Check moment feasibility* (13), (37), (38)

Compute $U(c, x, t)$ using (13).

If

$$Q(c,x,t) > -\eta_Q \text{ and } U(c,x,t) > -\eta_U$$

or

$$Q(c,x,t) < \eta_Q \text{ and } U(c,x,t) < \eta_U$$

discard c and stop the execution of CHECK.

Step 7 *Check perimeter-Manning feasibility*

Compute $W(c, x, t)$ by means of (46).

If (47) (which depends of η_{ming}) does not hold or (48) (which depends of η_{maxg}) does not hold, discard c and stop the execution of CHECK.

Step 8 *Check admissibility conditions*

If (x, t) is involved in admissibility constraints and (34) does not hold, discard c and stop the execution of CHECK.

Step 9 *Compute Manning coefficient and perimeter*

Compute $n_g(c, x, t)$ and $P(c, x, t)$ as indicated in Lemma 4.1 Stop.

6 Optimization Procedures

Given $x \in [x_{min}, x_{max}], t \in [t_{min}, t_{max}], c \in \mathbb{R}^n$, and the tolerances $\eta_h, \eta_1, \eta_2, \eta_Q, \eta_U$, $\eta_{maxng}, \eta_{ming}$, Algorithm CHECK may stop at steps 2, 4, 6, 7, or 8 and discard c, or it can stop at Step 9. The latter possibility characterizes a success in terms of satisfying the fundamental inequalities by the parameters c at (x, t), and the first 5 possibilities show that the trial c is not suitable for defining a solution of the of the Saint-Venant equations satisfying admissibility conditions.

The trial parameters c should be chosen in a region of $\mathbb{R}^n$, which is generally defined by a box $c_{min} \leq c_i \leq c_{max}$ for all $i = 1, \ldots, n$.

If, for some choice of c, a sufficiently large random sample of points $(x, t) \in [x_{min}, x_{max}] \times [t_{min}, t_{max}]$ is given such that CHECK stops at step 9, we consider that an acceptable solution of the Saint-Venant equations has been obtained. Formally speaking, a solution of our feasibility problem is characterized by the fulfillment of the basic inequalities and the admissibility conditions in a sufficiently large sequence of independent identically distributed (i.i.d.) data points.

On the other hand, if a point (x, t) is found such that CHECK stops discarding c, we must compute a new trial vector c or declare that no solution of the Saint-Venant equations with the admissibility conditions can be obtained with the given vector of parameters.

It is important to emphasize that the problem cannot be considered solved if, for a given sufficiently large i.i.d. sample of (x, t) points, parameters c that satisfy all the inequalities are found. The reason is that this fulfillment the inequalities for a single sample could characterize a situation in which overfitting (with respect to the given sample) occurs. In other words, the parameters could be satisfactory with respect to the sample that was used for their estimation, but could be poor for other independent large samples.

Instead, a set of parameters c can be considered to be a solution if the inequalities are satisfied for all points of an i.i.d. sample that have nothing to do with the procedure used to find c. This is a crucial difference between our problem and standard feasibility problems.

This procedure is formalized below by the algorithm FINDC, which is a generalization of Algorithm BUSCASOL, introduced in Martínez (2024). Unlike the case of Martínez (2024), we do not assume a rigid boundary condition for the inflow $Q(x_{min}, t)$. Furthermore, we do not assume that an analytical expression is proposed primarily for the cross section. Instead, the model $A(c, x, t)$ is derived from the model $z(c, x, t)$ in the form described in Sect. 3.

Algorithm FINDC

We assume that $x_{min} < x_{max}$, $t_{min} < t_{max}$, and $c_i^{min} < c_i^{max}$, for all $i = 1, \ldots, n$, $m >> 1, 0 < p < 1$, and $\ell_{max} >> m$. We assume that $c \in \mathbb{R}^n$ is an initial estimation of the model parameters which will be used only when $\ell = 0$.

Step 1

Set $\ell \leftarrow 0$.

Step 2

If $\ell = \ell_{max}$, stop without achieving the objectives.

Step 3

If $\ell > 0$, compute c_i uniformly random in $[c_i^{min}, c_i^{max}]$ for $i = 1, \ldots, n$.

Step 4

Set $n_s \leftarrow 0$.

Step 5
If $n_s = m$, stop declaring that $c \in \mathbb{R}^n$ is a satisfactory solution.
Step 6
With probability p, consider the set of points (x, t) where admissibility constraints are involved and randomly choose one point in this set. Otherwise, compute (x, t) uniformly random in $[x_{min}, x_{max}] \times [t_{min}, t_{max}]$.
Step 7
Call Algorithm CHECK.
If CHECK finishes at Step 9, set $n_s \leftarrow n_s + 1$ and go to Step 5.

If FINDC fails to encounter a satisfactory solution we try to find acceptable parameters c by means of Continuous Optimization algorithms. For this purpose we express our objective as a box-constrained nonlinear least-squares problem. More precisely, assume that we have a sample of points (x_ν, t_ν) for $\nu = 1, \ldots, m$. Then, we wish to find $c \in [c_{min}, c_{max}]^n$ such that, for all $\nu = 1, \ldots, m$,

$$z(c, x_\nu, t_\nu) - z_b(c, x_\nu) \geq \eta_h, \tag{53}$$

$$2\tilde{s}(c, x_\nu, t_\nu)\eta_1 \leq q(c, x_\nu, t_\nu) \leq 2\tilde{s}(c, x_\nu, t_\nu)\eta_2, \tag{54}$$

$$Q(c, x_\nu, t_\nu)U(c, x_\nu, t_\nu) \leq -\eta_Q \eta_U, \tag{55}$$

$$-\frac{U(c, x_\nu, t_\nu)A(c, x_\nu, t_\nu)^{7/3}}{Q(c, x_\nu, t_\nu)|Q(c, x_\nu, t_\nu)|} \geq \eta_{ming}^2 (2[D(c, x_\nu) + \sqrt{(\tilde{s}(c, x_\nu, t_\nu) - D(c, x_\nu))^2 + h(c, x_\nu, t_\nu)^2}])^{\frac{4}{3}}, \tag{56}$$

$$-\frac{U(c, x_\nu, t_\nu)A(c, x_\nu, t_\nu)^{7/3}}{Q(c, x_\nu, t_\nu)|Q(c, x_\nu, t_\nu)|} \leq \eta_{maxng}^2 (2[\tilde{s}(c, x_\nu, t_\nu) + h(c, x_\nu, t_\nu)])^{\frac{4}{3}}, \tag{57}$$

and

$$Z_j([Q, A, z, h, Q_x, Q_t, A_x, A_t, z_x, z_t, h_x, h_t](c, x_\nu, t_\nu)) \leq 0 \text{ for } j = 1, \ldots, d. \tag{58}$$

The inequations (53)–(58) can be expressed in the form

$$g_i(c_1, \ldots, c_n) \leq 0, i = 1, n_{ineq}, \tag{59}$$

where $n_{ineq} = m(d + 5)$ is the number of inequations is large and, roughly speaking, $n \leq 50$.

Therefore, we seek a solution of the minimization problem

$$\text{Minimize} \sum_{i=1}^{n_{ineq}} \max\{0, g_i(c_1, \ldots c_n)\}^2. \tag{60}$$

To solve this problem, the algorithm SESEM, introduced in Birgin and Martínez (2022), may be recommended due to its remarkable performance on PDE-constrained Saint-Venant problems. SESEM is a nonlinear least squares solver inspired by the Sequential Secant method, so that at each iteration only one evaluation of the objective function is used to compute the search direction. SESEM incorporates several ad hoc procedures that provide global convergence properties.

Moreover, assuming that the last n_{admis} components g_i represent admissibility conditions, it is interesting to consider the constrained optimization problem

$$\text{Minimize} \sum_{i=n_{ineq}-n_{admis}+1}^{n_{admis}} \max\{0, g_i(c)\}^2, \text{ subject to } g_i(c) \leq 0, i = 1, n_{ineq} - n_{admis}. \tag{61}$$

The problem (61) can be solved using a Safeguarded Augmented Lagrangian (Algencan) algorithm (Birgin and Martínez 2014). It is important to observe that, at each outer iteration, Algencan solves, approximately, a nonlinear least squares problem (Martínez 2023). Therefore, at each outer iteration of Algencan, one can use the SESEM algorithm. Algencan has been improved many times since its introduction in Andreani et al. (2008). See Andreani et al. (2008, 2011, 2022, 2016), Birgin and Martínez (2008, 2020).

The algorithm LOKAL, introduced below is an alternative to SESEM and Algencan. For this purpose, given the feasible parameters $c \in \mathbb{R}^n$, we define

$$f(c) = \sum_{i=1}^{n_{ineq}} \max\{0, g_i(c_1, \ldots c_n)\}^2. \tag{62}$$

LOKAL is a local variation random search algorithm. Its special feature is that the functional evaluations $f(c)$ are preceded by the computation of a random sample (x_ν, t_ν). Therefore, strictly speaking, $f(c)$ is determined by both c and the sample (x_ν, t_ν).

Algorithm LOKAL

Assume that the tolerances $\eta_h > 0, \eta_1 > 0, \eta_2 > 0, \eta_Q > 0, \eta_U > 0, \eta_{maxng} > 0$, $\eta_{ming} \geq 0$ are given as well as an initial feasible initial parameters c. Assume that $max_{trials} >> 1$ is given and set $k_{trial} \leftarrow 1$. Let $m \geq 1$ be the size of the random samples $(x_1, t_1), \ldots, (x_m, t_m)$ employed in this algorithm.

Step 1 *Initial evaluation of f*

Compute a random i.i.d. sample $(x_1, t_1), \ldots, (x_m, t_m)$ respecting the rule given at Step 6 of FINDC with respect to the inclusion of points that involve admissibility constraints, and define $f(c)$ by (62). For all $\nu = 1, \ldots, m$, set $(x, t) \leftarrow (x_\nu, t_\nu)$ and call Algorithm CHECK. If CHECK stops at Step 9 for all $\nu = 1, \ldots, m$, stop the execution of Algorithm LOKAL, declaring that a satisfactory solution has been found.

Step 2 *Stop if the number of evaluations exceeds* max_{trials}.

If $k_{trial} \geq max_{trials}$ stop declaring that maximum number of allowed trials was exhausted.

Step 3 *Compute a random variation of the current set of parameters.*

Compute c_i^{trial} uniformly random in $[\max\{c_i^{min}, c_i - 0.1|c_i|\}, \min\{c_i^{max}, c_i + 0.1|c_i|\}]$ for $i = 1, \ldots, n$.

Step 4 *Compute objective function at trial set of parameters*

Update $k_{trial} \leftarrow k_{trial} + 1$. Compute a random i.i.d. sample $(x_1, t_1), \ldots, (x_m, t_m)$ involving admissibility conditions as in Step 1, and define $f(c^{trial})$ by (62). For all $\nu = 1, \ldots, m$, set $(x, t) \leftarrow (x_\nu, t_\nu)$ and call Algorithm CHECK. If CHECK stops at Step 9 for all $\nu = 1, \ldots, m$, set $c \leftarrow c^{trial}$ and stop the execution of Algorithm LOKAL, declaring that a satisfactory solution has been found.

Step 5 Updating current parameters

If $f(c^{trial}) < f(c)$, set $c \leftarrow c^{trial}$ and go to Step 2.

These algorithms can be combined in different ways. Algorithm COMBI describes a particular combination that produced good results in several experiments.

Given tolerances $\eta_h, \eta_1, \eta_2, \eta_Q, \eta_U, \eta_{maxng}, \eta_{ming}$, the objective of COMBI is to find coefficients $c_i \in [c_i^{min}, c_i^{max}]$, $i = 1, \ldots, n$, such that (53)–(58) holds for a large i.i.d. sample of points (x, t) such that $x \in [x_{min}, x_{max}]$ and $t \in [t_{min}, t_{max}]$.

Algorithm COMBI

Assume that $m_{big} \geq m_0 >> 1$ and $\ell_{max} >> m$, $\nu_{max} \in \{1, 2, 3, \ldots\}$ and the tolerances $\eta_h > 0, \eta_1 > 0, \eta_2 > 0, \eta_Q > 0, \eta_U > 0, \eta_{maxng} > 0, \eta_{ming} \geq 0$ are given. Set $m \leftarrow m_0$. Let $c \in \mathbb{R}^n$ be an initial estimation for the parameters of the model.

Step 1 *Using FINDC*

Call Algorithm FINDC.

If FINDC finishes at Step 5, and $m \geq m_{big}$, stop declaring that a satisfactory solution has been found.

If FINDC finishes at Step 5, but $m < m_{big}$, define c as the solution found by FINDC and go to Step 6.

Step 2 *Relax tolerances*

Multiply each of the tolerances $\eta_h, \eta_1, \eta_2, \eta_Q, \eta_U, \eta_{maxng}, \eta_{ming}$ by 10 and call FINDC with the new tolerances. If FINDC finishes at Step 5 (declaring satisfactory solution with relaxed tolerances), go to Step 4.

Step 3 *Test failure*

If Step 2 has been executed more than ν_{max} consecutive times stopping with failure of FINDC, stop the execution of COMBI, declaring that no solution of the set of desired inequalities has been found. Otherwise, go to Step 2.

Step 4 *Improvement using Optimization*

Select a uniform random sample $x_\nu \in [x_{min}, x_{max}], t_\nu \in [t_{min}, t_{max}], \nu = 1, \ldots, m$. Recover the original tolerances $\eta_h, \eta_1, \eta_2, \eta_Q, \eta_U, \eta_{maxng}, \eta_{ming}$. Define n_{ineq} as in (59) and consider the optimization problems (60) and (61).

Go to Step 5, Step 6, or Step 7.

Step 5 *Using LOKAL*

Call Algorithm LOKAL, described above in the present section.

Step 6 *Using SESEM*

Call SESEM with its default algorithmic parameters (Birgin and Martínez 2022), and the parameters $c_1, \ldots, c_n$ found at Step 2 by FINDC as the first approximation required by SESEM. If inequalities (53)–(58) are satisfied by the output of SESEM, define c as the solution found by SESEM and go to Step 8.

Step 7 *Using Algencan*

Call Algencan with its default algorithmic parameters (Birgin and Martínez 2014), and the parameters $c_1, \ldots, c_n$ found at Step 2 by FINDC as the first approximation required by Algencan. If inequalities (53)–(58) are satisfied by the output of Algencan, define c as the solution found by Algencan and go to Step 8.

Step 8 *Using Inexact Restoration*

Increase the sample size m using inexact restoration (Birgin et al. 2018; Martínez and Santos 2023) and go to Step 1.

Since the cost of each iteration of the optimization algorithms is proportional to the sample size m, it is recommendable to use procedures that work with moderate values of m and increase the sample size only when necessary. For this purpose we can use the Inexact Restoration (IR) procedures introduced in Martínez and Santos (2023), Martínez (2024) and references therein.

The design of COMBI deserves a few comments. Note that the only place where COMBI returns declaring success is at Step 1, after a call to FINDC with a sample of at least m_{big} points, which ends at Step 5 of FINDC. This fact guarantees that, for large number of randomly chosen i.i.d. points (x, t), the Saint-Venant equations and the admissible constraints have been satisfied. We consider that this is a serious indication that the Saint-Venant equations and the admissible constraints hold at every point $(x, t) \in [x_{min}, x_{max}] \times [t_{min}, t_{max}]$. If FINDC stops at Step 5, but $m < m_{big}$, there is still no reason to declare success, but the parameters c deserve to be tested with larger samples, as indicated in Step 6.

If FINDC stops at Step 2 and does not meet the objectives, we relax the tolerances and try to find a set of them such that FINDC succeeds. This procedure is repeated several times until a suitable set of tolerances is found, or a maximum number of increases is reached. In the latter case, COMBI returns with a failure message. However, if FINDC was successful with relaxed tolerances we try to improve this result using LOKAL, SESEM, or Algencan. Obviously, a solution computed by LOKAL, SESEM or Algencan that satisfies (53)–(58) with $m \geq m_{big}$ is a strong candidate for a COMBI solution, but this fact must be confirmed by an independently chosen random sample, as indicated in Step 6 of COMBI.

7 Numerical Experiments

There are many reasons why it may be necessary to model natural channels. However, no two cases are the same.

Each case requires a careful definition of the purposes, which may include data completion problems (Caubet et al. 2019; Guellouz and Khayat 2022), data compression, creation of synthetic data, comparison of PDE solvers (Martínez 2024), prediction of the evolution of flood events in short time (Birgin and Martínez 2023), confirmation of Saint Venant models (Chalfen and Niemiec 1986; Ferreira et al. 2017), estimation of Manning coefficients (Agresta et al. 2021; Askar and Al-jumaily 2008; Ayvaz 2013; Marcus et al. 1992), reliability tests of optimization techniques (Guta and Prasad 2018), elaboration of analytical models (Chalfen and Niemiec 1986; Martínez 2024), study of extreme events (Lai and Khan 2012a), assessment of uncertainties and risks (Birgin et al. 2024; Pappenberger et al. 2005), testing of innovative PDE schemes (Ying et al. 2004), test of different scenarios and normative models, and many others.

In the presence of each particular situation one most collect the pertinent (both theoretical and empirical) information and one must build appropriate models before fitting their parameters.

The goal of our experiments is to show that the procedures proposed in this paper are reliable, in the sense that they can be implemented with moderate space and time resources, and that the results obtained in a few synthetic cases are plausible. Application to individual interesting real-world situations requires further work and should be reported elsewhere.

The experiments were run using a computer with a 5.2 GHz 12th Gen Intel(R) Core(TM) i9- 12900K processor and 16GB 3200 MHz DDR4 RAM memory, running Linux Mint 21.1. The code was compiled by the GFortran compiler of GCC (version 11.3.0) with the -O4 optimization directive enabled.

7.1 Experiment Without Admissibility Conditions

There are no admissibility conditions in this experiment. In particular, neither initial nor boundary conditions are imposed. In these circumstances it can be expected that many different solutions of the Saint-Venant equations are possible. On the other hand, this "large number of degrees of freedom" is reduced due to the imposition of model expressions for several quantities which are not present in the Saint-Venant system.

We will assume that $B(c, x)$, $M(c, x)$, $D(c, x)$, and $E(c, x)$ vary linearly between x_{min} and x_{max}. Therefore,

$$B(c,x) = a_B \frac{x_{max} - x}{x_{max} - x_{min}} + b_B \frac{x - x_{min}}{x_{max} - x_{min}}, \; M(c,x) = a_M \frac{x_{max} - x}{x_{max} - x_{min}} + b_M \frac{x - x_{min}}{x_{max} - x_{min}}, \tag{63}$$

$$D(c,x) = a_D \frac{x_{max} - x}{x_{max} - x_{min}} + b_D \frac{x - x_{min}}{x_{max} - x_{min}}, \; E(c,x) = a_E \frac{x_{max} - x}{x_{max} - x_{min}} + b_E \frac{x - x_{min}}{x_{max} - x_{min}}. \tag{64}$$

The unknown coefficients $a_B, b_B, a_M, b_M, a_D, b_D, a_E$, and b_E will be subject to bounds that must be established taking into account their geometrical meaning (Fig. 1) and available topographic information.

The analytical model $z_b(c, x)$ takes the form of a linear function plus an oscillatory function whose coefficients need to be determined. The coefficients of both the linear function and the trigonometric polynomial are a subset of the coefficients $c \in \mathbb{R}^n$. Namely,

$$z_b(c,x) = b_0^{z_b}(x - x_{min}) + a_0^{z_b} + \sum_{k=1}^{n_{z_b}} a_k^{z_b} \cos\left[\frac{2\pi(x - x_{min})k}{x_{max} - x_{min}}\right] + b_k^{z_b} \sin\left[\frac{2\pi(x - x_{min})k}{x_{max} - x_{min}}\right]. \tag{65}$$

The coefficient $a_0^{z_b}$ can be eliminated as only the derivative of z_b appears in the Saint-Venant equations. Different values of a^{z_b} merely modify the distance of $z_b(x)$ with respect to an arbitrary reference datum.

Reasonable bounds for the coefficients $a_k^{z_b}$ and $b_k^{z_b}$ are given below:

$$b_0^{z_b} \in [-0.001, 0]$$

$$a_k^{z_b} \in [-10^{-4}, 10^{-4}] \text{ and } b_k^{z_b} \in [-10^{-4}, 10^{-4}] \text{ for all } k = 1, \ldots n_{z_b}.$$

In addition, n_{z_b} is also an (integer) coefficient that needs to be determined. We consider that a reasonable value of which is 5.

The analytical model $Q(c, x_{min}, t)$ also takes the form of a linear function plus an oscillatory function on the variable t. Namely,

$$Q(c, x_{min}, t) = a_0^Q + b_0^Q (t - t_{min}) + \sum_{k=1}^{n_Q} a_k^Q \cos\left[\frac{2\pi(t - t_{min})k}{t_{max} - t_{min}}\right] + b_k^Q \sin\left[\frac{2\pi(t - t_{min})k}{t_{max} - t_{min}}\right]. \tag{66}$$

In order to suggest reasonable bounds for the coefficients that define $Q(c, x_{min}, t)$, we assume:

$$a_0^{Q_b} \in [5, 15],\ b_0^{Q_b} \in [0, 2/3600,$$

$$a_k^{Q_b} \in [0, 0.001] \text{ and } b_k^{Q_b} x \in [0, 0.001] \text{ for all } k = 1, \ldots n_{Q_b}.$$

In addition, set $n_Q = 5$.

We will postulate that $z(c, x, t)$ is a quadratic function that depends of x and $Q(c, x_{min}, t)$. Namely,

$$z(c, x, t) = c_0 + c_x \frac{x - x_{min}}{x_{max} - x_{min}} + c_Q Q(c, x_{min}, t) + c_{xx}\left(\frac{x - x_{min}}{x_{max} - x_{min}}\right)^2 + c_{QQ} Q(c, x_{min}, t)^2$$

$$+ c_{xQ} \frac{x - x_{min}}{x_{max} - x_{min}} Q(c, x_{min}, t). \tag{67}$$

Reasonable assumptions in order to define bounds for the coefficients that define $z(c, x, t)$ are:

$$|c_0| \le 5,\ |c_x| \le 0.1,\ |c_Q| \le 1,\ |c_{xx}|, |c_{xq}|, |c_{QQ}| \le 0.01.$$

For the lateral contribution we will assume the following model:

$$q(c, x, t) = c_0^q + \sum_{k=1}^{n_q} a_k^q \cos\left[\frac{2\pi(t - t_{min})k}{t_{max} - t_{min}}\right] + b_k^q \cos\left[\frac{2\pi(x - x_{min})k}{x_{max} - x_{min}}\right] \tag{68}$$

The unknown coefficients in (68) are subject to the following bound constraints:

$$c_q^0 \in [-1/3600, 1/3600],\ a_q^k \in [-1/3600, 1/3600],\ b_q^k \in [-1/3600, 1/3600].$$

We will also postulate that $n_q = 5$.

According to these model definitions, the number of parameters to be estimated are

$$n = 49. \tag{69}$$

Recall that time is measured in seconds and distances in meters. We used $x_{min} = 0$, $x_{max} = 1000$, $t_{min} = 0$, $t_{max} = 86400$ seconds, which corresponds to one day. This experiment do not include admissibility conditions (3).

Table 1 Values found for D, B, M, and E at x_{min} and x_{max}

	$x = x_{min}$	$x = x_{max}$
D	2.41	9.88
B	1.74	6.56
M	49.2	33.9
E	207	306

The results can be summarized as follows.

1. The values of the cross area parameters $B(c, x)$, $D(c, x)$, $M(c, x)$, $E(c, x)$ at x_{min} and x_{max} are given in Table 1. These parameters vary linearly between x_{min} and x_{max} and are independent of time. According to Table 1, the width of the river at the dry level goes from 4.82 m if $x = x_{min}$ to 19.76 m if $x = x_{max}$ and, at the surface level goes from 5.20 m to 20.0 m.
2. For each $t \in [t_{min}, t_{max}]$, $h(c, x, t)$ varies linearly as a function of x with $h(c, x_{min}, t) = 2.79$ m and $h(c, x_{max}, t) = 3.20$ m.
3. For each $t \in [t_{min}, t_{max}]$, $z(c, x, t)$ varies linearly as a function of x with $z(c, x_{min}, t) = 2.79$ m and $z(c, x_{max}, t) = 2.78$ m.
4. For each $t \in [t_{min}, t_{max}]$, $A(c, x, t)$ varies linearly as a function of x with $A(c, x_{min}, t) = 14.2\ \text{m}^2$ and $A(c, x_{max}, t) = 63.7\ \text{m}^2$.
5. $Q(c, x_{min}, t)$ varies linearly as a function of t with $Q(c, x_{min}, t_{min}) = 14\ \text{m}^3/\text{s}$ and $Q(c, x_{min}, t_{max}) = 17\ \text{m}^3/\text{s}$.
6. The outlet discharge $Q(x_{max}, t)$ varies as shown in Fig. 2 with a minimum of $11.2\ \text{m}^3/\text{s}$ and a maximum of $18.6\ \text{m}^3/\text{s}$.
7. For each $t \in [t_{min}, t_{max}]$, $Q(c, x, t)$ varies almost linearly as a function of x from $Q(x_{min}, t)$ to $Q(x_{max}, t)$.
8. The Yen-Manning coefficients oscillate between 0.07 and 0.09.

7.2 Experiments with Two Flow-Rate Observations

In these experiments we adopted the models given in (63)–(68). In addition, we introduced admissibility assumptions defined by two observations of the flow-rate. For this purpose two points (x, t) were randomly chosen in the space-temporal domain and a random flow-rate value between 10 and 30 m^3/s was assumed to be observed at those points. The data for seven cases are given in Table 2. The results are given in Tables 3, 4, 5, 6, 7, 8 and 9. In all the cases the observations were reproduced with a relative error less than 0.1. In these tables "Width" denotes the width of the river at the surface level.

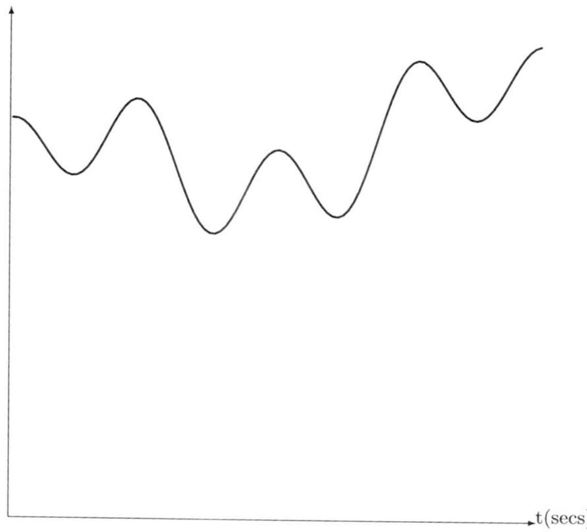

Fig. 2 Outlet discharge $Q(x_{max}, t)$. Minimum: 11.2 m^3/s, Maximum: 18.6 m^3/s

Table 2 Data for experiments with 2 flow-rate observations

Case	x_1	t_1	$Q(x_1, t_1)$	x_2	t_2	$Q(x_2, t_2)$
1	811	54541	21.2	865	81535	22.7
2	63	61860	18.5	534	52094	22.7
3	895	82191	14.0	929	59014	26.1
4	303	2593	16.2	965	25809	23.4
5	60	14666	10.0	445	30474	11.2
6	632	19252	12.0	362	78333	25.3
7	583	84578	20.4	926	4385	29.7

Table 3 Results of experiments with 2 flow-rate observations: Case 1

Point	A	z	h	Q	V	Width	Manning
(x_{min}, t_{min})	47.6	4.62	4.61	25.0	0.53	10.9	0.089
(x_{max}, t_{min})	97.9	4.57	5.57	29.2	0.30	17.9	0.16
(x_{min}, t_{max})	48.7	4.72	4.72	22.9	0.47	10.4	0.096
(x_{max}, t_{max})	99.9	4.68	5.68	27.0	0.27	17.9	0.18

Table 4 Results of experiments with 2 flow-rate observations: Case 2

Point	A	z	h	Q	V	Width	Manning
(x_{min}, t_{min})	1.04	0.12	0.12	18.4	17.6	8.88	0.013
(x_{max}, t_{min})	6.85	0.079	0.36	22.9	3.35	19.0	0.017
(x_{min}, t_{max})	0.71	0.079	0.080	19.5	27.5	8.86	0.011
(x_{max}, t_{max})	6.13	0.041	0.32	24.1	3.93	19.0	0.016

Table 5 Results of experiments with 2 flow-rate observations: Case 3

Point	A	z	h	Q	V	Width	Manning
(x_{min}, t_{min})	31.5	1.76	1.76	22.1	0.70	18.0	0.05
(x_{max}, t_{min})	35.3	1.70	2.59	19.7	0.56	13.9	0.069
(x_{min}, t_{max})	26.2	1.47	1.47	15.1	0.58	17.9	0.055
(x_{max}, t_{max})	31.2	1.40	2.29	12.8	0.41	13.8	0.088

Table 6 Results of experiments with 2 flow-rate observations: Case 4

Point	A	z	h	Q	V	Width	Manning
(x_{min}, t_{min})	8.46	0.92	0.92	17.7	2.09	9.27	0.034
(x_{max}, t_{min})	26.0	0.83	1.58	17.3	0.66	16.5	0.063
(x_{min}, t_{max})	9.05	0.98	0.98	22.4	2.48	9.28	0.033
(x_{max}, t_{max})	27.0	0.89	1.64	22.0	0.81	16.5	0.056

Table 7 Results of experiments with 2 flow-rate observations: Case 5

Point	A	z	h	Q	V	Width	Manning
(x_{min}, t_{min})	96.9	5.17	5.17	10.0	0.10	18.9	0.50
(x_{max}, t_{min})	90.5	5.12	5.68	15.7	0.17	16.2	0.29
(x_{min}, t_{max})	97.8	5.22	5.22	10.9	0.11	18.9	0.46
(x_{max}, t_{max})	91.2	5.17	5.72	16.6	0.18	16.2	0.28

Table 8 Results of experiments with 2 flow-rate observations: Case 6

Point	A	z	h	Q	V	Width	Manning
(x_{min}, t_{min})	4.96	0.27	0.27	13.0	2.62	18.4	0.022
(x_{max}, t_{min})	17.4	0.22	1.07	20.4	1.17	16.4	0.031
(x_{min}, t_{max})	5.04	0.28	0.27	21.4	4.25	18.4	0.023
(x_{max}, t_{max})	17.5	0.22	1.07	28.8	1.65	16.4	0.028

Table 9 Results of experiments with 2 flow-rate observations: Case 7

Point	A	z	h	Q	V	Width	Manning
(x_{min}, t_{min})	19.4	2.81	2.81	29.0	1.49	6.97	0.013
(x_{max}, t_{min})	21.9	2.81	3.22	28.4	1.30	7.07	0.020
(x_{min}, t_{max})	15.4	2.34	2.33	22.4	1.46	6.95	0.014
(x_{max}, t_{max})	17.8	2.24	2.64	21.9	1.22	6.98	0.018

Table 10 Results with several flow-rate observations

n_{dat}	MRE	AE (m^3/s)	ARE
3	0.078	4.7	0.077
10	0.13	4.8	0.071
25	0.14	5.2	0.083
50	0.17	5.5	0.085
100	0.17	5.5	0.084
200	0.21	5.1	0.078
500	0.22	5.9	0.090
1000	0.22	5.1	0.078

7.3 Experiments with Several Flow-Rate Observations

In these experiments we employed observations of flow-rates at n_{dat} points (x_ν, t_ν), randomly chosen in the region $[x_{min}, x_{max}] \times [t_{min}, t_{max}]$. For each point, the observed flow-rate Q_ν^{dat} was chosen randomly in the interval $[55, 75]$. The results are given in Table 10, where, for each n_{dat} we report:

$$\text{Maximal Relative Error} = \text{MRE} \; = \max\{|Q(x(x_\nu, t_\nu) - Q_\nu^{dat}|/|Q_\nu^{dat}|, \nu = 1, \ldots, n_{dat}\},$$

$$\text{Average Error} \; = AE = \frac{1}{n_{dat}} \sum_{\nu=1}^{n_{dat}} |Q(x(x_\nu, t_\nu) - Q_\nu^{dat}|,$$

and

$$\text{Average Relative Error} = \text{ARE} = \frac{1}{n_{dat}} \sum_{\nu=1}^{n_{dat}} |Q(x(x_\nu, t_\nu) - Q_\nu^{dat}|/|Q_\nu^{dat}|.$$

8 Conclusions

Feasibility problems are at the heart of practical optimization. On the one hand, in many constrained and unconstrained optimization problems, users can define, with appropriate tolerances, the conditions under which they will be satisfied by approximate solutions. The mathematical expressions of these conditions define feasibility problems. On the other hand, feasibility formulations come from the natural way in which users express their data and requirements.

In this paper, we have proposed a tool for dealing with channel flow simulations. The feasibility problem comes from the directive: Write down all you know and all you want for your problem, including empirical and theoretical information, and, then, proceed with the technical calculations.

In principle, our feasibility problems has infinitely many equations and inequations. Therefore, we rely on sufficiently large representative samples to handle them.

The theoretical information concerning the channel flow problem was given by the Saint-Venant equations. Therefore, the procedures introduced in this paper can be considered as ways of solving these equations with constraints derived from data, additional information, conjectures, or project design.

Each feasibility component is defined by a particular analytical model that depends on a set of parameters. The feasibility problem is to find the parameters for which the feasibility inequations hold for a sufficiently large sample. The choice of the analytical models that define the feasibility components is a critical task for the problem designer. When the number of parameters is very large, the chances of satisfying the problem inequalities increase, but the chances of satisfying the inequalities for points not present in the sample generally decrease. Therefore, if the number of unknown parameters is large, the size of the samples which must be employed to ensure optimality may become prohibitively large.

The tools presented in this paper exhibit remarkable flexibility, allowing the user to ignore the satisfaction of the PDE equations in regions where he or she believes the equations do not hold, as well as to include sedimentation phenomena through time variation of the bed slope and cross section coefficients. In addition, the methodology is compatible with semiempirical approaches, where a simplified physical model is accompanied by an empirical model correction of the physical error.

Future research includes the adaptation of the techniques presented in this paper to other PDE models related to fluid mechanics and the application of the main tools to different cases concerning real rivers.

Acknowledgements This work was supported by FAPESP (grants 2013/07375-0, 2016/01860-1, and 2018/24293-0) and CNPq (grants 302538/2019-4 and 302682/2019-8).

Declarations

Conflict of interest. The author declares that there is no conflict of interest associated with this publication.

Data availability statement No external data were used.

References

Agresta, A., Baioletti, M., Biscarini, C., Caraffini, F., Milani, A., Santucci, V.: Using optimisation meta-heuristics for the roughness estimation problem in river flow analysis. Appl. Sci. **11**, 10575 (2021)

Andreani, R., Birgin, E.G., Martínez, J.M., Schuverdt, M.L.: On augmented Lagrangian methods with general lower-level constraints. SIAM J. Optim. **18**, 1286–1309 (2008)

Andreani, R., Haeser, G., Martínez, J.M.: On sequential optimality conditions for smooth constrained optimization. Optimization **60**, 627–641 (2011)

Andreani, R., Haeser, G., Schuverdt, M.L., Secchin, L.D., Silva, P.J.S.: On scaled stopping criteria for a safeguarded augmented Lagrangian method with theoretical guarantees. Math. Progr. Comput. **14**, 121–146 (2022)

Andreani, R., Martínez, J.M., Ramos, A., Silva, P.J.S.: A cone-continuity constraint qualification and algorithmic consequences. SIAM J. Optim. **26**, 96–110 (2016)

Antil, H., Kouri, D.P., Lacasse, M.-D., Ridzal, D. (eds.): Frontiers in PDE-Constrained Optimization. The IMA Volumes in Mathematics and its Applications. Springer (2019)

Ascher, U.M., Petzold, L.R.: Computer Methods for Ordinary Differential Equations and Differential-Algebraic Equations. Society for Industrial and Applied Mathematics, Philadelphia (1998)

Askar, M.K., Al-jumaily, K.K.: A nonlinear optimization model for estimating Manning's roughness coefficient. In: Proceedings of the Twelfth International Water Technology Conference, IWTC12, Alexandria, Egypt, pp. 1299–1306 (2008)

Ayvaz, M.T.: A linked simulation-optimization model for simultaneously estimating the Manning's surface roughness values and their parameter structures in shallow water flows. J. Hydrol. **500**, 183–199 (2013)

Birgin, E.G., Correa, M.R., González-López, V.A., Martínez, J.M., Rodrigues, D.S.: Randomly supported models for the prediction of flows in channels, Technical Report of CRIAB-Mathematical Engineering, to appear in J. Hydr. Eng. (2024). https://doi.org/10.1061/JHEND8/HYENG-13748

Birgin, E.G., Krejić, N., Martínez, J.M.: On the employment of Inexact Restoration for the minimization of functions whose evaluation is subject to errors. Math. Comput. **87**, 1307–1326 (2018)

Birgin, E.G., Martínez, J.M.: Improving ultimate convergence of an Augmented Lagrangian method. Optim. Methods Softw. **23**, 177–195 (2008)

Birgin, E.G., Martínez, J.M.: Practical Augmented Lagrangian Methods for Constrained Optimization. Fundamentals of Algorithms, vol. 10. Society for Industrial and Applied Mathematics, Philadelphia, PA (2014)

Birgin, E.G., Martínez, J.M.: Complexity and performance of an Augmented Lagrangian algorithm. Optim. Methods Softw. **35**, 885–920 (2020)

Birgin, E.G., Martínez, J.M.: Accelerated derivative-free nonlinear least-squares applied to the estimation of Manning coefficients. Comput. Optim. Appl. 689–715 (2022)

Birgin, E.G., Martínez, J.M.: A PDE-informed optimization algorithm for river flow predictions. Numer. Algor. 1–16 (2023). https://doi.org/10.1007/s11075-023-01647-1

Caubet, F., Dardé, J., Godoy, M.: On the data completion problem and the inverse obstacle problem with partial Cauchy data for Laplace's equation. Fabien Caubet, Jérémi Dardé, Matías Godoy ESAIM: COCV **25**, 30 (2019). https://doi.org/10.1051/cocv/2017056

Chalfen, M., Niemiec, A.: Analytical and numerical solution of Saint-Venant equations. J. Hydrol. **86**, 1–13 (1986)

Chaudhry, M.H.: Open Channel Flow, 3rd edn. Springer (2022)

Ding, Y., Jia, Y., Wang, S.S.Y.: Identification of Manning's roughness coefficients in shallow water flows. J. Hydr. Eng. 501–510 (2004)

Ding, Y., Wang, S.S.Y.: Identification of Manning's roughness coefficients in channel network using adjoint analysis. Int. J. Comput. Fluid Dyn. **19**, 3–13 (2005)

Emmett, W.W., Myrick, W.W., Meade, R.H.: Field data describing the movement and storage of sediment in the East Fork River, Wyoming, Part 1. River Hydraulics and Sediment Transport, Report No. 1 (1979)

Ferreira, D.M., Fernandes, C.V.S., Gomes, J.: Verification of Saint-Venant equations solution based on the Lax diffusive method for flow routing in natural channels. Brazil. J. Water Res. RBRH **22**, e25 (2017). http://dx.doi.org/10.1590/2318-0331.011716104

Gouyet, J.-F.: Physics and Fractal Structures. Masson Springer, Paris/New York (1996)

Guellouz, L., Khayat, F.: A data completion method for identifying pollution intrusion in aquifers. Sci. Rep. **12**, 1 (2022). https://doi.org/10.1038/s41598-022-20131-9

Guta, K., Prasad, K.S.H.: Estimation of open channel flow parameters by using optimization techniques. Int. J. Sci. Res. **6**, 1295–1304 (2018)

Iserles, A.: A First Course in the Numerical Analysis of Differential Equations. Cambridge University Press (1996)

Lai, W., Khan, A.A.: Discontinuous Galerkin method for 1D shallow water flows in natural rivers. Eng. Appl. Comput. Fluid Mech. **6**, 74–86 (2012a)

Lai, W., Khan, A.A.: Modeling dam-break flood over natural rivers using discontinuous Galerkin method. J. Hydrodyn. **24**, 467–478 (2012b)

Lai, W., Khan, A.A.: Discontinuous Galerkin method for 1D shallow water flow in nonrectangular and non-prismatic channels. J. Hydraul. Eng. **138**, 285–296 (2012c)

LeVeque, R.J.: Numerical Methods for Conservation Laws. Lectures in Mathematics. ETH Zürich, Birkäuser (1992)

Marcus, W.A., Roberts, K., Harvey, L., Tackman, G.: An evaluation of methods for estimating Manning's n in small mountain streams. Mt. Res. Dev. **12**, 227–239 (1992)

Martínez, J.M.: Levenberg-Marquardt revisited and parameter tuning of river regression models. Comput. Appl. Math. (2023). https://doi.org/10.1007/s40314-023-02535-z

Martínez, J.M.: Obtaining analytical solutions to Saint-Venant equations using optimization tools, submitted (2024)

Martínez, J.M., Santos, L.T.: Inexact-restoration modelling with monotone interpolation and parameter estimation. Optim. Eng. 1–20 (2023). https://doi.org/10.1007/s11081-023-09861-5

Meade, R.H., Myrick, W.W., Emmett, W.W.: Field data describing the movement and storage of sediment in the East Fork River, Wyoming, Part 2. River Hydraulics and Sediment Transport, Report No. 2 (1979)

Pappenberger, F., Beven, K., Horrit, M., Blazkova, S.: Uncertainty in the calibration of effective roughness parameters in HEC-RAS using inundation and downstream level observations. J. Hydrol. **302**, 46–69 (2005)

Porto, R.M.: Hidráulica Básica. EESC-USP, São Paulo (2000)

Raissi, M., Perdikaris, P., Karniadakis, G.E.: Physics-informed neural networks: a deep learning framework for solving forward and inverse problems involving nonlinear partial differential equations. J. Comput. Phys. **378**, 686–707 (2019)

Saint-Venant, A.J.C.: Théorie du mouvement non-permanent des eaux, avec application aux crues des rivière at à l'introduction des marées dans leur lit. Comptes Rendus des Séances de Académie des Sciences **73**, 147–154 (1871)

Simões, A.L., Schulz, H.E., Porto, R.M.: Métodos Computacionais em Hidráulica. Edufba, Bahia (2017)

Wang, Y., Chen, S., Suqin, Wu, X.: A rational spectral collocation method for solving a class of parameterized singular perturbation problems. J. Comput. Appl. Math. **233**, 2652–2660 (2009)

Wright, S.J.: Optimization in machine learning and data science. SIAM News **56**, 1–2 (2023)

Yen, B.B.: Dimensionally homogeneous Manning's formula. J. Hydraul. Eng. **118**(9) (1992)

Ying, X., Khan, A.A., Wang, S.S.Y.: Upwind conservative scheme for the Saint Venant equations. J. Hydraul. Eng. **130**, 977–987 (2004)

Open Access This chapter is licensed under the terms of the Creative Commons Attribution 4.0 International License (http://creativecommons.org/licenses/by/4.0/), which permits use, sharing, adaptation, distribution and reproduction in any medium or format, as long as you give appropriate credit to the original author(s) and the source, provide a link to the Creative Commons license and indicate if changes were made.

The images or other third party material in this chapter are included in the chapter's Creative Commons license, unless indicated otherwise in a credit line to the material. If material is not included in the chapter's Creative Commons license and your intended use is not permitted by statutory regulation or exceeds the permitted use, you will need to obtain permission directly from the copyright holder.

Statistical Accuracy of Approximate Filtering Methods

J. A. Carrillo, F. Hoffmann, A. M. Stuart, and U. Vaes

Abstract Estimating the statistics of the state of a dynamical system, from partial and noisy observations, is both mathematically challenging and finds wide application. Furthermore, the applications are of great societal importance, including problems such as probabilistic weather forecasting (Kalnay 2003) and prediction of epidemics (Keeling and Eames 2005). Particle filters provide a well-founded approach to the problem, leading to provably accurate approximations of the statistics (Doucet et al. 2001). However these methods perform poorly in high dimensions (Bickel et al. 2008; Snyder et al. 2008). In 1994 the idea of ensemble Kalman filtering was introduced (Evensen 1994) leading to a methodology that has been widely adopted in the geophysical sciences (van Leeuwen et al. 2019) and also finds application to quite general inverse problems (Iglesias et al. 2013). However, ensemble Kalman filters have defied rigorous analysis of their statistical accuracy, except in the linear Gaussian setting (Le Gland et al. 2011; Mandel et al. 2011). In this article we describe recent work which takes first steps to analyze the statistical accuracy of ensemble Kalman filters beyond the linear Gaussian setting (Carrillo et al. 2022). The subject is inherently technical, as it involves the evolution of probability measures according to a nonlinear and nonautonomous dynamical system; and the approximation of this evolution. It can nonetheless be presented in a fairly accessible fashion, understandable with basic knowledge of dynamical systems, numerical analysis and probability. We undertake such a presentation here.

J. A. Carrillo
Mathematical Institute, University of Oxford, Oxford, UK
e-mail: carrillo@maths.ox.ac.uk

F. Hoffmann · A. M. Stuart (✉)
Department of Computing and Mathematical Sciences, Caltech, Pasadena, CA, USA
e-mail: astuart@caltech.edu

F. Hoffmann
e-mail: franca.hoffmann@caltech.edu

U. Vaes
MATHERIALS Project-Team, Inria Paris, Paris, France
e-mail: urbain.vaes@inria.fr

CERMICS,École des Ponts, Champs-sur-Marne, France

© The Author(s) 2026

S. Oishi et al. (eds.), *Recent Developments in Industrial and Applied Mathematics*, ICIAM2023 Springer Series 1, https://doi.org/10.1007/978-981-95-1446-5_14

1 Filtering Problem

Consider a dynamical system for state $\{v_n\}_{n\in\mathbb{Z}^+}$ evolving in $\mathbb{R}^d$, partially and noisily observed through data $\{y_n\}_{n\in\mathbb{N}}$ in $\mathbb{R}^K$, with state and data determined by the following system, holding for $n \in \mathbb{Z}^+$:

$$\begin{aligned} \textbf{State}: \quad & v_{n+1} = \Psi(v_n) + \xi_n\,, \\ \textbf{Data}: \quad & y_{n+1} = h(v_{n+1}) + \eta_{n+1}\,. \end{aligned}$$

Here $\mathbb{N} := \{1, 2, \ldots\}$ and $\mathbb{Z}^+ = \{0, 1, 2, \ldots\}$. We assume that the initial state of the system is a Gaussian random variable $v_0 \sim \mathcal{N}(m_0, C_0)$. Furthermore we assume that $\xi_n \sim \mathcal{N}(0, \Sigma)$ is the mean-zero noise affecting the state evolution and $\eta_{n+1} \sim \mathcal{N}(0, \Gamma)$ is the mean-zero noise entering the data acquisition process. We assume that the state evolution and data acquisition noise sequences are i.i.d. and that the following independence assumptions hold:

$$v_0 \perp\!\!\!\perp \{\xi_n\}_{n\in\mathbb{Z}^+} \perp\!\!\!\perp \{\eta_{n+1}\}_{n\in\mathbb{Z}^+}.$$

The objective of filtering is to determine the probability distribution on v_n given all the data acquired up to that point. To this end we define, for a given realization of the data, denoted by a dagger †,

$$Y_n^\dagger = \{y_\ell^\dagger\}_{\ell=1}^n, \qquad v_n | Y_n^\dagger \sim \mu_n.$$

With this notation we can state the objective of filtering more precisely: it is to determine the *filtering distribution* or *true filter* μ_n and update it sequentially in n. As we now show μ_n evolves according to a nonautonomous and nonlinear dynamical system on the space of probability measures.

To determine this dynamical system, and various approximations of it that follow, it is helpful to define

- $\mathcal{P}(\mathbb{R}^r)$: all probability measures on $\mathbb{R}^r$.
- $\mathcal{G}(\mathbb{R}^r)$: all Gaussian probability measures on $\mathbb{R}^r$.

For simplicity we will use the same symbol for measures and their densities throughout this article. Before writing the dynamical system for μ_n we first determine the evolution of the measures π_n, $\mathfrak{r}_n$ defined by

$$v_n \sim \pi_n, \qquad (v_n, y_n) \sim \mathfrak{r}_n.$$

These evolve according to

$$\begin{aligned} \pi_{n+1} &= \mathsf{P}\pi_n, \\ \mathfrak{r}_{n+1} &= \mathsf{Q}\pi_{n+1} \end{aligned}$$

Here $\mathsf{P}: \mathcal{P}(\mathbb{R}^d) \to \mathcal{P}(\mathbb{R}^d)$ is the linear operator

$$\mathsf{P}\pi(u) = \frac{1}{\sqrt{(2\pi)^d \det \Sigma}} \int \exp\left(-\frac{1}{2}|u - \Psi(v)|_\Sigma^2\right) \pi(v)\, \mathrm{d}v.$$

whilst $\mathsf{Q}: \mathcal{P}(\mathbb{R}^d) \to \mathcal{P}(\mathbb{R}^d \times \mathbb{R}^K)$, which is also a linear operator, is determined by

$$\mathsf{Q}\pi(u, y) = \frac{1}{\sqrt{(2\pi)^K \det \Gamma}} \exp\left(-\frac{1}{2}\left|y - h(u)\right|_\Gamma^2\right) \pi(u).$$

The evolution for π_n, which describes the probability of the state v_n, is determined by a Markov process on $\mathbb{R}^d$ defined via the linear operator P. The linear operator Q lifts π_n to the joint space of state and data (v_n, y_n). It is worth highlighting that, by moving from the evolution of $(v_n, y_n) \in \mathbb{R}^d \times \mathbb{R}^K$ (finite dimensions) to the evolution of $(\pi_n, \mathfrak{r}_n) \in \mathcal{P}(\mathbb{R}^d \times \mathbb{R}^K)$ (infinite dimensions) we have converted a nonlinear stochastic problem into a linear autonomous one.

However, the dynamical system for μ_n is nonlinear and nonautonomous. To determine the dynamical system for μ_n we need to introduce a nonlinear operator on probability measures. Specifically, let $\mathsf{B}(\bullet; y^\dagger): \mathcal{P}(\mathbb{R}^d \times \mathbb{R}^K) \to \mathcal{P}(\mathbb{R}^d)$ describe conditioning of a joint random variable (v, y) on observation $y = y^\dagger$:

$$\mathsf{B}(\rho; y^\dagger)(u) = \frac{\rho(u, y^\dagger)}{\int_{\mathbb{R}^d} \rho(u, y^\dagger)\, \mathrm{d}u}.$$

Armed with this we define $v_{n+1}|Y_n^\dagger \sim \widehat{\mu}_{n+1}$ and observe that

$$\begin{aligned}
\widehat{\mu}_{n+1} &= \mathsf{P}\mu_n, && v_{n+1}|Y_n^\dagger \sim \widehat{\mu}_{n+1}\\
\rho_{n+1} &= \mathsf{Q}\widehat{\mu}_{n+1}, && (v_{n+1}, y_{n+1})|Y_n^\dagger \sim \rho_{n+1}\\
\mu_{n+1} &= \mathsf{B}(\rho_{n+1}; y_{n+1}^\dagger), && \text{conditioning on } y_{n+1} = y_{n+1}^\dagger.
\end{aligned}$$

Thus we have the nonlinear and nonautonomous dynamical system

$$\mu_{n+1} = \mathsf{B}(\mathsf{QP}\mu_n; y_{n+1}^\dagger), \qquad \mu_0 = \mathcal{N}(m_0, C_0).$$

This evolution may be thought of in terms of sequential application of Bayes theorem: $\mathsf{P}\mu_n$ is prior prediction; $\mathsf{L}(\bullet; y^\dagger) := \mathsf{B}(\bullet; y^\dagger) \circ \mathsf{Q}$ maps prior to posterior according to Bayes theorem. This leads to the following equivalent formulation of the evolution:

$$\mu_{n+1} = \mathsf{L}(\mathsf{P}\mu_n; y_{n+1}^\dagger), \qquad \mu_0 = \mathcal{N}(m_0, C_0).$$

The resulting evolution is an infinite dimensional problem since μ_n is a probability density over $\mathbb{R}^d$. Approximating this evolution is thus a significant challenge.

2 Particle Filter

During the last quarter of the last century, the particle filter became a widely adopted methodology for solving the filtering problem in moderate dimensions d; the methodology is overviewed in Doucet et al. (2001).

To describe the basic form of the method we introduce a random map on the space of probability measures; this map encapsulates Monte Carlo sampling. To this end let Ω denote an abstract probability space (encapsulating the sampling underlying the Monte Carlo method) and define $\mathsf{S}^J : \mathcal{P}(\mathbb{R}^r) \times \Omega \to \mathcal{P}(\mathbb{R}^r)$ to be the empirical approximation operator:

$$\mathsf{S}^J \mu = \frac{1}{J} \sum_{j=1}^{J} \delta_{v_j}, \qquad v_j \overset{\text{i.i.d.}}{\sim} \mu.$$

Here Ω is a probability space capturing the process through which independent draws (from μ) are made. We suppress explicit dependence on $\omega \in \Omega$ in the expression $\mathsf{S}^J \mu$ for notational convenience. Then, for large J and fixed $\omega \in \Omega$, S^J is a random approximation of the identity on $\mathcal{P}(\mathbb{R}^r)$, a statement made precise in the remarks that follow Theorem 1.

Since $\mathsf{S}^J \approx I$, we may introduce the approximation $\mu_n^{\mathrm{PF}} \approx \mu_n$ evolving according to the map

$$\mu_{n+1}^{\mathrm{PF}} = \mathsf{L}(\mathsf{S}^J \mathsf{P} \mu_n^{\mathrm{PF}}; y_{n+1}^{\dagger}), \qquad \mu_0^{\mathrm{PF}} = \mu_0.$$

In the preceding expression it is implicit that each application of S^J is based on an independent realization of $\omega \in \Omega$. It then follows that, for $n \in \mathbb{N}$,

$$\mu_n^{\mathrm{PF}} = \sum_{j=1}^{J} w_n^{(j)} \delta_{\widehat{v}_n^{(j)}}$$

where the particles $\widehat{v}_n^{(j)}$ and weights $w_n^{(j)}$ evolve according to

$$\begin{aligned}
\widehat{v}_{n+1}^{(j)} &= \Psi\big(v_n^{(j)}\big) + \xi_n^{(j)}, \qquad v_n^{(j)} \overset{\text{i.i.d.}}{\sim} \mu_n^{\mathrm{PF}}, \\
\ell_{n+1}^{(j)} &= \exp\left(-\frac{1}{2}\big|y_{n+1}^{\dagger} - h\big(\widehat{v}_{n+1}^{(j)}\big)\big|_{\Gamma}^2\right), \\
w_{n+1}^{(j)} &= \ell_{n+1}^{(j)} \Big/ \Big(\sum_{m=1}^{J} \ell_{n+1}^{(m)}\Big).
\end{aligned}$$

Here $\xi_n^{(j)}$ are $\mathcal{N}(0, \Sigma)$ Gaussians that are i.i.d. with respect to both n and j.

Systemization of the analysis of this method may be found in the work of Del Moral (1997), with the significant extension to analysis over long time-intervals in Del

Moral and Guionnet (2001) and to continuous time in Crişan et al. (1999). We describe a prototypical theoretical result, based on the formulation underpinning Rebeschini and van Handel (2015); for precise statement of conditions under which the theorem holds see Law et al. (2015, Theorem 4.5). The theorem deploys a metric $\mathrm{d}(\bullet, \bullet)$ on random probability measures. This is defined as follows. First let μ, ν be elements of the space of random probability measures defined over a probability space Ω and let μ_ω (resp. ν_ω) denote the specific probability measure that arises when random probability measure μ (resp. ν) is evaluated at $\omega \in \Omega$. Let $\mathbb{E}^\omega$ denote expectation with respect to $\omega \in \Omega$. Then

$$\mathrm{d}(\mu, \nu)^2 = \sup_{|f|\le 1} \mathbb{E}^\omega \big|\mu_\omega[f] - \nu_\omega[f]\big|^2,$$

where $\mu_\omega[f] = \int f(u)\, \mu_\omega(\mathrm{d}u)$ and similarly for ν_ω.

The metric reduces to TV distance if applied to deterministic probability measures. In the application that follows Ω is defined to include the randomness that arises from the successive empirical approximations using S^J over multiple steps of the particle filter.

Theorem 1 (Law et al. 2015, Theorem 4.5) *It holds that*

$$\sup_{0\le n\le N} \mathrm{d}(\mu_n, \mu_n^{\mathrm{PF}}) \le \frac{C(N)}{\sqrt{J}}.$$

Remark 1 The following remarks explain and interpret the theorem:

- The theorem states that a distance between the particle filter and the true filter is bounded above by an error which decreases at the Monte Carlo rate with respect to the number of particles J.
- The proof follows a typical numerical analysis structure: Consistency + Stability Implies Convergence.
- Consistency: $\mathrm{d}(\mathsf{S}^J\mu, \mu) \le \frac{1}{\sqrt{J}}$. This shows that S^J approximates the identity.
- Stability: P, L are Lipschitz in $\mathrm{d}(\bullet, \bullet)$.
- The theorem assumes upper and lower bounds on the n-dependent likelihood arising from sequential application of Bayes theorem; the constant C depends on these bounds. Furthermore C grows exponentially with N.

In practice, especially in high dimensions d, the method often suffers from weight collapse (Bickel et al. 2008; Snyder et al. 2008). This refers to the phenomenon where one of the weights $\{w_n^{(j)}\}_{j=1}^J$ is close to 1 and the others are therefore necessarily close to 0. When this happens the method is of little value because the effective number of particles approximating μ_n is 1. This leads us to consider the ensemble Kalman filter which, by design, has equal weights.

3 Ensemble Kalman Filter

The ensemble Kalman filter, introduced by Evensen (1994), is overviewed in the texts (Evensen 2009; Evensen et al. 2022). We will state the basic particle form of the algorithm, from Evensen (1994). We will then show how it may be derived from a mean-field perspective highlighted in Calvello et al. (2025). Finally we describe analysis of this mean-field ensemble Kalman filter, and its relationship to the true filter (Carrillo et al. 2022). This analysis of the mean field method is a first step to eventually obtaining an error estimate between the ensemble Kalman filter and the true filter in the finite-ensemble setting.

To describe the particle form of the method we first introduce key notation relating to covariances. Specifically we write covariance under $\rho \in \mathcal{P}(\mathbb{R}^d \times \mathbb{R}^K)$ as:

$$\mathrm{cov}(\rho) = \begin{pmatrix} C^{vv}(\rho) & C^{vy}(\rho) \\ C^{vy}(\rho)^\top & C^{yy}(\rho) \end{pmatrix}.$$

We use similar notation (mean_ρ) for the mean under ρ. The ensemble Kalman filter from Evensen (1994) then has the form, for $n \in \mathbb{Z}^+$,

$$\begin{aligned}
\widehat{v}_{n+1}^{(j)} &= \Psi\big(v_n^{(j)}\big) + \xi_n^{(j)}, \\
\widehat{y}_{n+1}^{(j)} &= h(\widehat{v}_{n+1}^{(j)}) + \eta_{n+1}^{(j)}, \\
v_{n+1}^{(j)} &= \widehat{v}_{n+1}^{(j)} + C^{vy}\big(\rho_{n+1}^{\mathrm{EK},J}\big)C^{yy}\big(\rho_{n+1}^{\mathrm{EK},J}\big)^{-1}\big(y_{n+1}^\dagger - \widehat{y}_{n+1}^{(j)}\big), \\
\rho_{n+1}^{\mathrm{EK},J} &= \frac{1}{J}\sum_{j=1}^{J} \delta_{\left(\widehat{v}_{n+1}^{(j)}, \widehat{y}_{n+1}^{(j)}\right)}.
\end{aligned}$$

Here $\xi_n^{(j)} \sim \mathcal{N}(0, \Sigma)$ i.i.d. with respect to both n and j and $\eta_n^{(j)} \sim \mathcal{N}(0, \Gamma)$ i.i.d. with respect to both n and j. Furthermore the set of $\{\xi_n^{(j)}\}$ is independent of the set of $\{\eta_n^{(j)}\}$.

Remark 2 Note that the components of $\mathrm{cov}(\rho_{n+1}^{\mathrm{EK},J})$ involve empirical learning of the covariances Σ and Γ of the $\{\xi_n^{(j)}\}_{j=1}^J$ and $\{\eta_{n+1}^{(j)}\}_{j=1}^J$ respectively. It is possible, and indeed often desirable, to directly input the matrices Σ and Γ, only using empirical estimation for the covariance based on the $\{\Psi(v_n^{(j)})\}_{j=1}^J$ and $\{h(\widehat{v}_{n+1}^{(j)})\}_{j=1}^J$. In particular to use

$$C^{vy}(\rho_{n+1}^{\mathrm{EK},J}) \approx C^{vh,J}, \tag{2a}$$

$$C^{yy}(\rho_{n+1}^{\mathrm{EK},J}) \approx C^{hh,J} + \Gamma \tag{2b}$$

where $C^{vh,J}$ is the empirical cross-covariance between the $\{\Psi(v_n^{(j)})\}_{j=1}^J$ and the $\{h(\widehat{v}_{n+1}^{(j)})\}_{j=1}^J$, whilst $C^{hh,J}$ is the empirical covariance of the $\{h(\widehat{v}_{n+1}^{(j)})\}_{j=1}^J$. Note that this leads to a definition of $C^{yy}(\rho_{n+1}^{\mathrm{EK},J})$ which differs from the original one, which employs an empirical approximation Γ_{n+1} of the covariance Γ, formed from $\{\eta_{n+1}^{(j)}\}_{j=1}^{J}$.

From the particles defined by this algorithm we may define an empirical measure

$$\mu_n^{\mathrm{EK},J} = \frac{1}{J}\sum_{j=1}^{J}\delta_{v_n^{(j)}},$$

noting that for $n = 0$ we choose $v_0^{(j)} \sim \mu_0$ i.i.d. Practitioners like this methodology because it assigns equal weights to the particles and cannot suffer from the weight collapse arising in the particle filter. Furthermore, in the setting where Ψ, h are both linear, and μ_n is Gaussian, $\mu_n^{\mathrm{EK},J}$ converges to μ_n as $J \to \infty$, at the Monte Carlo rate (Le Gland et al. 2011; Mandel et al. 2011). In this linear setting, with Gaussian noise and initial condition, the problem can be solved explicitly by the Kalman filter (Kalman 1960); this is a nonautonomous and nonlinear dynamical system for the mean and covariance of the (in this setting) Gaussian μ_n. However forming and propagating $d \times d$ covariances, when $d \gg 1$, is impractical; ensemble methods, on the other hand, operate by computing a $J \times d$ low-rank approximation of the covariance and can be used when $d \gg 1$. Thus particle methods are of some practical value even in this linear Gaussian setting, avoiding the need to work with large covariances when $d \gg 1$.

We now outline recent new work aimed at determining the statistical accuracy of the mean field ensemble Kalman filter, beyond the linear Gaussian setting. The mean-field limit of the particle system, found by letting $J \to \infty$, is determined by the map

$$\begin{aligned}
\widehat{v}_{n+1} &= \Psi(v_n) + \xi_n,\\
\widehat{y}_{n+1} &= h(\widehat{v}_{n+1}) + \eta_{n+1},\\
v_{n+1} &= \widehat{v}_{n+1} + C^{vy}\big(\rho_{n+1}^{\mathrm{EK}}\big)C^{yy}\big(\rho_{n+1}^{\mathrm{EK}}\big)^{-1}\big(y_{n+1}^{\dagger} - \widehat{y}_{n+1}\big),\\
(\widehat{v}_{n+1}, \widehat{y}_{n+1}) &\sim \rho_{n+1}^{\mathrm{EK}}.
\end{aligned}$$

Once again $\xi_n \sim \mathcal{N}(0, \Sigma)$ i.i.d. with respect to n and $\eta_n \sim \mathcal{N}(0, \Gamma)$ i.i.d. with respect to n; and the set of $\{\xi_n\}$ is independent of the set of $\{\eta_n\}$.

Remark 3 The covariances C^{vy} and C^{yy} can be calculated exactly from the covariance of $\widehat{v}_{n+1}$ with $h(\widehat{v}_{n+1})$, from the covariance of $h(\widehat{v}_{n+1})$ with itself and from Γ. The formulae are the $J \to \infty$ limit of those given in (2).

Note that the map $v_n \mapsto v_{n+1}$ is nonlinear, stochastic and nonautonomous in the sense that it depends on the observed data. But, furthermore, the map depends on the law of v_n since knowledge of this law is required to define ρ_{n+1}^{EK}. Thus we now focus on finding an evolution equation for the law of v_n, which we denote by μ_n^{EK}.

With this goal we define $\mathfrak{T}$ as follows:

$$\mathfrak{T}(\bullet, \bullet; \rho, y^\dagger) \colon \mathbb{R}^d \times \mathbb{R}^K \to \mathbb{R}^d;$$
$$(v, y) \mapsto v + C^{vy}(\rho)C^{yy}(\rho)^{-1}\big(y^\dagger - y\big),$$

noting that this map is linear for any given pair $(\rho, y^\dagger) \in \mathcal{P}(\mathbb{R}^d \times \mathbb{R}^K) \times \mathbb{R}^K$. Recalling the notation $F_\sharp \nu$ for pushforward of a probability measure ν under a map F, we define the operator

$$\mathsf{T}(\rho; y^\dagger) = \big(\mathfrak{T}(\bullet, \bullet; \rho, y^\dagger)\big)_\sharp \rho.$$

Notice that T, like B, is nonlinear as a map on probability measures. With the definition of T it then follows (Calvello et al. 2025) that

$$\mu_{n+1}^{\mathrm{EK}} = \mathsf{T}(\mathsf{QP}\mu_n^{\mathrm{EK}}; y_{n+1}^\dagger), \quad \mu_0^{\mathrm{EK}} = \mu_0.$$

The mean-field map $v_n \mapsto v_{n+1}$ is now well-defined by coupling it to this nonlinear evolution equation for μ_n^{EK}, the law of v_n. We refer to it as a *mean-field map* precisely because of this dependence on its own law.

The key question we wish to address is the relationship between μ_n^{EK} and μ_n. To focus on this question we write their evolution equations in parallel:

$$\mu_{n+1}^{\mathrm{EK}} = \mathsf{T}(\mathsf{QP}\mu_n^{\mathrm{EK}}; y_{n+1}^\dagger),$$
$$\mu_{n+1} = \mathsf{B}(\mathsf{QP}\mu_n; y_{n+1}^\dagger).$$

From this it is clear that the key is to understand when $\mathsf{T} \approx \mathsf{B}$. In fact $\mathsf{T} \equiv \mathsf{B}$ on the set of Gaussian measures $\mathcal{G}(\mathbb{R}^d \times \mathbb{R}^K)$ (Calvello et al. 2025). This is related to the fact that, in the Gaussian case, the map T effects an exact transport from the joint distribution of state and data to the conditional distribution of state given data. To obtain an error bound between μ_n^{EK} and μ_n it is helpful to define

$$\mathsf{G} \colon \mathcal{P} \to \mathcal{G},$$
$$\mathsf{G}\pi = \arg\min_{\mathfrak{p} \in \mathcal{G}} d_{\mathrm{KL}}(\pi \| \mathfrak{p}),$$

where $d_{\mathrm{KL}}(\pi \| \mathfrak{p})$ is the Kullback–Leibler (KL) divergence of π from $\mathfrak{p}$, defined as

$$d_{\mathrm{KL}}(\pi \| \mathfrak{p}) = \int_{\mathbb{R}^d} \frac{\mathrm{d}\pi}{\mathrm{d}\mathfrak{p}} \log\left(\frac{\mathrm{d}\pi}{\mathrm{d}\mathfrak{p}}\right) \mathrm{d}\mathfrak{p}. \tag{3}$$

It then follows that $\mathsf{G}\pi = \mathcal{N}(\mathrm{mean}_\pi, \mathrm{cov}_\pi)$ (Sanz-Alonso et al. 2023, Theorem 4.7).

Remark 4 The map T is not an *optimal* transport, in the sense of transporting one measure into another at minimal cost (Villani 2009), but it is a transport that is well-adapted to numerical implementation. Links between transport and data assimilation were pioneered by Sebastian Reich; see Reich and Cotter (2015) and the references therein.

When analyzing the particle filter we used a metric on random probability measures which reduces to the TV distance in the non-random case. Here we use a different metric; we have no need to consider random probability measures, but standard TV alone does not allow us to control first and second moments. Control of these moments is useful for the analysis because of the Gaussian approximations underlying the use of T rather than B. We thus use a weighted TV metric, with weight $g(v) = 1 + |v|^2$. Specifically we define

$$\mathrm{d}_g(\mu, \nu) = \sup_{|f| \le g} \big|\mu[f] - \nu[f]\big|,$$

with $\mu[f]$ and $\nu[f]$ defined as before. In order to state our theorem about closeness of μ_n and μ_n^{EK} we introduce the following measure of how close the true filter $\{\mu_n\}$ is to being Gaussian, in the lifted space of state and data:

Definition 1 We define the closeness between the filtering distribution μ_n, lifted to the joint space of state and data, and its projection onto Gaussians, over N steps:

$$\varepsilon := \sup{}_{0 \le n \le N}\ \mathrm{d}_g(\mathsf{GQP}\mu_n, \mathsf{QP}\mu_n).$$

For precise statement and proof of the following theorem, and in particular details of the conditions under which it holds, see Carrillo et al. (2022).

Theorem 2 *Let* $\mu_0^{\mathrm{EK}} = \mu_0$ *and assume that* $\|\Psi\|_{L^\infty}$, $\|h\|_{L^\infty}$ *and* $|h|_{C^{0,1}}$ *are bounded by* r. *Then there is* $C := C(N, r) \in (0, \infty)$ *such that*

$$\sup_{0 \le n \le N} \mathrm{d}_g(\mu_n, \mu_n^{\mathrm{EK}}) \le C\varepsilon.$$

Remark 5 The following remarks explain and interpret the theorem:

- The theorem states that a distance between the mean-field ensemble Kalman filter and the true filter is bounded above by a quantity which measures how close the true filter is to being Gaussian. This is natural because T and B are identical on Gaussians.
- As for the particle filter, the proof follows a typical numerical analysis structure: Consistency + Stability Implies Convergence.
- Consistency: in Carrillo et al. (2022) a class of problems is identified within which there are sequences of problems along which $\varepsilon \to 0$. This demonstrates that there are problem classes within which the mean-field ensemble Kalman filter accurately approximates the true filter; at $\varepsilon = 0$ a Gaussian problem is obtained and so the result concerns a class of problems in a small neighbourhood of Gaussians.
- Stability: P, Q are Lipschitz in $\mathrm{d}_g(\bullet, \bullet)$. Whilst B, T are not Lipschitz, stability bounds can be proved in $\mathrm{d}_g(\bullet, \bullet)$, given certain information such as moment bounds and lower bounds on covariances.
- The theorem assumes upper bounds on the vector fields Ψ, h defining the filtering problems, and the constant C depends on these bounds. Furthermore C grows exponentially with N.

4 Conclusions

Ensemble Kalman filters are widely used, yet many open problems remain concerning their properties. In this article we have concentrated on understanding the sense in which they approximate the true filtering distribution. This is important for understanding the sense in which probabilistic forecasts made using ensemble Kalman filters predict accurate statistics. In the context of weather forecasting, an example is determining the probability of rain at levels that will cause flooding; in the context of forecasting epidemics, an example is determining the probability of an infection peak that will overwhelm health services, in the absence of interventions. The work described in Theorem 2 is a first step to build theory in this area, for non-Gaussian problems. Many challenges lie ahead to develop this theory so that it applies under more complex and realistic conditions.

Acknowledgements This article is based on an invited lecture delivered by AMS at ICIAM 2023 in Tokyo. JAC was supported by the Advanced Grant Nonlocal-CPD (Nonlocal PDEs for Complex Particle Dynamics: Phase Transitions, Patterns and Synchronization) of the European Research Council Executive Agency (ERC) under the European Union's Horizon 2020 research and innovation programme (grant agreement No. 883363). JAC was also partially supported by the Engineering and Physical Sciences Research Council (EPSRC) under grants EP/T022132/1 and EP/V051121/1. FH is supported by start-up funds at the California Institute of Technology. FH was also supported by the Deutsche Forschungsgemeinschaft (DFG, German Research Foundation) via project 390685813-GZ 2047/1-HCM. The work of AMS is supported by a Department of Defense Vannevar Bush Faculty Fellowship, and by the SciAI Center, funded by the Office of Naval Research (ONR), under Grant Number N00014-23-1-2729. UV is partially supported by the European Research Council (ERC) under the European Union's Horizon 2020 research and innovation programme (grant agreement No 810367), and by the Agence Nationale de la Recherche under grants ANR-21-CE40-0006 (SINEQ) and ANR-23-CE40-0027 (IPSO). The authors are grateful to Eviatar Bach, Ricardo Baptista, Edo Calvello and an anonymous referee for helpful suggestions relating to the exposition.

Competing Interests The authors have no conflicts of interest to declare that are relevant to the content of this chapter.

References

Bickel, P., Li, B., Bengtsson, T.: Sharp failure rates for the bootstrap particle filter in high dimensions. In: Pushing the Limits of Contemporary Statistics: Contributions in Honor of Jayanta K. Ghosh. Inst. Math. Stat. (IMS) Collect., vol 3 (2008)

Calvello, E., Reich, S., Stuart, A.M.: Ensemble Kalman methods: a mean field perspective. Acta Numer. (2025)

Carrillo, J.A., Hoffmann, F., Stuart, A.M., Vaes, U.: The Mean Field Ensemble Kalman Filter: Near-Gaussian Setting. arXiv:2212.13239 (2022)

Crişan, D., Moral, P., Lyons, T.J.: Interacting particle systems approximations of the Kushner-Stratonovitch equation. Adv. Appl. Probab. **31**(3), 819–838 (1999)

Del Moral, P.: Nonlinear filtering: interacting particle resolution. C. R. Acad. Sci. Paris Ser. I Math. **325**(6), 653–658 (1997)

Del Moral, P., Guionnet, A.: On the stability of interacting processes with applications to filtering and genetic algorithms. Ann. Inst. H. Poincaré Probab. Statist. **37**(2), 155–194 (2001)

Doucet, A., de Freitas, N., Gordon, N. (eds.): Sequential Monte Carlo methods in Practice. Statistics for Engineering and Information Science. Springer-Verlag, New York (2001)

Evensen, G., Vossepoel, F.C., van Leeuwen, P.J.: Data assimilation fundamentals: a unified formulation of the state and parameter estimation problem. In: Springer Textbooks in Earth Sciences, Geography and Environment. Springer, Cham (2022)

Evensen, G.: Data Assimilation: The Ensemble Kalman Filter, 2nd ed. Springer, Berlin (2009)

Evensen, G.: Sequential data assimilation with a nonlinear quasi-geostrophic model using Monte Carlo methods to forecast error statistics. J. Geophys. Res. Oceans **99**(C5), 10143–10162 (1994)

Iglesias, M.A., Law, K.J.H., Stuart, A.M.: Ensemble Kalman methods for inverse problems. Inverse Probl. **29**(4), 045001, 20 (2013)

Kalman, R.E.: A new approach to linear filtering and prediction problems. Trans. ASME Ser. D. J. Basic Eng. **82**(1), 35–45 (1960)

Kalnay, E.: Atmospheric Modeling, Data Assimilation and Predictability. Cambridge University Press (2003)

Keeling, M.J., Eames, K.T.: Networks and epidemic models. J. R. Soc. Interface **2**(4), 295–307 (2005)

Law, K., Stuart, A., Zygalakis, K.: Data Assimilation. Texts in Applied Mathematics, vol. 62. Springer, Cham (2015)

Le Gland, F., Monbet, V., Tran, V.-D.: Large sample asymptotics for the ensemble Kalman filter. In: The Oxford Handbook of Nonlinear Filtering, pp. 598–631. Oxford University Press, Oxford (2011)

Mandel, J., Cobb, L., Beezley, J.D.: On the convergence of the ensemble Kalman filter. Appl. Math. **56**(6), 533–541 (2011)

Rebeschini, P., van Handel, R.: Can local particle filters beat the curse of dimensionality? Ann. Appl. Probab. **25**(5), 2809–2866 (2015)

Reich, S., Cotter, C.: Probabilistic Forecasting and Bayesian Data Assimilation. Cambridge University Press, New York (2015)

Sanz-Alonso, D., Stuart, A., Taeb, A.: Inverse Problems and Data Assimilation. LMS Student Lecture Notes, vol. 107. Cambridge University Press (2023)

Snyder, C., Bengtsson, T., Bickel, P., Anderson, J.: Obstacles to high-dimensional particle filtering. Mon. Weather Rev. **136**(12), 4629–4640 (2008)

van Leeuwen, P.J., Künsch, H.R., Nerger, L., Potthast, R., Reich, S.: Particle filters for high-dimensional geoscience applications: a review. Q. J. R. Meteorol. Soc. **145**(723), 2335–2365 (2019)

Villani, C.: Optimal Transport. Grundlehren der Mathematischen Wissenschaften [Fundamental Principles of Mathematical Sciences], vol. 338. Springer, Berlin (2009). Old and new

Open Access This chapter is licensed under the terms of the Creative Commons Attribution 4.0 International License (http://creativecommons.org/licenses/by/4.0/), which permits use, sharing, adaptation, distribution and reproduction in any medium or format, as long as you give appropriate credit to the original author(s) and the source, provide a link to the Creative Commons license and indicate if changes were made.

The images or other third party material in this chapter are included in the chapter's Creative Commons license, unless indicated otherwise in a credit line to the material. If material is not included in the chapter's Creative Commons license and your intended use is not permitted by statutory regulation or exceeds the permitted use, you will need to obtain permission directly from the copyright holder.

MIX
Papier aus verantwortungsvollen Quellen
Paper from responsible sources
FSC® C105338

If you have any concerns about our products,
you can contact us on
ProductSafety@springernature.com

In case Publisher is established outside the EU,
the EU authorized representative is:
Springer Nature Customer Service Center GmbH
Europaplatz 3, 69115 Heidelberg, Germany

Printed by Libri Plureos GmbH
in Hamburg, Germany